BIRKHÄUSER

Operator Theory: Advances and Applications

Vol. 188

Spectral Theory in Inner Product Spaces and Applications

6th Workshop on Operator Theory in Krein Spaces and Operator Polynomials, Berlin, December 2006

Jussi Behrndt
Karl-Heinz Förster
Heinz Langer
Carsten Trunk
Editors

Birkhäuser
Basel · Boston · Berlin

Editors:

Jussi Behrndt
Technische Universität Berlin
Institut für Mathematik
Straße des 17. Juni 136, MA 6–4
10623 Berlin, Germany
e-mail: behrndt@math.tu-berlin.de

Heinz Langer
Technische Universität Wien
Institut für Analysis und Scientific Computing
Wiedner Hauptstraße 8
1040 Wien, Austria
e-mail: hlanger@mail.zserv.tuwien.ac.at

Karl-Heinz Förster
Technische Universität Berlin
Institut für Mathematik, MA 6–4
Straße des 17. Juni 136
10623 Berlin, Germany
e-mail: foerster@math.tu-berlin.de

Carsten Trunk
Technische Universität Ilmenau
Institut für Mathematik
Postfach 100565
98684 Ilmenau, Germany
e-mail: carsten.trunk@tu-ilmenau.de

2000 Mathematical Subject Classification: Primary 47-06; Secondary 47B50, 47Axx, 34Bxx, 15Axx, 12D10

Library of Congress Control Number: 2008935488

Bibliographic information published by Die Deutsche Bibliothek.
Die Deutsche Bibliothek lists this publication in the Deutsche Nationalbibliografie; detailed bibliographic data is available in the Internet at http://dnb.ddb.de

ISBN 978-3-7643-8910-9 Birkhäuser Verlag AG, Basel - Boston - Berlin

Basel · Boston · Berlin
P.O. Box 133, CH-4010 Basel, Switzerland
Part of Springer Science+Business Media
Printed on acid-free paper produced from chlorine-free pulp. TCF ∞
Printed in Germany

ISBN 978-3-7643-8910-9 e-ISBN 978-3-7643-8911-6

9 8 7 6 5 4 3 2 1 www.birkhauser.ch

Contents

Preface

This volume contains papers written by participants of the 6th Workshop on Operator Theory in Krein Spaces and Operator Polynomials, which was held at the Technische Universität Berlin, Germany, December 14 to 17, 2006. This workshop was attended by 67 participants from 14 countries.

The lectures covered topics from spectral and perturbation theory of linear operators in inner product spaces and from operator polynomials. They included the theory of generalized Nevanlinna and Schur functions, differential operators, singular perturbations, de Branges spaces, scattering problems, block numerical ranges, nonnegative matrices and relations. All these topics are reflected in the present volume. Besides, it contains an after dinner speech from an earlier workshop, which we think may be of interest for the reader, as well as a speech on the occasion of the retirement of Peter Jonas.

It is a pleasure to acknowledge the substantial financial support received from the

- Deutsche Forschungsgemeinschaft (DFG),
- Berlin Mathematical School (BMS),
- DFG-Forschungszentrum MATHEON "Mathematik für Schlüsseltechnologien",
- Institute of Mathematics of the Technische Universität Berlin.

We would also like to thank Petra Grimberger for her great help in the organisation. Without her assistance the workshop might not have taken place.

The Editors

Lectures Held at the 6th Workshop on Operator Theory in Krein Spaces

1. *Vadim Adamyan*
   Non-negative perturbations of non-negative self-adjoint operators
2. *Daniel Alpay*
   Generalized Nevanlinna functions in the Banach space setting
3. *Elena Andreisheva*
   Approximation of generalized Schur functions
4. *Yury Arlinskii*
   Matrix representations of contractions with rank one defects and inverse spectral problems
5. *Tomas Azizov*
   Pontryagin theorem and an analysis of spectral stability of solitons
6. *Jussi Behrndt*
   On the eigenvalues of non-canonical self-adjoint extensions
7. *Paul Binding*
   Darboux transformations and commutation in Pontryagin space
8. *Marina Chugunova*
   Diagonalization of the coupled-mode system
9. *Branko Curgus*
   Positive operators in Krein spaces similar to self-adjoint operators in Hilbert spaces
10. *Mikhail Denisov*
    Spectral function for some product of self-adjoint operators
11. *Maxim Derevyagin*
    A criterion for the resolvent set of generalized Jacobi operators acting in Krein spaces
12. *Vladimir Derkach*
    Coupling method in the theory of generalized resolvents of symmetric operators

13. *Aad Dijksma*
    The Schur transformation for generalized Nevanlinna functions centered at a point in $\mathbb{C}^+$
14. *Aurelian Gheondea*
    A Krein space interpretation of Dirac operators
15. *Uwe Günther*
    A toy model of PT-symmetric quantum mechanics, the squire equation and UV-IR-duality
16. *Niels Hartanto*
    Block numerical ranges of nonnegative matrices
17. *Rostyslav Hryniv*
    What spectra can non-self-adjoint Sturm-Liouville operators have?
18. *Birgit Jacob*
    Location of the spectrum of operator matrices which are associated to second order equations
19. *Marinus Kaashoek*
    The inverse problem for Krein orthogonal entire matrix functions
20. *Ilya Karabash*
    The abstract kinetic equation on a half-line
21. *Victor Khatskevich*
    On the structure of semigroups of operators acting in spaces with indefinite metric
22. *Aleksey Kostenko*
    Indefinite Sturm-Liouville operators with the singular critical point zero
23. *Pavel Kurasov*
    Operator models for singular ordinary differential expressions
24. *Ekaterina Lopushanskaya*
    On the representation of generalized Caratheodory functions in the area Ω_ν
25. *Annemarie Luger*
    On non-canonical extensions with finitely many negative squares
26. *Mark Malamud*
    Completeness of root vectors for arbitrary boundary value problems for Sturm-Liouville equations
27. *Alexander Markus*
    On convexity of ranges of quadratic forms on various sets
28. *Hagen Neidhardt*
    Wigner's R-matrix and Weyl functions
29. *Vyacheslav Pivovarchik*
    On spectra of a certain class of quadratic operator pencils with one-dimensional linear part

30. *Andre C.M. Ran*
On the relation between $XX^{[*]}$ and $X^{[*]}X$ in an indefinite inner product space
31. *Leiba Rodman*
On factorization of matrix functions in the Wiener algebra
32. *Adrian Sandovici*
Composition of Dirac structures for infinite-dimensional systems
33. *Igor Sheipak*
Indefinite Sturm-Liouville operators with singular self-similar weights
34. *Andrei Shkalikov*
Remarks on global stability for inverse Sturm-Liouville problems
35. *Yuri Shondin*
On approximation of a boundary value problem at 'regular singular' endpoint
36. *Henk de Snoo*
A Lebesgue decomposition for nonnegative forms
37. *Vladimir Strauss*
A test for commutative J-symmetric families of D_κ^+-class
38. *Lyudmila I. Sukhocheva*
On linearizators of hyperbolic type pencils
39. *Franciszek H. Szafraniec*
Rows versus columns
40. *Matej Tusek*
Influence of the curvature on the impurity spectrum in a quantum dot
41. *Carsten Trunk*
Normal and hyponormal matrices in inner product spaces
42. *Kresimir Veselic*
On eigenvalue bounds for indefinite self-adjoint operators
43. *Henrik Winkler*
Kato decompositions for quasi-Fredholm relations
44. *Michal Wojtylak*
Domination in Krein spaces
45. *Harald Woracek*
Admissible majorants for de Branges spaces of entire functions

Operator Theory:
Advances and Applications, Vol. 188, xiii–xv

After Dinner Speech on the Inventor of the Equality Sign

Aad Dijksma

The day after tomorrow I will have the pleasure of having attended all four Berlin workshops on Operator Theory in Krein spaces. At the first workshop there were four speakers: Tomas Azizov, Hagen Neidhardt, Yuri Shondin, and myself. Since then the number of speakers has increased steadily to 41. Moreover, the number of days has increased. The first workshop was held for one day in November, now it is spread out over three days in December. What has not changed is that each speaker is allotted the same amount of time for his or her lecture: one hour each at the first workshop and 25 minutes each now. Indeed, it is a characteristic and nice feature of these Berlin workshops that we are all treated equally. For this reason I would like to say a few words about equality, or more precisely about the equality sign. Although this may seem a sudden digression to you, for me it is not as I will explain.

This summer I attended workshops in Newcastle and Cardiff, and Myra, my wife, accompanied me. After Cardiff we spent a week camping in South Wales along the coast near Tenby, now a tourist center with a very picturesque colorful harbor. Centuries ago it was a major port. One day we were hiking along the Pembroke coast when we met an English couple. We started talking and when I told them that I was a mathematician they said, "Oh, but then you are in the right place, because the inventor of the equality sign is born right here in Tenby. There is a commemorative plaque in St Mary's Church." The next day we went to the church and found the shield on a pillar honoring Robert Recorde as the inventor of the equality sign. He was born in Tenby in 1510, almost 500 years ago.

I do not know how it is with you, but before this event I had never really much thought about who invented which sign and was a bit sceptical. But that changed after I spoke with my colleague at the University of Groningen Jan van Maanen, who is a specialist in the History of Mathematics. He knew the name of Robert

This speech was delivered at the fourth Workshop on Operator Theory in Krein spaces in Berlin, Friday, December 18, 2004.

Recorde quite well and to learn more about such inventions he recommended Florian Cajori's *A history of mathematical notations* published in two large volumes in 1928–29; now available as a Dover pocket edition [1].

Anyway, I became interested in this Robert Recorde. The following information is taken from the website [2] of the St. Andrews University on the History of Mathematics. Robert Recorde studied medicine first in Cambridge and then in Oxford. He was highly educated and taught in both places. In 1545 he moved to London where he became the court physician to King Edward VI and Queen Mary. In 1549 he was appointed as the controller of the Bristol mint and later he was placed in charge of the silver mines in Wexford, Ireland and was made the technical supervisor of the Dublin mint. In this capacity he introduced the silver crown of five shillings which was the first English coin to have a date written in Arabic numerals rather than in Roman numerals. Robert Recorde was not always very tactical in his dealings with the intrigues of the court and sometimes lost favour. In 1556 he tried to charge his arch-enemey at court, the Earl of Pembroke, with misconduct in order to regain his position. Recorde may have had a valid case, but the Earl would not allow a minor servant like Recorde to get the better of him. He countered Recorde's charges by suing him for libel. Recorde lost and was ordered to pay Pembroke 1000 pounds, which he could or would not pay and subsequently he was put in jail in 1557. It would be very ironic if he could not pay, because he was owed exactly this amount for his services in Ireland. He received these 1000 pounds in 1570, but this was of no use to him at all (only to his next of kin) as he had died in the King's Bench prison in Southwark twelve years earlier.

Robert Recorde published four books. The first three have the clear titles *Grounde of Artes*, *Pathwaie to Knowledge*, and *Castle of Knowledge*. The fourth book, published a year before he died, bears the criptic title: *The Whetstone of Witte*[1]. In this book he introduces the equality sign (two parallel horizontal lines, about 1 cm long, longer than we are used to now) and justifies it by writing: "bicause noe 2 thynges can be moare equalle." The symbol was not immediately popular. Two vertical lines were used by some and *æ* (or *œ*) from the word "aequalis" meaning equal was widely used into the 1700's. The latter symbol was also used in astronomy as an abbreviation for the planet Saturnus, which was rather convenient for the printers, so that they did not have to make a new symbol in lead. The title of the book is a clever pun by Recorde which nowadays needs an explanation: "Cosa" is Latin for "thing" and "thing" was used in algebra to denote

---

[1]Jan van Maanen reminded me that the Dutch daily evening newspaper NRC-Handelsblad in its own advertisements calls itself the "whetstone of the mind" and suggested I write a note to tell that almost 450 years earlier this slogan was already used as the title of a book. An adapted version [3] of the dinner speech was subsequently published. I sent a copy of it to Jack and Margaret O'Higgins, the English couple alluded to in the speech, and they encouraged me to submit a note to the Tenby Times to explain how it came about that Tenby was mentioned on the back page of a leading Dutch newspaper: not because of its picturesque harbor but because of a chance meeting, a dinner speech, and a slogan. A short report [4] appeared in the Tenby Times five months later.

the unknown; now, following Descartes, we often use the letter x. Algebraists were called cossists and algebra the cossic art. The word "cos" is Latin for whetstone, a stone for sharpening razors and tools. Hence the pun: it was an algebra book on which to sharpen one's mathematical wit.

I would like to make a toast to the organizers, but this too reminds me of a holiday in England about thirty five years ago. Before camping there, we visited Prof. Felix M. Arscot in Reading, who died 8 years ago. He was an expert in differential equations but also knew a lot about English history. He drove us around and took us to the Winchester Cathedral Graveyard, where he showed us Private Thomas Thetcher's grave who died in 1776. On the tombstone it was written as a warning to all who drink cold alcohol, in particular, beer:

Here sleeps in peace a
Hampshire grenadier
who caught his death by
drinking cold small beer.
Soldiers be warned by his
untimely fall
and when you're hot
drink strong or not at all.

My speech has been long enough to bring our drinks to an agreeable temperature and so ideal for toasting the organizers for a memorable meeting. I am sure I speak on behalf of all present when I thank them for all the work they have done for the workshop and for making it such a pleasant event. We are looking forward to the fifth.

References

[1] Florian Cajori, *A history of mathematical notations* ; Vol. I: *Notations in elementary mathematics*, Vol. II: *Notations mainly in higher mathematics*, Dover Publications, 1993.

[2] For a biography on Robert Recorde and other mathematicians, see the website: http://www-gap.dcs.st-and.ac.uk/ history/Mathematicians.

[3] Aad Dijksma, *Slijpsteen voor de geest*, NRC-Handelsblad, Donderdag 23 maart 2006, page 30 (Achterpagina).

[4] *Dutch professor's "Recorde" of Tenby*, Tenby Times and diaries, Issue 65, August 2006, page 10.

Aad Dijksma
Department of Mathematics
University of Groningen
P.O. Box 407
9700 AK Groningen, The Netherlands
e-mail: dijksma@math.rug.nl

Operator Theory:
Advances and Applications, Vol. 188, xvii–xx

Speech on the Occasion of the Retirement of Peter Jonas

Heinz Langer

Dear friends and colleagues,

On the fourth workshop in 2004, Aad Dijksma gave an after dinner speech on some question from the history of mathematics: he taught us that the equality sign in mathematics was introduced by Robert Recorde from Tenby (Wales, UK) in the first half of the 16th century. The organizers decided to continue this tradition of after dinner speeches. Since 'after dinner' last night was rather late and I could not be sure to attract your attention, it was transformed into an 'after coffee break speech'.

I will also start with a historical question which is close to the title of our workshop: Who invented the name 'Krein space'? Perhaps some of you know the answer: it was Janos Bognár in his book, which appeared in 1974. I remember that Mark Krein was quite surprised when he came to know about this. Being a very modest person he had some doubts that this name was justified. But if this would have been the case, 'Krein space' would not have found such rapid acceptance. A lecture of Janos Bognár, which my supervisor of the diploma and PhD thesis, Professor P.H. Müller, had attended in 1959 in Balatonföldvar, was also the reason that I became interested in these spaces. Before Janos Bognár had introduced this name, these spaces were called 'J-spaces', or 'spaces with indefinite metric', 'spaces with indefinite inner product' or with 'indefinite bilinear form', and 'Nevanlinna spaces' in a paper by Ilppo Simo Louhivaara. In fact, Rolf Nevanlinna was perhaps the first to introduce such spaces axiomatically. At the end of the 50ies and beginning of the 60ies of the last century there appeared quite a few papers about general linear or topological linear spaces with indefinite inner product. However, the axiomatic introduction of Pontryagin spaces goes back to a paper by I.S.Iokhvidov and M.G.Krein, and the definition of a Krein space as linear space $\mathcal{K}$ with an indefinite inner product $[\cdot,\cdot]$ that admits an orthogonal decomposition $\mathcal{K} =$

This speech was delivered at the sixth Workshop on Operator Theory in Krein spaces in Berlin, Saturday, December 16, 2006.

$\mathcal{K}_+[\dot{+}]\mathcal{K}_-$, where $(\mathcal{K}_\pm, \pm[\cdot,\cdot])$ are Hilbert spaces, was first given by Mark Krein. In the 1950ies and 60ies, the interest in operators in these spaces was concentrated in Odessa, in Finland (by Rolf Nevanlinna and his students Louhivaara and Pesonen), in Hungary, in the US by Ralph Philips, later in Dresden, and in the middle of the 60ies also in Paris (by Laurent Schwarz). Between some of these places there was hardly any connection at the beginning. I met Professor Akilov in Leningrad in 1962 (we had translated the book 'Functional Analysis' by L.V. Kantorovič and G.P. Akilov into German and there some mathematical questions arose which I wanted to settle with him), and he asked me about my field of interest. When I said 'operators in spaces with indefinite inner product' his reaction was 'eto takaja moda na Ukraine' ('there is such a fashion in the Ukraina').

The situation has changed, and in the 1980ies spaces with indefinite inner product even became an item in the AMS classification. Strangely enough, even now the two related items have different names for these spaces (probably meant to be more general than Krein spaces): 'Spaces with indefinite inner products' (46C20) and 'Operators in spaces with an indefinite metric' (47B50).

Nowadays the notion 'Krein space' is well established, at least in some parts of the world. This workshop on Krein spaces takes place here in Berlin already for the 6th time, and I think it is remarkable that a yearly conference finds such a growing interest: from a one day meeting in the year 2001 with five lectures, where two of them were given by Aad Dijksma, it has grown into a four days conference with a very dense program of 45 lectures and 67 participants from 14 countries this year. I would like to thank the organizers, Professor Karl-Heinz Förster, Dr. Jussi Behrndt, Dr. Peter Jonas, Dr. Christian Mehl, Dr. Carsten Trunk, and, last but not least, Frau Petra Grimberger very much for their efforts.

The question arises: how did Krein spaces find their way to Berlin? To answer this question we have to go back for almost 50 years: In 1959 Peter Jonas began to study mathematics at the Technische Hochschule in Dresden. I probably met him the same year, 47 years ago, since I was giving the problem classes in analysis for these students' group. But this I do not really remember. However, 4–5 years later, in 1964, two students of this group wrote their diploma theses with me, and one of them was Peter Jonas. These two were indeed my very first diploma students! The other one, Volker Nollau, retired this summer at the TU Dresden. In his diploma thesis Peter considered a question of stability for infinite-dimensional Hamiltonian systems, close to a paper by Dergusov. After he got the diploma, if I remember this well, he did not want to stay in Dresden in Saxonia, but he wanted to return to Berlin instead, where he had spent his childhood after the war (he was born in 1941 in Memel, in Eastern Prussia at that time, nowadays in Lithuania). So in 1964 he got a pre-doc position at the Institut für Reine Mathematik der Akademie der Wissenschaften der DDR. His supervisor was Professor Josef Naas, who at that time was an influential person for mathematics in the GDR. Naturally Peter had to change the field, and in 1969 he got his PhD with a thesis on questions of differential geometry. His publications in this period had titles like 'Über Folgen von Bogenlängenkurven zu distributionentheoretisch konvergenten Kurvenfolgen'.

Then for a while he worked on elliptic PDE's and quasiconformal mappings, a well established area at this institute; finally, in the 1970ies he returned to the theory of self-adjoint and unitary operators in Krein spaces and their applications in the theory of linear Hamiltonian differential equations. Having learnt a lot about distributions during his PhD work in Berlin, he used them for an intensive study of the spectral function or spectral distribution of definitizable operators in Krein spaces.

During all these years, from 1965 to 1989, or even a little longer, Peter was employed by the Karl-Weierstraß-Institute, a research institute for mathematics at the Academy of Sciences of the GDR. For an extended period he worked in the group of Professor Baumgärtel, who let Peter the freedom to follow his own interests. Under the influence of this group Peter worked on perturbation and scattering theory for operators in Krein spaces. In 1987 he wrote his habilitation thesis 'The trace formula for some classes of unitary and self-adjoint operators in Krein space'. However, at this institute, Peter was always somehow mathematically isolated, and there were no students. Moreover, there was often some danger that the direction of research he was interested in would be closed and that he would have to work in more applied areas. He tried to compensate this isolation partly by establishing contacts to colleagues in the East: he travelled to Kiev, Moscow, Budapest, Warsow, Iasi, Leningrad, Odessa, Krakow, Voronesh, Bucuresti, Dubna, Tscheljabinsk and other places. He developed close contacts with Aurelian Gheondea, Tomas Azizov, Tsembeltsogt Bajasgalan, and many others. And Peter's apartment in Berlin was always open for visitors, also for a stay over night.

Then came the changes of 1989, which meant greater freedom but also new difficulties. Peter's position became rather uncertain. With the help of Professor Ilppo Simo Louhivaara he became somehow associated to the Free University and, after a positive evaluation of his work at the Karl-Weierstraß-Institute and still in the group of Professor Baumgärtel, he was temporarily enrolled at the University of Potsdam. Finally, and this was most important for Peter and for us, in 1991 Professor Karl-Heinz Förster gave Peter the possibility to jointly organize at the Technical University here in Berlin the 'Colloquium Operator Theory'. It seems to me that now gradually and with the support of Karl-Heinz, here at the TU Berlin Peter finally has had the chance to fully realize his plans and ideas. He could give lectures and seminars, and started to work with students and graduate students. During the last 15 years he wrote as many papers as during the 25 years before, and has supervised 10 diploma students, and two of them, Carsten and Jussi, also got their PhD with Peter. Let me mention some of the topics of his research:

- Stability criteria for positive perturbations in Krein spaces,
- Compact perturbations and the trace formula in Krein spaces,
- Spectral functions and spectral distributions for definitizable operators,
- Locally definitizable operators and functions,
- Applications to the Klein Gordon equation and to spectral problems with eigenvalue depending boundary conditions.

As you perhaps all know, since 1990, Peter has travelled not only to the East but to conferences and workshops all over the world including Caracas in Venecuela, Rehovot and Beer Sheva in Israel, Bellingham in Washington, and Seoul in South Korea. Also in his private journeys he now prefers exotic countries like Thailand, Laos, Burma and Kambodcha. Peter, having reached the age of 65 on the 18th of July this year, officially retired by the end of July 2006. But he is still very active in mathematics: at the moment he supervises three diploma students and one PhD student. Peter, I wish you good health and all the best for many years to come!

Added in June 2008: Sadly, none of these wishes came true. Only seven months after this speech was given Peter Jonas died of brain cancer.

Heinz Langer
Institut für Analysis und Scientific Computing
Technische Universität Wien
Wiedner Hauptstr. 8
A-1040 Wien, Austria
e-mail: hlanger@mail.zserv.tuwien.ac.at

Operator Theory:
Advances and Applications, Vol. 188, 1–30

Augmented Schur Parameters for Generalized Nevanlinna Functions and Approximation

D. Alpay, A. Dijksma and H. Langer

Abstract. Schur parameters of a Schur function are a well-known concept in Schur analysis. Here we define the Schur transformation and the sequence of Schur parameters for a Nevanlinna function and a generalized Nevanlinna function. They are applied to approximate the Nevanlinna function or the generalized Nevanlinna function by rational ones.

Mathematics Subject Classification (2000). Primary 41A20, 47A57, 46C20; Secondary 30C80, 47B32, 30D30.

Keywords. Schur transform, Schur algorithm, Schur parameter, generalized Nevanlinna function, $(-\mathrm{i}J)$-unitary matrix function, approximation.

1. Introduction

A function s is called a *Schur function* if it is a holomorphic function on the open unit disk $\mathbb{D}$ with $|s(z)| \leq 1$, $z \in \mathbb{D}$. When s is a Schur function and not identically equal to a unimodular complex number, its *Schur transform* $\widehat{s}$ is defined by

$$\widehat{s}(z) = \frac{1}{z}\frac{s(z)-s(0)}{1-s(z)s(0)^*}, \quad z \in \mathbb{D}.$$

The mapping $s \mapsto \widehat{s}$ is called the *Schur transformation.* It maps the set of Schur functions, which are not identically equal to a unimodular number, into the set of Schur functions. If $|\widehat{s}(0)| < 1$ then $\widehat{s}$ is not identically equal to a unimodular number and the Schur transform of $\widehat{s}$ can be defined. Repeating this procedure, we obtain what is called the *Schur algorithm*:

$$s_0 := s,\ s_1 := \widehat{s}_0,\ s_2 := \widehat{s}_1, \ \text{ in general } s_j := \widehat{s}_{j-1},$$

D. Alpay thanks the Earl Katz family for endowing the chair which supported his research. The research of A. Dijksma and H. Langer was supported in part by the Center for Advanced Studies in Mathematics (CASM) of the Department of Mathematics, Ben–Gurion University.

and the sequence of *Schur parameters of the Schur function* s:

$$\rho_0 := s_0(0),\ \rho_1 := s_1(0),\ \rho_2 := s_2(0),\ \text{ in general } \rho_j := s_j(0),$$

where $j = 1, 2, \dots$ if all the numbers $\rho_j, j = 1, 2, \dots$, have modulus < 1, and $j = 1, 2, \dots, m$ if m is the smallest integer ≥ 1 such that $|\rho_m| = 1$; in the latter case the algorithm stops after m steps with s_m. There is a vast literature about Schur functions and its Schur parameters, see [11] and the references given therein. Part of this has been generalized in [10] to *generalized Schur functions* using the Schur transformation from [9], see also [4]. In this (indefinite) generalization finitely many of the Schur parameters, which are complex numbers in the above situation, had to be replaced by augmented Schur parameters which carry more information than just complex numbers.

In what follows $\mathbb{R}$ denotes the real line, $\mathbb{C}$ the complex plane, and $\mathbb{C}^+$, $\mathbb{C}^-$ the open upper and lower half-plane, respectively. In this note we consider the Schur transformation for Nevanlinna functions and generalized Nevanlinna functions as it was introduced in [3], [5], and [4]. Recall that a complex function n is a *Nevanlinna function* if $\operatorname{hol}(n)$, its domain of holomorphy, contains $\mathbb{C} \setminus \mathbb{R}$, n is symmetric with respect to $\mathbb{R}$, that is, $n(z^*) = n(z)^*$, $z \in \operatorname{hol}(n)$, and satisfies

$$\frac{\operatorname{Im} n(z)}{\operatorname{Im} z} \geq 0, \quad z \in \mathbb{C} \setminus \mathbb{R}.$$

An equivalent definition is obtained if the latter inequality is replaced by the condition that the Nevanlinna kernel

$$L_n(z, w) = \begin{cases} \dfrac{n(z) - n(w)^*}{z - w^*}, & w \neq z^*, \\ n'(z), & w = z^*, \end{cases} \tag{1.1}$$

is nonnegative on $\mathbb{C} \setminus \mathbb{R}$. We denote the class of Nevanlinna functions by $\mathbf{N}_0$. A function n is called a *generalized Nevanlinna function with κ negative squares*, if it is meromorphic on $\mathbb{C} \setminus \mathbb{R}$, satisfies $n(z^*) = n(z)^*$ for all $z \in \operatorname{hol}(n)$, and the kernel $L_n(z, w)$ has κ negative squares. The set of such functions will be denoted by $\mathbf{N}_\kappa$. If n belongs to $\mathbf{N}_\kappa$ then n has at most κ poles in $\mathbb{C}^+$. Clearly, $\mathbf{N}_\kappa$ with $\kappa = 0$ coincides with the class of Nevanlinna functions defined above. The functions from the class $\mathbf{N} = \cup_{\kappa \geq 0} \mathbf{N}_\kappa$ are called *generalized Nevanlinna functions.* If $n \in \mathbf{N}$, then by $\kappa_-(n)$ we denote the number of negative squares of n. Evidently, if $n \in \mathbf{N}$ is constant, then, because of the condition $n(z^*) = n(z)^*$ for all $z \in \operatorname{hol}(n)$, this constant must be real.

Let $z_1 \in \mathbf{C}^+$ be fixed. If n is a Nevanlinna function ($n \in \mathbf{N}_0$), the Schur transform $\widehat{n}$ of n, centered at z_1, is defined as follows:

$$\widehat{n}(z) = \begin{cases} \dfrac{\beta(z)n(z) - |n(z_1)|^2}{n(z) - \alpha(z)}, & \text{if } n \text{ is not linear, that is, } n \neq \alpha, \\ \infty, & \text{if } n \text{ is linear,} \end{cases} \tag{1.2}$$

where

$$\alpha(z) = n(z_1) + \frac{n(z_1)-n(z_1)^*}{z_1-z_1^*}(z-z_1), \quad \beta(z) = n(z_1)^* - \frac{n(z_1)-n(z_1)^*}{z_1-z_1^*}(z-z_1).$$

In the quotient in (1.2), the numerator and denominator have simple zeros in z_1 which cancel, and $\widehat{n}$ turns out to be again a Nevanlinna function or ∞, see [3, Subsection 2 of Section 2]. In the special case when n is rational for the degrees of n and $\widehat{n}$ we have $\deg \widehat{n} = \deg n - 1$, see [5, Theorem 4.2]; recall that the *degree* $\deg f$ of a rational complex function f is the sum of all the multiplicities of all its poles (including ∞). If $n_1 := \widehat{n}$ is not constant or ∞, its Schur transform $n_2 := \widehat{n}_1$ can be considered, etc. As in the case of Schur functions above, through this *Schur algorithm* we obtain either an infinite sequence of functions $n_j := \widehat{n}_{j-1}$, $j = 1, 2, \ldots$, or, if some function n_m is ∞ or real at z_1 (and hence constant), a finite sequence with $j = 1, 2, \ldots, m$. The corresponding sequence (ρ_j) of *Schur parameters* is then defined as the finite or infinite sequence of the values $\rho_j = n_j(z_1)$. For the more complicated definition of the Schur transform centered at z_1 of a *generalized* Nevanlinna function and the corresponding sequence of *augmented Schur parameters* we refer to Section 4.

As to the main results, in the present paper we characterize the sequences of Schur parameters or of augmented Schur parameters and show that they determine the given functions n from $\mathbf{N}_0$ or $\mathbf{N}$ uniquely. Moreover, we define a sequence of 'rational' functions $r^{[\ell]}$, which depend only on the first $\ell + 1$ Schur parameters or augmented Schur parameters of n, and which converge locally uniformly on $\mathbb{C}^+$ if $n \in \mathbf{N}_0$ and locally uniformly on $\mathbb{C}^+ \cap \operatorname{hol}(n)$ if $n \in \mathbf{N}$. These main results are proved in Section 3 for Nevanlinna functions and in Section 4 for generalized Nevanlinna functions. In Section 3 we also derive an explicit formula which expresses the Schur parameter ρ_j of the function $n \in \mathbf{N}_0$ by the first $j + 1$ Taylor coefficients of n at z_1; in Section 2 we collect some definitions and preliminary results.

We thank the referee for very carefully reading the manuscript and valuable suggestions.

2. Preliminaries

For a 2×2 matrix

$$M = \begin{pmatrix} \alpha & \beta \\ \gamma & \delta \end{pmatrix}$$

we define the *linear fractional transformation* $\mathcal{T}_M$ by

$$\mathcal{T}_M(u) = \frac{\alpha u + \beta}{\gamma u + \delta} \quad \text{if } u \in \mathbb{C}, \qquad \mathcal{T}_M(\infty) = \frac{\alpha}{\gamma},$$

if only these quotients make sense. If M is a 2×2 matrix function and u is a scalar function then $\mathcal{T}_M(u)$ is the scalar function with values $(\mathcal{T}_M(u))(z) = \mathcal{T}_{M(z)}(u(z))$ for all z for which the right-hand side is defined.

In the sequel we set

$$J = \begin{pmatrix} 0 & 1 \\ -1 & 0 \end{pmatrix},$$

hence the matrix $-\mathrm{i}J$ is hermitian as well as unitary. A 2×2 matrix function Θ is called $(-\mathrm{i}J)$-*contractive on* $\mathcal{D}$, a nonempty connected open subset of $\mathbb{C}$, if it is meromorphic on $\mathcal{D}$ and satisfies:

$$\Theta(z)\big(-\mathrm{i}J\big)\Theta(z)^* \leq -\mathrm{i}J, \quad z \in \mathrm{hol}\,(\Theta) \cap \mathcal{D}.$$

The matrix function Θ is called $(-\mathrm{i}J)$-inner if it is $(-\mathrm{i}J)$-contractive on $\mathbb{C}^+$ and the nontangential boundary values on $\mathbb{R}$ satisfy

$$\Theta(z)(-\mathrm{i}J)\Theta(z)^* = (-\mathrm{i}J), \quad \text{almost all } z \in \mathbb{R}. \tag{2.1}$$

In the paper we consider rational 2×2 matrix functions Θ, only Lemma 2.1 is concerned with a more general situation. If Θ is rational and (2.1) holds for all $z \in \mathrm{hol}\,(\Theta) \cap \mathbb{R}$, then Θ is called $(-\mathrm{i}J)$-*unitary on* $\mathbb{R}$. Thus if Θ is rational and $(-\mathrm{i}J)$-inner on $\mathbb{C}^+$, then it is $(-\mathrm{i}J)$-unitary on $\mathbb{R}$. If Θ is rational and $(-\mathrm{i}J)$-unitary on $\mathbb{R}$ then

$$\Theta(z)J\Theta(z^*)^* = J, \quad z \in \mathrm{hol}\,(\Theta), \tag{2.2}$$

and hence

$$\Theta(z^*)^* = \frac{1}{\det \Theta(z)}\,\Theta(z)^\top, \quad z \in \mathrm{hol}\,(\Theta). \tag{2.3}$$

Let z_1 be a fixed chosen point in $\mathbb{C}^+$. A rational 2×2 matrix function Θ has a unique pole in z_1^* (and, in particular, is analytic at infinity) if and only if it is of the form

$$\Theta(z) = \sum_{i=0}^{q} \frac{T_i}{(z - z_1^*)^i}, \tag{2.4}$$

where q is an integer ≥ 1 and T_i, $i = 0, 1, \ldots, q$, are complex 2×2 matrices of which the last q are not all zero. If we denote by $\deg \Theta$ the McMillan degree of Θ and if Θ has the form (2.4), then

$$\deg \Theta = \mathrm{rank} \begin{pmatrix} T_q & 0 & \cdots & 0 & 0 \\ T_{q-1} & T_q & \cdots & 0 & 0 \\ \vdots & \vdots & \ddots & \vdots & \vdots \\ T_2 & T_3 & \cdots & T_q & 0 \\ T_1 & T_2 & \cdots & T_{q-1} & T_q \end{pmatrix}; \tag{2.5}$$

for the definition of the McMillan degree and the relation (2.5) we refer to [8].

We denote by $\mathcal{U}^{z_1}$ the class of all rational 2×2 matrix functions which are constant or have a unique pole at z_1^* and are $(-\mathrm{i}J)$-unitary on $\mathbb{R}$, and by $\mathcal{I}^{z_1}$ the set of all $\Theta \in \mathcal{U}^{z_1}$ which are $(-\mathrm{i}J)$-inner. If we set

$$\mathrm{b}(z) = \frac{z - z_1}{z - z_1^*},$$

then for every $\Theta \in \mathcal{U}^{z_1}$ there is a real number θ such that

$$\det \Theta(z) = e^{i\theta}\, b(z)^{\deg \Theta}, \tag{2.6}$$

see [4, Theorem 3.6].

For $\Theta \in \mathcal{U}^{z_1}$ the kernel

$$K_\Theta(z,w) = \frac{J - \Theta(z)J\Theta(w)^*}{2\pi(z - w^*)}$$

has finitely many positive and finitely many negative squares because the matrix function Θ is rational and satisfies (2.2). Hence (see [6, Theorem 8.1]) the reproducing kernel space $\mathcal{P}(\Theta)$ with reproducing kernel K_Θ is a finite-dimensional Pontryagin space of dimension

$$\dim \mathcal{P}(\Theta) = \deg \Theta,$$

and we have

$$\mathcal{P}(\Theta) = \mathbf{H}_{2,J} \ominus \Theta\, \mathbf{H}_{2,J},$$

where $\mathbf{H}_{2,J}$ stands for the Krein space with reproducing kernel

$$H_J(z,w) = \frac{J}{2\pi(z - w^*)}.$$

By definition the *number* $\kappa_-(\Theta)$ *of negative squares of* $\Theta \in \mathcal{U}^{z_1}$ is the number of negative squares of the kernel K_Θ and this number coincides with the negative index of the Pontryagin space $\mathcal{P}(\Theta)$ it generates. If $\Theta \in \mathcal{I}^{z_1}$, then the kernel K_Θ is positive definite, $\mathcal{P}(\Theta)$ is a Hilbert space, and $\kappa_-(\Theta) = 0$.

Both sets $\mathcal{U}^{z_1}$ and $\mathcal{I}^{z_1}$ are closed under matrix multiplication and this operation is *minimal*, that is, the degrees add up:

$$\deg \Theta_1\Theta_2 = \deg \Theta_1 + \deg \Theta_2, \quad \Theta_1, \Theta_2 \in \mathcal{U}^{z_1}. \tag{2.7}$$

This readily follows from (2.6) and the kernel decomposition

$$K_{\Theta_1\Theta_2}(z,w) = K_{\Theta_1}(z,w) + \Theta_1(z)K_{\Theta_2}(z,w)\Theta_1(w)^*, \tag{2.8}$$

which imply

$$\mathcal{P}(\Theta_1\Theta_2) = \mathcal{P}(\Theta_1) \oplus \Theta_1\mathcal{P}(\Theta_2), \tag{2.9}$$

see [3, Theorem 6.1] and [4, Theorem 3.13]. Note that the last equality implies that also the numbers of negative squares add up:

$$\kappa_-(\Theta_1\Theta_2) = \kappa_-(\Theta_1) + \kappa_-(\Theta_2), \quad \Theta_1, \Theta_2 \in \mathcal{U}^{z_1}. \tag{2.10}$$

The following lemma is well known, see for example [12, Corollary 2.10 (i)]. We give a proof for completeness.

Lemma 2.1. *Let*

$$\Theta = \begin{pmatrix} a & b \\ c & d \end{pmatrix} \tag{2.11}$$

be a nonconstant 2×2 *matrix function which is holomorphic and* $(-iJ)$*-contractive on some nonempty connected open subset* $\mathcal{D}$ *of* $\mathbb{C}^+$. *Then:*

(1) *At most one of the entries a, b, c, d vanishes identically on $\mathcal{D}$, and the functions*

$$b/a,\ -a/b,\ d/c,\ -c/d \tag{2.12}$$

have a nonnegative imaginary part on their domain of holomorphy in $\mathcal{D}$ if only the denominator does not vanish identically; the same holds for the functions

$$a/c,\ -c/a,\ b/d,\ -d/b. \tag{2.13}$$

if, additionally, $\det\Theta \not\equiv 0$.

(2) *If Θ is rational and $(-\mathrm{i}J)$-inner on $\mathbb{C}^+$, then the functions in* (2.12) *and in* (2.13) *are Nevanlinna functions if only the denominator does not vanish identically.*

(3) *If $\Theta \in \mathcal{I}^{z_1}$ then a, b, c, d do not vanish at z_1 and all the functions in* (2.12) *and* (2.13) *are Nevanlinna functions.*

Proof. (1) First we observe that with Θ also the matrix functions

$$J\Theta = \begin{pmatrix} c & d \\ -a & -b \end{pmatrix},\quad \Theta J = \begin{pmatrix} -b & a \\ -d & c \end{pmatrix},\quad J\Theta J = \begin{pmatrix} -d & c \\ b & -a \end{pmatrix}$$

are $(-\mathrm{i}J)$-contractive on $\mathcal{D}$. Since Θ is $(-\mathrm{i}J)$-contractive, we have for $z \in \mathcal{D}$ that

$$\mathrm{i}\begin{pmatrix} a(z)b(z)^* - a(z)^*b(z) & a(z)d(z)^* - b(z)c(z)^* - 1 \\ b(z)c(z)^* - d(z)a(z)^* + 1 & c(z)d(z)^* - d(z)c(z)^* \end{pmatrix} \geq 0. \tag{2.14}$$

If $a = d = 0$ this inequality can only hold if $b(z)c(z)^* = 1$, $z \in \mathcal{D}$. Since $b(z)$ is holomorphic in z and $c(z)^*$ is holomorphic in z^*, both functions must be constants which was excluded by assumption. The proofs for $a = c = 0$ and $a = b = 0$ are similar; for other combinations we can use the twisted matrix functions above. From (2.14) it also follows that

$$\mathrm{i}(a(z)b(z)^* - b(z)a(z)^*) \geq 0,\quad \mathrm{i}(c(z)d(z)^* - d(z)c(z)^*) \geq 0,\qquad z \in \mathcal{D},$$

and hence

$$\operatorname{Im}\frac{b(z)}{a(z)} \geq 0,\quad \operatorname{Im}\frac{d(z)}{c(z)} \geq 0,$$

for all z in the domain of holomorphy of these functions if only the denominator does not vanish identically. The claims for the functions in (2.13) follow in the same way from the fact that

$$\Theta(z)^{-1} = (\det\Theta(z))^{-1}\begin{pmatrix} d(z) & -b(z) \\ -c(z) & a(z) \end{pmatrix}$$

is $(\mathrm{i}J)$-contractive:

$$\Theta(z)^{-1}(\mathrm{i}J)\Theta(z)^{-*} \leq \mathrm{i}J.$$

(2) If Θ is rational and $(-\mathrm{i}J)$-inner on $\mathbb{C}^+$ then it is $(-\mathrm{i}J)$-contractive on $\mathbb{C}^+$ and $(-\mathrm{i}J)$-unitary on $\mathbb{R}$, which, by (2.2), implies that $\det\Theta(z) \neq 0$, $z \in \operatorname{hol}(\Theta)$. Hence the conclusions of part (1) hold. By (2.3),

$$a(z^*)^* = \frac{a(z)}{\det\Theta(z)},$$

and corresponding relations hold for the other entries of $\Theta(z)$. This implies that the functions in (2.12) and in (2.13) are symmetric with respect to the real axis and hence Nevanlinna functions.

(3) Assume $\Theta \in \mathcal{I}^{z_1}$. If $c(z_1) = 0$, then

$$\mathrm{i}\begin{pmatrix} a(z_1)b(z_1)^* - a(z_1)^*b(z_1) & a(z_1)d(z_1)^* - 1 \\ -d(z_1)a(z_1)^* + 1 & 0 \end{pmatrix} \geq 0,$$

and hence $a(z_1)d(z_1)^* = 1$. This implies that $\det \Theta(z_1) = a(z_1)d(z_1) \neq 0$, contradicting (2.6) according to which $\det \Theta(z_1) = 0$. Hence $c(z_1) \neq 0$; using the twisted matrix functions the same follows for the other entries of $\Theta(z)$. □

Now let n be a Nevanlinna function and write its Taylor expansion at z_1 as

$$n(z) = \sum_{j=0}^{\infty} \nu_j (z - z_1)^j.$$

We denote by $\mathbf{f}_0, \mathbf{f}_1, \ldots$ the rational 2×1 vector functions

$$\mathbf{f}_0(z) = \frac{\begin{pmatrix} \nu_0^* \\ 1 \end{pmatrix}}{z - z_1^*}, \quad \mathbf{f}_j(z) = \frac{\mathbf{f}_{j-1}(z)}{z - z_1^*} + \frac{\begin{pmatrix} \nu_j^* \\ 0 \end{pmatrix}}{z - z_1^*}, \qquad j \geq 1, \tag{2.15}$$

and call it the *chain of functions associated with* n (at the point z_1). These functions play a role in the following lemma, which we apply in Section 3.

Lemma 2.2. *Let $\Theta \in \mathcal{I}^{z_1}$ with $\deg \Theta = s \geq 1$ be written in the form (2.11), let $\check{n}$ be a Nevanlinna function and assume that d/c is not constant, or let $\check{n}$ be identically equal to ∞. Then the function*

$$n = \mathcal{T}_\Theta(\check{n}) \tag{2.16}$$

is a Nevanlinna function and the space $\mathcal{P}(\Theta)$ is spanned by the first s elements $\mathbf{f}_0, \ldots, \mathbf{f}_{s-1}$ of the chain of functions associated with n.

Proof. Assume first $\check{n} \neq \infty$. The relation (2.16) with Θ as in (2.11) is equivalent to

$$\begin{pmatrix} 1 & -n \end{pmatrix} \Theta = \frac{\det \Theta}{c\check{n} + d} \begin{pmatrix} 1 & -\check{n} \end{pmatrix}. \tag{2.17}$$

If the numerator $c\check{n} + d$ vanishes at z_1, then $\check{n}(z_1) = -d(z_1)/c(z_1)$ and since the functions $\check{n}$ and d/c are Nevanlinna functions (see Lemma 2.1), $\operatorname{Im} d(z_1)/c(z_1) = 0$. The maximum modulus principle implies that d/c is equal to a real constant, and this was excluded by assumption. Hence $c\check{n} + d$ does not vanish at z_1 and the right-hand side of (2.17) has a zero of order s at z_1, see (2.6). By expanding the left-hand side in powers of $z - z_1$, we obtain the desired result as in the proof of [2, Proposition 6.4].

The proof is the same if $\check{n} = \infty$, because then $n = a/c$ and

$$\begin{pmatrix} 1 & -n \end{pmatrix} \Theta = (\det \Theta) \begin{pmatrix} 0 & -\frac{1}{c} \end{pmatrix}.$$

□

3. Schur parameters and approximation for Nevanlinna functions

3.1. Recall that z_1 is a fixed chosen point in $\mathbb{C}^+$, and let $n \in \mathbf{N}_0$ be nonconstant. We define the numbers

$$\nu_0 = n(z_1), \quad \mu = \frac{\nu_0 - \nu_0^*}{z_1 - z_1^*}$$

and the linear functions

$$\alpha(z) = \nu_0 + \mu(z - z_1), \quad \beta(z) = \nu_0^* - \mu(z - z_1).$$

Then $\operatorname{Im} \nu_0 = \operatorname{Im} n(z_1) > 0$ (otherwise, by the maximum modulus principle, n would be a real constant), hence $\mu > 0$, and the functions α and $-\beta$ are Nevanlinnna functions, in particular, they are symmetric with respect to the real axis. The *Schur transform of n, centered at z_1*, is the function $\widehat{n}$ defined by

$$\widehat{n}(z) = \begin{cases} \dfrac{\beta(z)n(z) - |\nu_0|^2}{n(z) - \alpha(z)} & \text{if } n \text{ is not linear,} \\ \infty & \text{if } n \text{ is linear.} \end{cases} \tag{3.1}$$

The quotient in this formula makes sense only for $z \in \mathbb{C} \setminus (\mathbb{R} \cup \{z_1, z_1^*\})$, and it can be continued analytically at least into the points z_1, z_1^*; in fact it was shown in [3, Subsection 2 of Section 2] that, if n is not linear, the function $\widehat{n}$ belongs to $\mathbf{N}_0$ and is therefore holomorphic in $\mathbb{C} \setminus \mathbb{R}$. Note that if n is linear and not constant, that is, if $n(z) = az + b$, $z \in \mathbb{C}$, with $a > 0$, $b \in \mathbb{R}$, then $n(z) = \alpha(z)$, and the denominator in the quotient on the right-hand side of (3.1) vanishes. The map $n \mapsto \widehat{n}$ is called the *Schur transformation on $\mathbf{N}_0$ centered at z_1*. It maps the set of nonconstant Nevanlinna functions into the set $\mathbf{N}_0 \cup \{\infty\}$.

The definition (3.1) immediately implies the following:

$$n(z) = \rho \in \mathbb{C}^+ \text{ for } z \in \mathbb{C}^+ (\text{and hence } n(z) = \rho^* \text{ for } z \in \mathbb{C}^-) \Longleftrightarrow \widehat{n} = n. \tag{3.2}$$

The relation (3.1) between n and $\widehat{n}$ can be written as

$$n(z) = \frac{\alpha(z)\widehat{n}(z) - |\nu_0|^2}{\widehat{n}(z) - \beta(z)} = \mathcal{T}_{\Theta(z)}(\widehat{n}(z)) \tag{3.3}$$

with

$$\Theta(z) = \frac{1}{\mu(z - z_1^*)} \begin{pmatrix} \alpha(z) & -|\nu_0|^2 \\ 1 & -\beta(z) \end{pmatrix}.$$

The factor $1/\mu(z - z_1^*)$ in the definition of Θ has been chosen so that Θ belongs to $\mathcal{I}^{z_1}$ and is normalized by $\Theta(\infty) = I_2$. This matrix function Θ is of the form

$$\Theta(z) = I_2 + (\mathrm{b}(z) - 1)\frac{\mathbf{u}\mathbf{u}^* J}{\mathbf{u}^* J \mathbf{u}}, \tag{3.4}$$

where

$$\mathrm{b}(z) = \frac{z - z_1}{z - z_1^*}, \quad \mathbf{u} = \begin{pmatrix} \nu_0^* \\ 1 \end{pmatrix}, \quad J = \begin{pmatrix} 0 & 1 \\ -1 & 0 \end{pmatrix}.$$

According to [3, Theorem 6.2], a matrix function Θ of the form (3.4) is elementary in $\mathcal{I}^{z_1}$, that is, it cannot be written as a nontrivial product of two functions from $\mathcal{I}^{z_1}$, see [3], [4].

Note that, with z_1 being fixed apriori, the 2×2 matrix function Θ is built from just one complex number $\nu_0 \in \mathbb{C}^+$. Hence in (3.3) n is determined by $\widehat{n}$ and the complex number ν_0; the latter turns out to be the value of n at z_1.

If the function $\widehat{n}$ in (3.1) is not a constant, we can consider its Schur transform, and so on. That is, with a function $n \in \mathbf{N}_0$, which is not constant, we can associate a finite or infinite sequence of Nevanlinna functions (n_j) by repeatedly applying the Schur transformation:

$$n_0 := n,\ n_1 := \widehat{n}_0,\ n_2 := \widehat{n}_1, \dots,\ \text{in general}\ n_j := \widehat{n}_{j-1}.$$

The sequence (n_j) is finite and terminates after m steps, $m \geq 1$, if the function n_m is constant or ∞ since in this case n_{m+1} is not defined; otherwise, if no function n_j is constant or ∞, the sequence (n_j) is infinite. This repeated application of the Schur transformation is called the *Schur algorithm.*

With the sequence (n_j), the *sequence* (ρ_j) *of Schur parameters of the Nevanlinna function* n is defined by

$$\rho_j = n_j(z_1), \quad j = 0, 1, \dots, m \text{ or } j = 0, 1, \dots.$$

Clearly, $\rho_0 = \nu_0$, and in the first case, when the sequences (n_j) and (ρ_j) are finite,

(a) $\operatorname{Im} \rho_j > 0$ for $j = 0, 1, \dots, m-1$, $\rho_m \in \mathbb{R} \cup \{\infty\}$,

in the second case

(b) $\operatorname{Im} \rho_j > 0$ for $j = 0, 1, \dots$.

In the following, any finite complex sequence (ρ_j) with the property (a) or infinite complex sequence (ρ_j) with the property (b) is called a *Schur sequence.* Further, with any number $\rho \in \mathbb{C}^+$ we define the number $\mu_\rho = \dfrac{\rho - \rho^*}{z_1 - z_1^*}$, the functions

$$\alpha_\rho(z) = \rho + \mu_\rho(z - z_1), \quad \beta_\rho(z) = \rho^* - \mu_\rho(z - z_1),$$

and the 2×2 matrix function

$$\Theta_\rho(z) = \frac{1}{\mu_\rho(z - z_1^*)} \begin{pmatrix} \alpha_\rho(z) & -|\rho|^2 \\ 1 & -\beta_\rho(z) \end{pmatrix}; \tag{3.5}$$

if ρ is a term ρ_j of a Schur sequence we write Θ_j instead of Θ_{ρ_j} etc. Since

$$\Theta_\rho(z_1) = \frac{1}{\mu_\rho(z_1 - z_1^*)} \begin{pmatrix} \rho & -|\rho|^2 \\ 1 & -\rho^* \end{pmatrix},$$

we have for any $w \in \mathbb{C}$

$$\mathcal{T}_{\Theta_\rho(z_1)}(w)) = \rho. \tag{3.6}$$

3.2. In this subsection we show that any Schur sequence, that is, any complex sequence (ρ_j) with the properties (a) or (b), is the sequence of Schur parameters of a unique Nevanlinna function. To this end, for any Schur sequence (ρ_j) we define the functions $r^{[\ell]}$:

$$r^{[0]}(z) = \begin{cases} \rho_0, & z \in \mathbb{C}^+, \\ \rho_0^*, & z \in \mathbb{C}^-, \end{cases} \qquad r^{[\ell]}(z) = \begin{cases} \mathcal{T}_{\Theta_0(z)\Theta_1(z)\cdots\Theta_{\ell-1}(z)}(\rho_\ell), & z \in \mathbb{C}^+, \\ \mathcal{T}_{\Theta_0(z)\Theta_1(z)\cdots\Theta_{\ell-1}(z)}(\rho_\ell^*), & z \in \mathbb{C}^-, \end{cases} \tag{3.7}$$

with $\ell = 1, 2, \ldots, m$ in case (a), and $\ell = 1, 2, \ldots$ in case (b).

Proposition 3.1. *For a given Schur sequence (ρ_j), the functions $r^{[\ell]}$ are Nevanlinna functions with the following properties*:

(1) *If $\rho_\ell \in \mathbb{C}^+$, then the restriction $r^{[\ell]}|_{\mathbb{C}^+}$ admits an analytic continuation to the complex plane $\mathbb{C}$ which is a rational function with poles only in the lower half-plane and whose imaginary part on the real axis is positive and rational; the sequence of Schur parameters of the function $r^{[\ell]}$ is*

$$\rho_0, \rho_1, \ldots, \rho_{\ell-1}, \rho_\ell, \rho_\ell, \rho_\ell, \ldots. \tag{3.8}$$

(2) *If $\rho_m \in \mathbb{R} \cup \{\infty\}$, then $r^{[m]}$ is a rational Nevanlinna function and the sequence of Schur parameters of the function $r^{[m]}$ is*

$$\rho_0, \rho_1, \ldots, \rho_{m-1}, \rho_m. \tag{3.9}$$

Proof. Let ρ_j be a term of a Schur sequence with $j \neq m$, that is, ρ_j is not the last term of a finite Schur sequence and $\operatorname{Im} \rho_j > 0$. Then for any Nevanlinna function f the denominator of the quotient

$$r(z) := \mathcal{T}_{\Theta_j(z)}(f(z)) = \frac{\alpha_j(z) f(z) - |\rho_j|^2}{f(z) - \beta_j(z)} \tag{3.10}$$

does not vanish on $\mathbb{C}^+$ since $\operatorname{Im}(-\beta_j(z)) > 0$ and $\operatorname{Im} f(z) \geq 0$ for $z \in \mathbb{C}^+$. The matrix function Θ_j is $(-\mathrm{i}J)$-contractive in $\mathbb{C}^+$, therefore we have for $z \in \mathbb{C}^+$

$$\begin{aligned} \frac{r(z) - r(z)^*}{2\mathrm{i}} &= \frac{\begin{pmatrix} 1 & -r(z) \end{pmatrix} J \begin{pmatrix} 1 \\ -r(z)^* \end{pmatrix}}{2\mathrm{i}} \\ &\geq \frac{\begin{pmatrix} 1 & -r(z) \end{pmatrix} \Theta_j(z) J \Theta_j(z)^* \begin{pmatrix} 1 \\ -r(z)^* \end{pmatrix}}{2\mathrm{i}}. \end{aligned}$$

With the equality

$$\begin{pmatrix} 1 & -r(z) \end{pmatrix} \Theta_j(z) = \mu_j (z - z_1^*) \frac{\det \Theta_j(z)}{f(z) - \beta_j(z)} \begin{pmatrix} 1 & -f(z) \end{pmatrix}$$

we obtain for $z \in \mathbb{C}^+$

$$\begin{aligned} \frac{r(z) - r(z)^*}{2\mathrm{i}} &\geq \left|\frac{\mu_j(z - z_1^*)\det\Theta_j(z)}{f(z) - \beta_j(z)}\right|^2 \frac{\begin{pmatrix}1 & -f(z)\end{pmatrix} J \begin{pmatrix}1 \\ -f(z)^*\end{pmatrix}}{2\mathrm{i}} \\ &= \left|\frac{\mu_j(z - z_1^*)\det\Theta_j(z)}{f(z) - \beta_j(z)}\right|^2 \operatorname{Im} f(z) \geq 0. \end{aligned} \tag{3.11}$$

Since the functions α_j, β_j and f are symmetric with respect to the real axis, so is the function r. This shows that r is a Nevanlinna function, and hence all the functions $r^{[\ell]}$ are Nevanlinna functions.

If the function f in (3.10) can be extended by holomorphy from the upper half-plane $\mathbb{C}^+$ to a rational function $\widetilde{f}$ on $\mathbb{C}$ with poles only in the lower half-plane and $\operatorname{Im}\widetilde{f}(x) > 0$ for $x \in \mathbb{R}$, then also the function r can be extended to a rational function $\widetilde{r}$ on $\mathbb{C}$ with poles only in $\mathbb{C}^-$, and the relation (3.11), applied for real $z = x$, yields that $\operatorname{Im}\widetilde{r}(x) > 0$ for $x \in \mathbb{R}$. This applies in particular to $f(z) = \rho_\ell$, $z \in \mathbb{C}^+$, with $\operatorname{Im}\rho_\ell > 0$, and in this case all the functions

$$\mathcal{T}_{\Theta_j(z)\Theta_{j+1}(z)\cdots\Theta_{\ell-1}(z)}(\rho_\ell),\ j = \ell-1, \ell-2, \ldots, 0, \tag{3.12}$$

can be extended to rational functions on $\mathbb{C}$ with poles only in $\mathbb{C}^-$ and a positive imaginary part on $\mathbb{R}$. If, on the other hand, we start with a real ρ_ℓ or $\rho_\ell = \infty$, that is, $\rho_\ell = \rho_m$ is the last element of a finite Schur sequence, then it is easy to see that all the functions in (3.12) are real on the real axis and hence are rational Nevanlinna functions.

To prove the relations (3.8) and (3.9) about the Schur parameters of $r^{[\ell]}$ we denote the Schur transforms of $r^{[\ell]}$ by $r_1^{[\ell]}, r_2^{[\ell]}, \ldots$. The relation

$$r^{[\ell]}(z) = \mathcal{T}_{\Theta_0(z)\Theta_1(z)\cdots\Theta_{\ell-1}(z)}(\rho_\ell) = \mathcal{T}_{\Theta_0(z)}\left(\mathcal{T}_{\Theta_1(z)\cdots\Theta_{\ell-1}(z)}(\rho_\ell)\right), \quad z \in \mathbb{C}^+,$$

and (3.6) imply $r^{[\ell]}(z_1) = \rho_0$. Further, the Schur transform $r_1^{[\ell]}$ of $r^{[\ell]}$ is obtained by applying the transformation $\mathcal{T}_{\Theta_0^{-1}}$ to the latter, that is,

$$r_1^{[\ell]}(z) = \mathcal{T}_{\Theta_1(z)}(\mathcal{T}_{\Theta_2(z)\cdots\Theta_{\ell-1}(z)}(\rho_\ell)), \quad r_1^{[\ell]}(z_1) = \rho_1,$$

etc. That in case (b) the sequence of Schur parameters becomes stationary follows from (3.2). □

Recall that an arbitrary Nevanlinna function f admits an integral representation

$$f(z) = az + b + \int_{\mathbb{R}} \left(\frac{1}{t - z} - \frac{t}{t^2 + 1}\right) d\sigma(t), \tag{3.13}$$

with $a \geq 0$, b real, and a nondecreasing function σ on $\mathbb{R}$ such that $\displaystyle\int_{\mathbb{R}} \frac{d\sigma(t)}{1 + t^2} < \infty$. The function σ is sometimes called the *spectral function of the Nevanlinna function f*. In part (1) of Proposition 3.1 the spectral function σ of the Nevanlinna function $r^{[\ell]}$ is absolutely continuous; if the analytic continuation of $r^{[\ell]}\big|_{\mathbb{C}^+}$ to the real axis is also denoted by $r^{[\ell]}$, then the density σ' becomes $\sigma'(x) = \frac{1}{\pi}\mathrm{Im}\, r^{[\ell]}(x)$, $x \in \mathbb{R}$, hence it is a rational and nonnegative function on $\mathbb{R}$.

For a rational Nevanlinna function f the representation (3.13) becomes

$$f(z) = az + b + \sum_{j=1}^{k} \frac{\sigma_j}{\lambda_j - z}, \tag{3.14}$$

with pairwise different real numbers λ_j and positive numbers σ_j, $j = 1, 2, \ldots, k$; the degree of this function f equals

$$\deg f = \begin{cases} k & \text{if } a = 0, \\ k+1 & \text{if } a > 0. \end{cases}$$

This degree of f also coincides with the dimension of the Hilbert space $\mathcal{H}(f)$ in any minimal self-adjoint realization of f, for the latter see [5]; here $\mathcal{H}(f)$ can also be replaced by the reproducing kernel Hilbert space $\mathcal{L}(f)$ generated by the Nevanlinna kernel L_f from (1.1).

In the following we show that any Schur sequence, that is, any sequence (ρ_j) with property (a) or property (b), is the sequence of Schur parameters of a unique Nevanlinna function n, and we give a procedure for the construction of n. We start with a finite Schur sequence $(\rho_j)_0^m$.

Theorem 3.2. *The sequence of Schur parameters of a rational Nevanlinna function n of degree m is a finite Schur sequence $(\rho_j)_0^m$ (as in case (a)); conversely, each finite Schur sequence $(\rho_j)_0^m$ is the sequence of Schur parameters of a unique Nevanlinna function n, which is rational and of degree m. This function n can be written as*

$$n(z) = \mathcal{T}_{\Theta_0(z)\Theta_1(z)\cdots\Theta_{m-1}(z)}(\rho_m) = r^{[m]}(z), \quad z \in \mathbb{C}^+, \tag{3.15}$$

where Θ_j is the matrix function Θ_{ρ_j} from (3.5), $j = 0, 1, \ldots, m-1$. Furthermore, n is holomorphic at ∞ (that is, $a = 0$ in (3.13)) if $\rho_m \in \mathbb{R}$, and n has a (simple) pole at ∞ if $\rho_m = \infty$; in the latter case all the Schur transforms n_j of n, $j = 1, 2, \ldots, m-1$, have a pole at ∞.

Proof. From the definition of the Schur transform it follows that for a rational Nevanlinna function n, which is not linear, the Schur transform $\widehat{n}$ is again a rational Nevanlinna function. Furthermore, according to [5, Theorem 4.2], $\dim \mathcal{H}(\widehat{n}) = \dim \mathcal{H}(n) - 1$, that is, which each step in the Schur algorithm the degree of the function decreases by one. Consequently, if we start with a rational Nevanlinna function of degree m after $m-1$ steps we get a function n_{m-1} which is either linear or of the form $n_{m-1}(z) = \frac{\sigma_0}{\lambda_0 - z}$ with $\sigma_0 > 0$, λ_0 real. In the first case we have $n_m = \infty$, in the second case n_m is a real constant.

If a finite Schur sequence $(\rho_j)_0^m$ is given, according to Proposition 3.1 (2) it is the sequence of Schur parameters of the function n in (3.15), and any function n with this sequence as sequence of Schur parameters is of this form.

Now consider a rational Nevanlinna function n which has a linear term as in (3.14) with $a > 0$ and which is not linear (that is, $k \geq 1$ in (3.14)). Then also its

Schur transform $\widehat{n}$ has a linear term $\widehat{a}z + \widehat{b}$ with $\widehat{a} > 0$; this follows easily from the definition of $\widehat{n}$ in (3.1). Therefore the second last step of the Schur algorithm gives a linear nonconstant function. □

Remark 3.3. If a Nevanlinna function n in its integral representation (3.13) has a nonconstant linear term ($a > 0$), then all the Nevanlinna functions n_j in the Schur algorithm for n have also an integral representation with a nonconstant linear term, except for the last one when the algorithm is finite. This can be proved as above where the function n is rational. It can also be seen from the minimal self-adjoint realization of n in the Hilbert space $\mathcal{H}(n)$:

$$n(z) = n(z_1)^* + (z - z_1^*)\big\langle\big(I + (z - z_1)(A - z)^{-1}\big)u, u\big\rangle_{\mathcal{H}(n)},$$

where $u \in \mathcal{H}(n)$, and A is a self-adjoint relation in $\mathcal{H}(n)$; recall that A has a nontrivial multi-valued part if and only if n has a linear term, that is, in the representation (3.13) of n we have $a > 0$. If A has a nontrivial multi-valued part, then the self-adjoint relation in the realization of any n_j also has a nontrivial multi-valued part, except for the last one when the algorithm is finite, see [5, Corollary 4.3].

The formula (3.15), written in the form

$$n(z) = \mathcal{T}_{\Theta_0(z)}(\mathcal{T}_{\Theta_1(z)}(\cdots(\mathcal{T}_{\Theta_{m-1}(z)}(\rho_m))\cdots)),$$

can be used to determine the rational Nevanlinna function n from its Schur parameters $\rho_0, \rho_1, \ldots, \rho_m$: with ρ_{m-1} we calculate $\mu_{m-1} = \dfrac{\rho_{m-1} - \rho_{m-1}^*}{z_1 - z_1^*}$ (> 0),

$$\alpha_{m-1}(z) = \rho_{m-1} + \mu_{m-1}(z - z_1), \quad \beta_{m-1}(z) = \rho_{m-1}^* - \mu_{m-1}(z - z_1),$$

$$\Theta_{m-1}(z) = \frac{1}{\mu_{m-1}(z - z_1^*)}\begin{pmatrix} \alpha_{m-1}(z) & -|\rho_{m-1}|^2 \\ 1 & -\beta_{m-1}(z) \end{pmatrix},$$

and

$$\mathcal{T}_{\Theta_{m-1}(z)}(\rho_m) = \begin{cases} \alpha_{m-1}(z) & \text{if } \rho_m = \infty, \\ \dfrac{\alpha_{m-1}(z)\rho_m - |\rho_{m-1}|^2}{\rho_m - \beta_{m-1}(z)} & \text{if } \rho_m \in \mathbf{R}. \end{cases}$$

Now we apply successively the transformations $\mathcal{T}_{\Theta_{m-2}(z)}, \ldots, \mathcal{T}_{\Theta_0(z)}$, which are determined by $\rho_{m-2}, \ldots, \rho_0$, respectively, and obtain $n(z)$. This finite sequence of functions consists of rational Nevanlinna functions of finite degree which increases by one with each step.

If the Nevanlinna function n is not rational, in the following subsection with the infinite sequence of Schur parameters of n we define a sequence of Nevanlinna functions which converges to n, locally uniformly on $\mathbb{C} \setminus \mathbb{R}$, that is, uniformly on compact subsets of $\mathbb{C} \setminus \mathbb{R}$.

3.3 We start with a lemma about the convergence of Nevanlinna functions.

Lemma 3.4.

(i) *Let* $n \in \mathbf{N}_0$ *and let* (f_j) *be a sequence in* $\mathbf{N}_0$ *such that the difference* $n - f_j$, $j = 1, 2, \ldots$, *has a zero of order* τ_j *at* z_1. *If* $\lim_{j\to\infty} \tau_j = \infty$, *then* $\lim_{j\to\infty} f_j = n$, *locally uniformly on* $\mathbb{C} \setminus \mathbb{R}$.

(ii) *Let* (f_j) *be a sequence in* $\mathbf{N}_0$ *such that the difference* $f_j - f_k$, $j, k = 1, 2, \ldots$, *has a zero of order* $\tau_{j,k}$ *at* z_1. *If* $\lim_{j,k\to\infty} \tau_{j,k} = \infty$, *then there exists a function* $n \in \mathbf{N}_0$ *such that* $\lim_{j\to\infty} f_j = n$, *locally uniformly on* $\mathbb{C} \setminus \mathbb{R}$.

Proof. A set of Nevanlinna functions, which at an interior point of $\mathbb{C}^+$ are uniformly bounded, is a normal family, see [1, Exercise 1, p 227]. By assumption, if (f_{j_ν}) is any subsequence which converges locally uniformly in $\mathbb{C}^+$, the limit function has the same value and derivatives at z_1 as the given function n, therefore it must coincide with n. Thus, every convergent subsequence of (f_j) has the same limit which means that the sequence (f_j) itself is convergent. The proof of the second statement is analogous. □

Theorem 3.5.

(1) *Assume* $n \in \mathbf{N}_0$ *is not a rational function, denote by* (ρ_j) *the (infinite) sequence of Schur parameters of* n, *and let* $r^{[\ell]}$, $\ell = 1, 2, \ldots$, *be the sequence of Nevanlinna functions as defined in* (3.7). *Then*

$$n = \lim_{\ell\to\infty} r^{[\ell]}, \quad \textit{locally uniformly on } \mathbb{C} \setminus \mathbb{R}. \tag{3.16}$$

(2) *Given any infinite Schur sequence* (ρ_j), *that is,* $\rho_j \in \mathbb{C}^+$, $j = 0, 1, 2, \ldots$. *Then there exists a unique function* $n \in \mathbf{N}_0$ *such that* (ρ_j) *is the sequence of Schur parameters of* n; *this function* n *is given by* (3.16) *with* $r^{[\ell]}$ *as defined in* (3.7).

Proof. (1) We claim that $n - r^{[\ell]}$ has a zero at z_1 of order $\geq \ell + 1$, $\ell = 1, 2, \ldots$. With

$$\Theta = \Theta_0 \Theta_1 \cdots \Theta_{\ell-1} = \begin{pmatrix} a & b \\ c & d \end{pmatrix}$$

we have

$$n = \mathcal{T}_\Theta(n_\ell), \quad r^{[\ell]} = \mathcal{T}_\Theta(\rho_\ell),$$

and

$$n - r^{[\ell]} = \frac{(n_\ell - \rho_\ell)\det\Theta}{(c\rho_\ell + d)(cn_\ell + d)}. \tag{3.17}$$

The claim follows from this equality, since $\det\Theta = \mathrm{b}^\ell$ (see (2.6)) has a zero at z_1 of order ℓ, the function $n_\ell - \rho_\ell$ has a zero at z_1 of order ≥ 1, and both functions $cn_\ell + d$ and $c\rho_\ell + d$ in the denominator do not vanish in z_1. To see the latter, observe that, according to Lemma 2.1, $c(z_1) \neq 0$ and d/c is a Nevanlinna function. Thus

$$c(z_1)\rho_\ell + d(z_1) = c(z_1)\left(\rho_\ell + \frac{d(z_1)}{c(z_1)}\right) \neq 0$$

since $\operatorname{Im}\rho_\ell > 0$, $\operatorname{Im}\frac{d(z_1)}{c(z_1)} \geq 0$; for $c(z_1)n_\ell(z) + d(z_1)$ this follows in the same way since $\operatorname{Im} n_\ell(z_1) > 0$. Part (1) now follows from the claim and Lemma 3.4 (i).

(2) Instead of (3.17) we consider now the relation

$$r^{[\ell]} - r^{[\ell+k]} = \mathcal{T}_{\Theta_0\cdots\Theta_{\ell-1}}(\rho_\ell) - \mathcal{T}_{\Theta_0\cdots\Theta_{\ell+k-1}}(\rho_{\ell+k})$$
$$= \mathcal{T}_{\Theta_0\cdots\Theta_{\ell-1}}(\rho_\ell) - \mathcal{T}_{\Theta_0\cdots\Theta_{\ell-1}}\big(r_\ell^{[\ell+k]}\big) = \frac{\big(\rho_\ell - r_\ell^{[\ell+k]}\big)\det\Theta}{(c\rho_\ell + d)\big(cr_\ell^{[\ell+k]} + d\big)},$$

where

$$r_\ell^{[\ell+k]} = \mathcal{T}_{\Theta_\ell\cdots\Theta_{\ell+k-1}}(\rho_{\ell+k}).$$

As in part (1) of this proof it can be shown that the zero at z_1 of the difference $r^{[\ell]} - r^{[\ell+k]}$ has order at least ℓ and hence by Lemma 3.4 (ii), there is a Nevanlinna function n which is the limit of the sequence $\big(r^{[\ell]}\big)$. The sequence of Schur parameters for n coincides with the given Schur sequence, see (3.8). To see uniqueness, if $\widetilde{n}$ is any function with the given Schur sequence (ρ_j) then according to part (1) of the theorem we have also $\widetilde{n} = \lim_{\ell\to\infty} r^{[\ell]}$, and hence $\widetilde{n} = n$. □

Remark 3.6. In (3.16) the definition of $r^{[\ell]}$ from (3.7) can be replaced by $r^{[0]} = f_0$ and

$$r^{[\ell]} = \mathcal{T}_{\Theta_0\Theta_1\cdots\Theta_{\ell-1}}(f_\ell), \quad \ell = 1, 2, \ldots,$$

where the f_ℓ are arbitrary Nevanlinna functions. If we choose each f_ℓ a real constant then the $r^{[\ell]}$ are rational Nevanlinna functions which converge locally uniformly on $\mathbb{C}\setminus\mathbb{R}$ to n.

As a consequence of Theorems 3.2 and 3.5 we have:

Corollary 3.7. *The Schur algorithm establishes a one-to-one correspondence between the class of nonconstant Nevanlinna functions and the class of Schur sequences.*

As an illustration we consider a linear Nevanlinna function $n(z) = az + b$ with $a > 0$, b real. Then the Schur sequence consists of the two terms $n(z_1), \infty$, and the real numbers a, b are determined by the complex value $n(z_1)$.

3.4. In this subsection, for a nonconstant Nevanlinna function besides the sequence (ρ_j) of its Schur parameters we consider also the sequence (ν_j) of its Taylor coefficients at z_1, that is, the coefficients of the expansion

$$n(z) = \sum_{j=0}^{\infty} \nu_j (z - z_1)^j. \tag{3.18}$$

We shall show that, for any integer $k \geq 1$, the first k Schur parameters are determined by the first k Taylor coefficients.

We first recall some notation from [4]. If $n \in \mathbf{N}_0$ the Nevanlinna kernel $L_n(z,w)$ is holomorphic in z and in w^* at $z = w = z_1$ and hence has a Taylor expansion of the form:

$$L_n(z,w) = \frac{n(z) - n(w)^*}{z - w^*} = \sum_{i,j=0}^{\infty} \gamma_{ij}(z - z_1)^i (w - z_1)^{*j}.$$

The coefficient matrix $\Gamma = \left(\gamma_{ij}\right)_{i,j=0}^{\infty}$ is called the *Pick matrix for n at z_1*. It can be expressed in terms of the Taylor coefficients ν_j of n as follows. Set

$$\Sigma = \begin{pmatrix} \nu_0 & 0 & 0 & 0 & \cdots \\ \nu_1 & \nu_0 & 0 & 0 & \cdots \\ \nu_2 & \nu_1 & \nu_0 & 0 & \cdots \\ \nu_3 & \nu_2 & \nu_1 & \nu_0 & \cdots \\ \vdots & \vdots & \vdots & \vdots & \ddots \end{pmatrix}$$

and define the matrix $\Gamma^0 = \left(\gamma_{ij}^0\right)_{i,j=0}^{\infty}$ by

$$\gamma_{ij}^0 = \frac{\partial^{i+j}}{\partial z^i \partial w^{*j}} \frac{2\,\mathrm{i}}{z - w^*}\bigg|_{z=w=z_1} = \binom{i+j}{i} \frac{2\,\mathrm{i}\,(-1)^i}{(z_1 - z_1^*)^{i+j+1}}, \qquad i,j = 0,1,\ldots,$$

then

$$\Gamma = \frac{1}{2\,\mathrm{i}} \left(\Sigma\,\Gamma^0 - \Gamma^0 \Sigma^*\right). \tag{3.19}$$

If we denote by A_j the principal $j \times j$ submatrix of a square matrix A, then it follows from (3.19) that

$$\Gamma_j = \frac{1}{2\,\mathrm{i}} \left(\Sigma_j\,\Gamma_j^0 - \Gamma_j^0 \Sigma_j^*\right).$$

To formulate the next theorem we introduce vectors $\mathbf{b}_j$ through the relation

$$\Gamma_{j+1} = \begin{pmatrix} \Gamma_j & \mathbf{b}_j \\ \mathbf{b}_j^* & \gamma_{j+1,j+1} \end{pmatrix},$$

and denote by $\mathbf{e}_j$ and $\mathbf{v}_j$ the $j \times 1$ vectors

$$\mathbf{e}_j = \begin{pmatrix} 1 & 0 & \cdots & 0 \end{pmatrix}^\top, \quad \mathbf{v}_j = \begin{pmatrix} \nu_0 & \nu_1 & \cdots & \nu_{j-1} \end{pmatrix}^\top.$$

Theorem 3.8. *Let the nonconstant Nevanlinna function n have the Taylor expansion (3.18) with the sequence (ν_j) of Taylor coefficients, and let (ρ_j) be the sequence of Schur parameters of n. Then $\rho_0 = \nu_0$ and*

$$\rho_j = \frac{\nu_j - \mathbf{b}_j^* \Gamma_j^{-1} \mathbf{v}_j}{\mathbf{b}_j^* \Gamma_j^{-1} \mathbf{e}_j}, \qquad j = 1,2,\ldots. \tag{3.20}$$

Proof. Let $\mathcal{M}$ be the span in $\mathbf{H}_{2,J}$ of the functions $\mathbf{f}_0, \mathbf{f}_1, \dots$ in (2.15). Then, according to [3, Theorem 4.1] and its proof, the map

$$\mathbf{f} \mapsto \sqrt{2\pi} \begin{pmatrix} 1 & -n \end{pmatrix} \mathbf{f} \tag{3.21}$$

is an isometry from $\mathcal{M}$ into the reproducing kernel Hilbert space $\mathcal{L}_+(n)$ with kernel $L_n(z,w)$ and z, w restricted to $\mathbb{C}^+$. For $j = 0, 1, \dots$, the matrix $2\pi\Gamma_j$ is the Gram matrix associated with the functions $\mathbf{f}_0, \dots, \mathbf{f}_{j-1}$ (see the chain of equalities on [3, p. 690]), and hence $\Gamma_j \geq 0$. Let

$$\mathcal{M}_j = \operatorname{span}\ \{\mathbf{f}_0, \dots, \mathbf{f}_{j-1}\}.$$

If $\Gamma_1 = \mu_0 = 0$, then ρ_0 is real, and $n \equiv \rho_0$, which is excluded. Assume now that $\Gamma_j > 0$ for some $j \geq 1$; then also $\Gamma_0, \Gamma_1, \dots, \Gamma_{j-1} > 0$. By Lemma 2.2 with $\Theta = \Theta_0 \cdots \Theta_{j-1}$ and $s = j$ we have

$$\mathcal{M}_j = \mathcal{P}(\Theta).$$

Further, [7, Theorem 3.1] implies that

$$\mathcal{M}_{j+1} = \mathcal{M}_j \oplus (\Theta_0 \cdots \Theta_{j-1})\mathcal{N}_j \tag{3.22}$$

with $\mathcal{N}_j$ spanned by a nonzero vector function of the form

$$\frac{\begin{pmatrix} \xi_j \\ \eta_j \end{pmatrix}}{z - z_1^*},$$

where ξ_j and η_j satisfy

$$\begin{pmatrix} \xi_j \\ \eta_j \end{pmatrix}^* (-\mathrm{i}J) \begin{pmatrix} \xi_j \\ \eta_j \end{pmatrix} \geq 0,$$

and are chosen such that

$$\mathbf{f}_j(z) = \begin{pmatrix} \mathbf{f}_0(z) & \cdots & \mathbf{f}_{j-1}(z) \end{pmatrix} \Gamma_j^{-1}\mathbf{b}_j + \Theta_0(z)\cdots\Theta_{j-1}(z)\frac{\begin{pmatrix} \xi_j \\ \eta_j \end{pmatrix}}{z - z_1^*}.$$

Multiplying both sides of this equality by z, letting z go to infinity, and taking into account the normalization of $\Theta_j(z)$ at infinity we get

$$\begin{pmatrix} \nu_j^* \\ 0 \end{pmatrix} = \begin{pmatrix} \mathbf{v}_j^* \\ \mathbf{e}_j^* \end{pmatrix} \Gamma_j^{-1}\mathbf{b}_j + \begin{pmatrix} \xi_j \\ \eta_j \end{pmatrix},$$

hence

$$\xi_j = \nu_j^* - \mathbf{v}_j^*\Gamma_j^{-1}\mathbf{b}_j, \quad \eta_j = -\mathbf{e}_j^*\Gamma_j^{-1}\mathbf{b}_j.$$

It remains to establish the connection between ξ_j, η_j and ρ_j. We claim that if $\eta_j \neq 0$, then $\rho_j = (\xi_j/\eta_j)^*$, which implies the formula in the theorem; $\eta_j = 0$ corresponds to $\rho_j = \infty$.

First consider the case where $\mathcal{M}_{j+1}$ is a strictly positive subspace. By (3.22) the space $\mathcal{N}_j$ is a Hilbert space, hence $\eta_j \neq 0$. Then, by for example [6, Theorem 6.9], there exists a normalized matrix function Ψ in $\mathcal{I}^{z_1}$ such that $\mathcal{M}_{j+1}$ coincides with the reproducing kernel Hilbert space $\mathcal{P}(\Psi)$ with kernel K_Ψ. Since $\mathcal{M}_j \subset$

$\mathcal{M}_{j+1}$ there exists a Θ_j of the form (3.5) such that $\Psi = \Theta\Theta_j = \Theta_0 \cdots \Theta_{j-1}\Theta_j$. Moreover, again by (3.22), $\mathcal{P}(\Theta_j)$ is spanned by the function

$$\frac{\mathbf{u}}{z - z_1^*}, \quad \mathbf{u} = \begin{pmatrix} \xi_j/\eta_j \\ 1 \end{pmatrix},$$

that is,

$$\Theta_j(z) = I_2 + \big(\mathrm{b}(z) - 1\big)\frac{\mathbf{u}\mathbf{u}^* J}{\mathbf{u}^* J\mathbf{u}}.$$

From the relation

$$L_n(z,w) = 2\pi \begin{pmatrix} 1 & -n(z) \end{pmatrix} K_\Psi(z,w) \begin{pmatrix} 1 \\ -n(w)^* \end{pmatrix} + R_\Psi(z,w)$$

with

$$R_\Psi(z,w) = \begin{pmatrix} 1 & -n(z) \end{pmatrix} \frac{\Psi(z) J \Psi(w)^*}{z - w^*} \begin{pmatrix} 1 \\ -n(w)^* \end{pmatrix}$$

and the fact that the map (3.21) is an isometry, it follows that $\mathcal{P}(\Psi)$ is isometrically included in the Hilbert space $\mathcal{L}(n)$, and hence the kernel R_Ψ is positive definite. Taking $z = w$ and writing $\Psi = (\psi_{ij})$ we obtain

$$\mathrm{i}\left((\psi_{11}^* - n^*\psi_{21}^*)(\psi_{12} - n\psi_{22}) - (\psi_{12}^* - n^*\psi_{22}^*)(\psi_{11} - n\psi_{21})\right) \geq 0. \tag{3.23}$$

If $n = \psi_{11}/\psi_{21}$ then

$$\mathcal{T}_{\Theta_0(z)\cdots\Theta_{j-1}(z)}(n_j(z)) = n(z) = \mathcal{T}_{\Psi(z)}(\infty)$$

and hence $n_j = \mathcal{T}_{\Theta_j}(\infty)$. It follows that $\rho_j = n_j(z_1) = \mathcal{T}_{\Theta_j(z_1)}(\infty) = (\xi_j/\eta_j)^*$. Now assume that $n \neq \psi_{11}/\psi_{21}$. Then (3.23) implies that

$$w = \frac{\psi_{21} - n\psi_{22}}{\psi_{11} - n\psi_{21}}$$

is a Nevanlinna function. Hence $n = \mathcal{T}_\Psi(w)$ and as above we find $\rho_j = (\xi_j/\eta_j)^*$.

Now assume that $\mathcal{M}_{j+1}$ is degenerate. Then the function

$$\Theta_0(z) \cdots \Theta_{j-1}(z) \frac{\begin{pmatrix} \xi_j \\ \eta_j \end{pmatrix}}{z - z_1^*}$$

is neutral in $\mathcal{M}_{j+1}$, and hence belongs to the kernel of the mapping (3.21):

$$\begin{pmatrix} 1 & -n(z) \end{pmatrix} \Theta_0(z) \cdots \Theta_{j-1}(z) \frac{\begin{pmatrix} \xi_j \\ \eta_j \end{pmatrix}}{z - z_1^*} \equiv 0.$$

Thus, writing

$$\Theta_0 \cdots \Theta_{j-1} = \begin{pmatrix} a & b \\ c & d \end{pmatrix},$$

we obtain

$$(a - nc)\,\xi_j = -(b - nd)\,\eta_j \tag{3.24}$$

If $\eta_j = 0$, then $\xi_j \neq 0$, and we have $n = a/c$; thus

$$\mathcal{T}_{\Theta_0\cdots\Theta_{j-1}}(n_j) = n = \mathcal{T}_{\Theta_0\cdots\Theta_{j-1}}(\infty),$$

and hence $\rho_j = \infty$. If $\eta_j \neq 0$, we have that $\xi_j/\eta_j \in \mathbb{R}$ and (3.24) implies that

$$\mathcal{T}_{\Theta_0\cdots\Theta_{j-1}}(n_j) = n = \mathcal{T}_{\Theta_0\cdots\Theta_{j-1}}(\xi_j/\eta_j),$$

which shows that $\rho_j = \xi_j/\eta_j$. □

4. Schur algorithm and approximation for generalized Nevanlinna functions

4.1 In this section we formulate and prove analogs of the theorems in the previous section in an indefinite setting. Recall that $\mathbf{N}$ is the set of all generalized Nevanlinna functions, z_1 is a fixed point in $\mathbb{C}^+$, and $\mathbf{N}^{z_1}$ is the set of all $n \in \mathbf{N}$ which are holomorphic at z_1. We write the Taylor expansion of the function $n \in \mathbf{N}^{z_1}$ as

$$n(z) = \sum_{j=0}^{\infty} \nu_j (z - z_1)^j, \tag{4.1}$$

in particular $\nu_0 = n(z_1)$.

If $n \in \mathbf{N}$ is not constant, we define its *Schur transform* $\widehat{n}$ *centered at* z_1 as follows.

(i) Assume $n \in \mathbf{N}^{z_1}$ and $\operatorname{Im} \nu_0 \neq 0$. Then we define

$$\widehat{n}(z) = \begin{cases} \dfrac{\beta(z)n(z) - |\nu_0|^2}{n(z) - \alpha(z)} & \text{if } n \text{ is not linear,} \\ \infty & \text{if } n \text{ is linear,} \end{cases}$$

where α and β are the same as in (3.1):

$$\alpha(z) = \nu_0 + \mu(z - z_1), \quad \beta(z) = \nu_0^* - \mu(z - z_1), \quad \mu = \frac{\nu_0 - \nu_0^*}{z_1 - z_1^*}.$$

(ii) Assume $n \in \mathbf{N}^{z_1}$ and $\operatorname{Im} \nu_0 = 0$. Then, since n is not constant, the function $1/(n - \nu_0)$ has poles at z_1 and z_1^* and can be written as

$$\frac{1}{n - \nu_0} = H_{z_1} + H_{z_1^*} + g, \tag{4.2}$$

where H_{z_1}, $H_{z_1^*}$ are the principal parts of the Laurent expansions of $1/(n-\nu_0)$ at z_1 and z_1^*, and g is a function which is holomorphic at z_1 and z_1^*. Since $n(z)^* = n(z^*)$, the orders of the poles at z_1 and z_1^* are the same and equal to the smallest integer $k \geq 1$ such that $\nu_k \neq 0$ (see (4.1)), and $H_{z_1^*}(z) = H_{z_1}(z^*)^*$. Thus

$$H_{z_1}(z) + H_{z_1^*}(z) = \frac{p(z)}{(z - z_1)^k (z - z_1^*)^k}, \tag{4.3}$$

where p is a real polynomial: $p(z)^* = p(z^*)$, of degree $\leq 2k-1$, and such that $p(z_1) \neq 0$. Now we define

$$\widehat{n}(z) = \begin{cases} \dfrac{\widetilde{\beta}(z)n(z) - \nu_0^2}{n(z) - \widetilde{\alpha}(z)} & \text{if } g \neq 0 \text{ in (4.2)}, \\ \infty & \text{if } g = 0 \text{ in (4.2)}, \end{cases}$$

where

$$\widetilde{\alpha}(z) = \nu_0 + \frac{(z-z_1)^k(z-z_1^*)^k}{p(z)}, \quad \widetilde{\beta}(z) = \nu_0 - \frac{(z-z_1)^k(z-z_1^*)^k}{p(z)}.$$

(iii) If n has a pole at z_1 then we define

$$\widehat{n} = n - h_{z_1} - h_{z_1^*},$$

where h_{z_1} and $h_{z_1^*}$ are the principal parts of the Laurent expansions of n at z_1 and z_1^*. As in case (ii), if k is the order of the pole at z_1, we have

$$h_{z_1}(z) + h_{z_1^*}(z) = \frac{p(z)}{(z-z_1)^k(z-z_1^*)^k}, \tag{4.4}$$

where p is a real polynomial of degree $\leq 2k-1$ such that $p(z_1) \neq 0$.

That the Schur transform is again a generalized Nevanlinna function follows from the next theorem, see [3, Theorem 7.3].

Theorem 4.1. *Let $n \in \mathbf{N}$ and assume it is not constant. For its Schur transform the following holds in the cases* (i), (ii), *and* (iii) *of the definition above.*

(i) $n \in \mathbf{N}_\kappa^{z_1} \Longrightarrow \widehat{n} \in \mathbf{N}_{\widehat{\kappa}}$ *with* $\widehat{\kappa} = \kappa$ *if* $\operatorname{Im} \nu_0 > 0$ *and* $\widehat{\kappa} = \kappa - 1$ *if* $\operatorname{Im} \nu_0 < 0$.
(ii) $n \in \mathbf{N}_\kappa^{z_1}$ *and* $n - \nu_0$ *has a zero at* z_1 *of order* $k \geq 1 \Longrightarrow k \leq \kappa$ *and* $\widehat{n} \in \mathbf{N}_{\kappa-k}$.
(iii) $n \in \mathbf{N}_\kappa$ *and* n *has a pole at* z_1 *of order* $k \geq 1 \Longrightarrow k \leq \kappa$ *and* $\widehat{n} \in \mathbf{N}_{\kappa-k}^{z_1}$.

The relation between n and its Schur transform $\widehat{n}$ can also be expressed via the inverse formula

$$n = \mathcal{T}_\Theta(\widehat{n}),$$

where with

$$J = \begin{pmatrix} 0 & 1 \\ -1 & 0 \end{pmatrix}, \quad \mathrm{b}(z) = \frac{z - z_1}{z - z_1^*},$$

Θ has the following form:

In case (i):

$$\begin{aligned} \Theta_{(\mathrm{i})}(z) &= \frac{1}{\mu(z-z_1^*)} \begin{pmatrix} \alpha(z) & -|\nu_0|^2 \\ 1 & -\beta(z) \end{pmatrix} \\ &= I_2 + (\mathrm{b}(z) - 1)\frac{\mathbf{u}\mathbf{u}^* J}{\mathbf{u}^* J \mathbf{u}}, \quad \mathbf{u} = \begin{pmatrix} \nu_0^* \\ 1 \end{pmatrix}, \end{aligned} \tag{4.5}$$

in case (ii):

$$\begin{aligned}\Theta_{(\mathrm{ii})}(z) &= \frac{p(z)}{(z-z_1^*)^{2k}}\begin{pmatrix}\widetilde{\alpha}(z) & -\nu_0^2\\ 1 & -\widetilde{\beta}(z)\end{pmatrix}\\ &= \mathrm{b}(z)^k I_2 - \frac{p(z)}{(z-z_1^*)^{2k}}\,\mathbf{u}\mathbf{u}^* J, \quad \mathbf{u} = \begin{pmatrix}\nu_0\\ 1\end{pmatrix},\end{aligned} \tag{4.6}$$

and in case (iii):

$$\begin{aligned}\Theta_{(\mathrm{iii})}(z) &= \mathrm{b}(z)^k\begin{pmatrix}1 & \dfrac{p(z)}{(z-z_1)^k(z-z_1^*)^k}\\ 0 & 1\end{pmatrix}\\ &= \mathrm{b}(z)^k I_2 + \frac{p(z)}{(z-z_1^*)^{2k}}\,\mathbf{e}\mathbf{e}^* J, \quad \mathbf{e} = \begin{pmatrix}1\\ 0\end{pmatrix}.\end{aligned} \tag{4.7}$$

In cases (i) and (ii) the Schur transform $\widehat{n}$ of n may have a pole at z_1, in case (iii) it is holomorphic at z_1. Criteria for the existence of a pole in cases (i) and (ii) and further details are given in [3] and [5]. Below, if a Nevanlinna function n has a pole at z_1, we set $n(z_1) = \infty$.

The *Schur algorithm* in the indefinite setting is a repeated application of the Schur transformation starting from the function $n \in \mathbf{N}$ which is not equal to a real constant:

$$n_0 := n,\ n_1 := \widehat{n}_0, \dots, n_j := \widehat{n}_{j-1}, \dots.$$

The sequence is finite and stops with n_m if and only if $n_m = r$ with $r \in \mathbb{R} \cup \{\infty\}$, for then the Schur transform of n_m is not defined. The following theorem holds, see [5, Theorem 7.1].

Theorem 4.2. *If $n \in \mathbf{N}$ is not constant and the Schur algorithm applied to n gives an infinite sequence (n_j), then there exists an index j_0 such that $n_j \in \mathbf{N}_0$ for all $j \geq j_0$.*

As in Section 3 with the finite or infinite sequence (n_j) of Nevanlinna functions we associate the finite or infinite sequence (ρ_j) with:

$$\rho_j = n_j(z_1), \quad j = 0, 1, \dots, m \text{ or } j = 0, 1, \dots.$$

These numbers are again called the *Schur parameters of n at z_1*. They are numbers from $\mathbb{C} \cup \{\infty\}$, and if the sequence (ρ_j) is finite and stops with ρ_m, then $\rho_m \in \mathbb{R} \cup \{\infty\}$. The Schur sequence has the property that

$$\rho_j \neq \rho_{j-1}^*, \quad j = 1, 2, \dots, \tag{4.8}$$

where $\infty^* = \infty$, $j = 1, 2, \dots, m$ if the sequence is finite and ends with ρ_m, and $j = 1, 2, \dots$ if the sequence is infinite. Indeed, for $\rho_{j-1} \neq \infty$ this follows from the basic interpolation theorems [5, Theorem 3.2 and Theorem 3.4] and if $\rho_{j-1} = \infty$, then n_{j-1} has a pole at z_1, therefore its Schur transform n_j belongs to $\mathbf{N}^{z_1}$ and hence $\rho_j = n_j(z_1) \neq \infty$. If the sequence (ρ_j) is infinite then, by Theorem 4.2, there is an integer $j_0 \geq 0$ such that for $j \geq j_0$ we have $\rho_j \in \mathbb{C}^+$, and then, evidently, $\rho_{j+1} \neq \rho_j^*$.

For a nonconstant generalized Nevanlinna function $n \in \mathbf{N}$, the sequence (ρ_j) of Schur parameters of n at z_1 need not contain enough information to determine n. Uniqueness will follow if (some of) the Schur parameters ρ_j are augmented with additional data in accordance with the definition of the Schur transform: The *augmented Schur parameter of* n_j in the Schur algorithm for n at z_1 will be denoted by $\widetilde{\rho}_j$, $j = 0, 1, \ldots$, and is defined as follows.

(I) If $n_j \in \mathbf{N}^{z_1}$ and $\operatorname{Im} n_j(z_1) \neq 0$, or if the sequence of Schur parameters ends with ρ_j, then $\widetilde{\rho}_j = \rho_j$.

(II) If $n_j \in \mathbf{N}^{z_1}$ and $\operatorname{Im} n_j(z_1) = 0$, then $\widetilde{\rho}_j$ is the *triple* $\widetilde{\rho}_j = \{\rho_j, k_j, p_j\}$, where the integer k_j and the polynomial p_j are as in the definition of the Schur transform of n_j according to formula (4.3) (hence $k_j \geq 1$ and p_j is a real polynomial of degree $\leq 2k_j - 1$ with $p_j(z_1) \neq 0$).

(III) If n_j has a pole of order $k_j \geq 1$ at z_1 then $\widetilde{\rho}_j$ is the *triple* $\widetilde{\rho}_j = \{\infty, k_j, p_j\}$, where p_j is the polynomial in the definition of the Schur transform of n_j according to formula (4.4) (hence p_j is a real polynomial of degree $\leq 2k_j - 1$ with $p_j(z_1) \neq 0$).

The sequence $(\widetilde{\rho}_j)$ will be called the *sequence of augmented Schur parameters of* n *centered at* z_1. Note that every such sequence has two or more entries and that the first entry is either a number from $\mathbb{C} \setminus \mathbb{R}$ or a triple described by (II) or (III).

4.2 A sequence $(\widetilde{\rho}_j)$ with two or more entries will be called an *augmented Schur sequence* (*centered at* z_1) if

(α) except for at most finitely many values of j, $\widetilde{\rho}_j$ is a number: $\widetilde{\rho}_j = \rho_j \in \mathbb{C}^+$,

(β) the exceptional $\widetilde{\rho}_j$ is either

  $(\beta)_1$ a number: $\widetilde{\rho}_j = \rho_j$ with $\rho_j \in \mathbb{C}^-$, and if the sequence is finite and ends with $\widetilde{\rho}_m$, then $m \geq 1$ and $\widetilde{\rho}_m = \rho_m$ with $\rho_m \in \mathbb{R} \cup \{\infty\}$, or

  $(\beta)_2$ a triple $\widetilde{\rho}_j = \{\rho_j, k_j, p_j\}$, where ρ_j is from $\mathbb{R} \cup \{\infty\}$, k_j is an integer ≥ 1 and p_j is a real polynomial of degree $\leq 2k_j - 1$ with $p_j(z_1) \neq 0$, and

(γ) the relation (4.8) holds.

Clearly, the sequence of augmented Schur parameters of a Nevanlinna function is an augmented Schur sequence. In the following we prove that the converse also holds.

Let $(\widetilde{\rho}_j)$ be an augmented Schur sequence. With each entry $\widetilde{\rho}_j$ we associate a 2×2 matrix function $\widetilde{\Theta}_j$ in accordance with formulas (4.5), (4.6), and (4.7):

If $\widetilde{\rho}_j = \rho_j \in \mathbb{C} \setminus \mathbb{R}$ we set:

$$\widetilde{\Theta}_j(z) = \left(I_2 + (\mathrm{b}(z) - 1) \frac{\mathbf{u}_j \mathbf{u}_j^* J}{\mathbf{u}_j^* J \mathbf{u}_j} \right), \quad \mathbf{u}_j = \begin{pmatrix} \rho_j^* \\ 1 \end{pmatrix},$$

if $\widetilde{\rho}_j = \{\rho_j, k_j, p_j\}$ with $\rho_j \in \mathbb{R}$ then:

$$\widetilde{\Theta}_j(z) = \left(\mathrm{b}(z)^{k_j} I_2 - \frac{p_j(z)}{(z - z_1^*)^{2k_j}} \mathbf{u}_j \mathbf{u}_j^* J \right), \quad \mathbf{u}_j = \begin{pmatrix} \rho_j \\ 1 \end{pmatrix},$$

and if $\widetilde{\rho}_j = \{\infty, k_j, p_j\}$:

$$\widetilde{\Theta}_j(z) = \left(\mathrm{b}(z)^{k_j} I_2 + \frac{p_j(z)}{(z - z_1^*)^{2k_j}} \, \mathbf{e}\mathbf{e}^* J \right), \quad \mathbf{e} = \begin{pmatrix} 1 \\ 0 \end{pmatrix}.$$

Remark 4.3. If the implication (4.8) is violated then the product of the two corresponding matrix functions $\widetilde{\Theta}_{j-1}$ and $\widetilde{\Theta}_j$ has a zero at z_1, which means that the linear transformation $\mathcal{T}_{\widetilde{\Theta}_{j-1}\widetilde{\Theta}_j}$ can be written as $\mathcal{T}_\Psi$ for some $\Psi \in \mathcal{U}^{z_1}$ with $\deg \Psi < \deg \widetilde{\Theta}_{j-1}\widetilde{\Theta}_j$. For example, consider

$$\widetilde{\Theta}_j(z) = \mathrm{b}(z)^{k_j} \begin{pmatrix} 1 & \dfrac{p_j(z)}{(z - z_1)^{k_j}(z - z_1^*)^{k_j}} \\ 0 & 1 \end{pmatrix},$$

corresponding to the augmented Schur parameter $\{\infty, k_j, p_j\}$ centered at z_1, $j = 1, 2$. Then, under the assumption that $k_1 \geq k_2 \geq 1$,

$$\widetilde{\Theta}_1(z)\widetilde{\Theta}_2(z) = \mathrm{b}(z)^{k_2+s}\Psi(z) \tag{4.9}$$

with

$$\Psi(z) = \mathrm{b}(z)^{k_1-s} \begin{pmatrix} 1 & \dfrac{q(z)}{(z - z_1)^{k_1}(z - z_1^*)^{k_1}} \\ 0 & 1 \end{pmatrix},$$

where $q(z) = p_1(z) + p_2(z)(z - z_1)^{k_1-k_2}(z - z_1^*)^{k_1-k_2}$ and s is the order of the zero at z_1 of the polynomial $q(z)$; if $q(z_1) \neq 0$, then $s = 0$. By (2.6)–(2.10), the following equalities hold:

$$\deg \widetilde{\Theta}_1\widetilde{\Theta}_2 = 2(k_1 + k_2), \quad \kappa_-(\widetilde{\Theta}_1\widetilde{\Theta}_2) = k_1 + k_2,$$
$$\deg \Psi = 2(k_1 - s), \qquad \kappa_-(\Psi) = k_1 - s;$$

the factor b^{k_2+s} in (4.9) makes up for the difference of the degrees and the difference of the numbers of negative squares: $b^{k_2+s}I_2 \in \mathcal{U}^{z_1}$ and

$$\deg b^{k_2+s}I_2 = 2(k_2 + s), \quad \kappa_-(b^{k_2+s}I_2) = k_2 + s.$$

With the augmented Schur sequence $(\widetilde{\rho}_j)$ we associate the functions $\widetilde{r}^{[\ell]}$ defined as follows

$$\widetilde{r}^{[0]}(z) = 0 \text{ if } \rho_0 \in \mathbb{R} \cup \{\infty\}, \quad \widetilde{r}^{[0]}(z) = \begin{cases} \pm\rho_0, & z \in \mathbb{C}^+, \\ \pm\rho_0^*, & z \in \mathbb{C}^-, \end{cases} \quad \text{if } \rho_0 \in \mathbb{C}^\pm,$$

and

$$\widetilde{r}^{[\ell]}(z) = \begin{cases} \mathcal{T}_{\widetilde{\Theta}_0(z)\widetilde{\Theta}_1(z)\cdots\widetilde{\Theta}_{\ell-1}(z)}(\rho'_\ell), & z \in \mathbb{C}^+, \\ \mathcal{T}_{\widetilde{\Theta}_0(z)\widetilde{\Theta}_1(z)\cdots\widetilde{\Theta}_{\ell-1}(z)}(\rho'^*_\ell), & z \in \mathbb{C}^-, \end{cases}$$

where $\ell = 1, 2, \ldots, m$ if the sequence $(\widetilde{\rho}_j)$ terminates with $\widetilde{\rho}_m$ $(m \geq 1)$ and $\ell = 1, 2, \ldots$ if $(\widetilde{\rho}_j)$ is an infinite sequence; here

$$\rho'_\ell = \rho_\ell \quad \text{if } \rho_\ell \in \mathbb{C}^+ \cup \mathbb{R} \cup \{\infty\}$$

and ρ'_ℓ is an arbitrarily chosen number from $\mathbb{C}^+\backslash\{\rho^*_{\ell-1}\}$ if $\rho_\ell \in \mathbb{C}^-$, hence $\rho'_\ell \neq \rho^*_{\ell-1}$ and

$$\rho_\ell \in \mathbb{C}^\pm \Rightarrow \rho'_\ell \in \mathbb{C}^+. \tag{4.10}$$

In the sequence $(\widetilde{\rho}_j)$ of augmented Schur parameters of a generalized Nevanlinna function and in an augmented Schur sequence $(\widetilde{\rho}_j)$ the integer k_j is defined only if $\widetilde{\rho}_j$ is a triplet. In what follows we set $k_j = 0$ if $\widetilde{\rho}_j$ is just a number.

Proposition 4.4.

(i) *The function* $\widetilde{r}^{[\ell]}$, $\ell \geq 0$, *is a generalized Nevanlinna function.*

(ii) $\widetilde{r}^{[\ell]}$ *is a rational generalized Nevanlinna function if and only if* $\rho_\ell \in \mathbb{R}\cup\{\infty\}$. *In this case and for* $\ell \geq 1$

$$\begin{aligned} \deg \widetilde{r}^{[\ell]} &= \#\{j : 0 \leq j < \ell, \operatorname{Im}\rho_j \neq 0\} + 2\textstyle\sum_{0\leq j<\ell} k_j\,, \\ \kappa_-(\widetilde{r}^{[\ell]}) &= \#\{j : 0 \leq j < \ell, \operatorname{Im}\rho_j < 0\} + \textstyle\sum_{0\leq j<\ell} k_j\,, \end{aligned}$$

and the sequence of augmented Schur parameters of $\widetilde{r}^{[\ell]}$ *is given by*

$$\widetilde{\rho}_0, \ldots, \widetilde{\rho}_{\ell-1}, \widetilde{\rho}_\ell = \rho_\ell.$$

(iii) *There is an integer* j_0 *such that for each* $\ell \geq j_0$ *the function* $\widetilde{r}^{[\ell]}$ *is a rational generalized Nevanlinna function with*

$$\begin{aligned} \deg \widetilde{r}^{[\ell]} &= \#\{j \geq 0 : \operatorname{Im}\rho_j \neq 0\} + 2\textstyle\sum_{j\geq 0} k_j\,, \\ \kappa_-(\widetilde{r}^{[\ell]}) &= \#\{j \geq 0 : \operatorname{Im}\rho_j < 0\} + \textstyle\sum_{j\geq 0} k_j\,. \end{aligned}$$

(iv) *If* $\rho_\ell \in \mathbb{C}^\pm$, *then*

$$\kappa_-(\widetilde{r}^{[\ell]}) = \#\{j : 0 \leq j < \ell, \operatorname{Im}\rho_j < 0\} + \sum_{0\leq j<\ell} k_j$$

and the sequence of augmented Schur parameters of $\widetilde{r}^{[\ell]}$ *is given by*

$$\widetilde{\rho}_0, \ldots, \widetilde{\rho}_{\ell-1}, \widetilde{\rho}_\ell = \rho'_\ell, \rho'_\ell, \rho'_\ell, \ldots.$$

Proof. We first prove the statement (i) and the formulas for the degrees and numbers of negative squares in (ii)–(iv). Then we give the proofs for the formulas of the augmented Schur sequences associated with $\widetilde{r}^{[\ell]}$ in (ii) and (iv).

1. First we assume $\ell \geq 1$ and $\rho_\ell \neq \infty$. We set

$$\widetilde{\Theta}_0 \cdots \widetilde{\Theta}_{\ell-1} = \Theta = \begin{pmatrix} a & b \\ c & d \end{pmatrix}.$$

Then

$$\begin{aligned} \frac{\widetilde{r}^{[\ell]}(z) - \widetilde{r}^{[\ell]}(w)^*}{z - w^*} &= 2\pi \begin{pmatrix} 1 & -\widetilde{r}^{[\ell]}(z) \end{pmatrix} K_\Theta(z,w) \begin{pmatrix} 1 \\ -\widetilde{r}^{[\ell]}(w)^* \end{pmatrix} \\ &\quad + \left(a(z) - c(z)\widetilde{r}^{[\ell]}(z)\right) \left(\frac{\rho'_\ell - \rho'^*_\ell}{z - w^*}\right) \left(a(w) - c(w)\widetilde{r}^{[\ell]}(w)\right)^*. \end{aligned}$$

The reproducing kernel space with Nevanlinna kernel

$$L_{\rho'_\ell}(z,w) = \frac{\rho'_\ell - \rho'^*_\ell}{z - w^*}$$

is an infinite-dimensional Hilbert space if $\rho_\ell \in \mathbb{C}^\pm$ (see (4.10)) and it is trivial if $\rho_\ell \in \mathbb{R}$. The statements (i)–(iv) and the formulas for the degrees and the number of negative squares now follow from Theorem 4.2 and from equating the dimension and the negative index of the Pontryagin space $\mathcal{P}(\Theta)$, see the remarks before Lemma 2.1.

2. Now assume $\ell \geq 2$ and $\rho_\ell = \infty$. Then with

$$\widetilde{\Theta}_0 \cdots \widetilde{\Theta}_{\ell-2} = \Theta = \begin{pmatrix} a & b \\ c & d \end{pmatrix} \text{ and } \widetilde{\Theta}_{\ell-1} = \begin{pmatrix} a_{\ell-1} & b_{\ell-1} \\ c_{\ell-1} & d_{\ell-1} \end{pmatrix}$$

we obtain the equality

$$\frac{\widetilde{r}^{[\ell]}(z) - \widetilde{r}^{[\ell]}(w)^*}{z - w^*} = 2\pi \begin{pmatrix} 1 & -\widetilde{r}^{[\ell]}(z) \end{pmatrix} K_\Theta(z,w) \begin{pmatrix} 1 \\ -\widetilde{r}^{[\ell]}(w)^* \end{pmatrix}$$
$$+ \left(a(z) - c(z)\widetilde{r}^{[\ell]}(z)\right) \frac{\dfrac{a_{\ell-1}(z)}{c_{\ell-1}(z)} - \left(\dfrac{a_{\ell-1}(w)}{c_{\ell-1}(w)}\right)^*}{z - w^*} \left(a(w) - c(w)\widetilde{r}^{[\ell]}(w)\right)^*.$$

Note that by (4.8) $\rho_{\ell-1} \neq \infty$, hence $c_{\ell-1} \neq 0$ and $a_{\ell-1}/c_{\ell-1}$ is well defined. The equality

$$\kappa_-(a_{\ell-1}/c_{\ell-1}) = \kappa_-(\widetilde{\Theta}_{\ell-1})$$

readily implies (ii) and (iii).

3. If $\ell = 1$ and $\rho_1 = \infty$, then the proof is similar because in this case

$$\frac{\widetilde{r}^{[1]}(z) - \widetilde{r}^{[1]}(w)^*}{z - w^*} = \frac{\dfrac{a_0(z)}{c_0(z)} - \left(\dfrac{a_0(w)}{c_0(w)}\right)^*}{z - w^*}.$$

4. We now prove the formulas for the augmented Schur sequences associated with $\widetilde{r}^{[\ell]}$. For that we consider the functions

$$f_j = T_{\Theta_j \cdots \Theta_{\ell-1}}(\rho'_\ell), \quad j = 0, 1, \ldots, \ell - 1.$$

By the reasoning used in the previous parts of the proof these functions are generalized Nevanlinna functions. We use induction to show that

$$f_j(z_1) = \rho_j, \quad j = \ell - 1, \ell - 2, \ldots, 0. \tag{4.11}$$

From formula (4.8) and (compare with(4.5)–(4.7))

$$f_{\ell-1}(z) = T_{\widetilde{\Theta}_{\ell-1}}(\rho'_\ell)(z) = \begin{cases} \rho_{\ell-1} + g_{\ell-1}(z) & \text{if } \rho_{\ell-1} \neq \infty, \rho_j = \infty, \\ \dfrac{(\rho_{\ell-1} + g_{\ell-1}(z))\rho'_\ell - |\rho_{\ell-1}|^2}{\rho'_\ell - \rho^*_{\ell-1} + g_{\ell-1}(z)} & \text{if } \rho_{\ell-1} \neq \infty, \rho_\ell \neq \infty, \\ \rho'_\ell + \dfrac{p_{\ell-1}(z)}{(z - z_1)^{k_{\ell-1}}(z - z_1^*)^{k_{\ell-1}}} & \text{if } \rho_{\ell-1} = \infty, \rho_\ell \neq \infty, \end{cases}$$

where $g_{\ell-1}$ is a rational function with $g_{\ell-1}(z_1) = 0$, we obtain $f_{\ell-1}(z_1) = \rho_{\ell-1}$. Now assume $f_j(z_1) = \rho_j$. Then

$$f_{j-1}(z) = T_{\Theta_{j-1}(z)}(f_j(z)) = \begin{cases} \dfrac{(\rho_{j-1} + g_{j-1}(z))f_j(z) - |\rho_{j-1}|^2}{f_j(z) - \rho_{j-1}^* + g_{j-1}(z)} & \text{if } \rho_{j-1} \neq \infty, \\ f_j(z) + \dfrac{p_{j-1}(z)}{(z-z_1)^{k_{j-1}}(z-z_1^*)^{k_{j-1}}} & \text{if } \rho_{j-1} = \infty, \end{cases}$$

where g_{j-1} is a rational function with $g_{j-1}(z_1) = 0$, and letting $z \to z_1$ we obtain $f_{j-1}(z_1) = \rho_{j-1}$. This proves (4.11). That f_1 is the Schur transform of $f_0 = \widetilde{r}^{[\ell]}$. readily follows from

$$f_1 = T_{\Theta_0^{-1}}(f_0),$$

the fact that

$$f_1(z_1) = \rho_1 \neq \rho_0^*,$$

and the following formulas:

(a) if $\rho_0 \in \mathbb{C} \setminus \mathbb{R}$, then, according to (4.11), $f_0(z_1) = \rho_0$,

(b) if $\rho_0 \in \mathbb{R}$, then

$$\frac{1}{f_0(z) - \rho_0} = -\frac{p_0(z)}{(z-z_1)^{k_0}(z-z_1^*)^{k_0}} + \frac{1}{f_1(z) - \rho_0},$$

and

(c) if $\rho_0 = \infty$, then

$$f_0(z) = f_1(z) + \frac{p_0(z)}{(z-z_1)^{k_0}(z-z_1^*)^{k_0}}.$$

These formulas also imply that the augmented Schur parameter for f_0 is the first entry $\widetilde{\rho}_0$ of the given augmented Schur sequence. In the same way one can now prove that f_j is the Schur transform of f_{j-1} and that the augmented Schur parameter for f_{j-1} is given by $\widetilde{\rho}_{j-1}$, $j = 1, 2, \dots, \ell - 1$. Finally, also in the same way one can show that

$$\widehat{f}_{\ell-1} = \rho'_\ell = \rho_\ell \quad \text{if } \rho_\ell \in \mathbb{R} \cup \{\infty\}$$

and that the augmented Schur sequence associated with $f_0 = \widetilde{r}^{[\ell]}$ stops with ρ_ℓ. This proves the last statement in (ii). The last statement in (iv) follows from (3.2) and the fact that

$$\widehat{f}_{\ell-1}(z) = \begin{cases} \rho'_\ell, & z \in \mathbb{C}^+, \\ \rho'^*_\ell, & z \in \mathbb{C}^-, \end{cases} \quad \text{if } \rho_\ell \in \mathbb{C} \setminus \mathbb{R}. \qquad \square$$

Theorem 4.5. *If $n \in \mathbf{N}$ is not constant, then the sequence $\left(\widetilde{\rho}_j\right)$ of augmented Schur parameters of n is finite if and only if n is a rational generalized Nevanlinna function. In this case we have*

$$\begin{aligned} \deg n \quad &= \#\{j \geq 0 : \operatorname{Im} \rho_j \neq 0\} + 2\textstyle\sum_{j\geq 0} k_j\,, \\ \kappa_-(n) \quad &= \#\{j \geq 0 : \operatorname{Im} \rho_j < 0\} + \textstyle\sum_{j\geq 0} k_j\,, \end{aligned}$$

where $k_j = 0$ *if* $\widetilde{\rho}_j$ *is just a number. If* $(\widetilde{\rho}_j)$ *stops with* $\widetilde{\rho}_m$*, then* $n = \widetilde{r}^{[m]}$ *and we have* $\rho_m \in \mathbb{R}$ *if and only if* $n(z) = \mathrm{o}(z)$ *for* $z = \mathrm{i}y$, $y \to \pm\infty$*, and otherwise* $\rho_m = \infty$.

Proof. The first part of the theorem follows from [5, Remark 7.5 (1)]. The second part can be proved by writing

$$\Theta = \widetilde{\Theta}_0 \cdots \widetilde{\Theta}_{m-1} = \frac{1}{(z - z_1^*)^s} \begin{pmatrix} a & b \\ c & d \end{pmatrix},$$

where $s = \deg \Theta$ and $a, b, c,$ and d are polynomials with, since $\Theta(\infty) = I_2$,

$$\deg a = \deg d = s, \quad \deg b < s, \quad \deg c < s.$$

It follows that if $\rho_m \in \mathbb{R}$, then

$$n(z) = \widetilde{r}^{[m]}(z) = \frac{a(z)\rho_m + b(z)}{c(z)\rho_m + d(z)} = \mathrm{o}(z),$$

whereas if $\rho_m = \infty$, then

$$n(z) = \widetilde{r}^{[m]}(z) = \frac{a(z)}{c(z)} \neq \mathrm{o}(z). \qquad \square$$

If the sequence $(\widetilde{\rho}_j)$ is infinite then n can be approximated by the functions $\widetilde{r}^{[\ell]}$ in the following sense.

Theorem 4.6. *If* $n \in \mathbf{N}$ *is not constant and the sequence* $(\widetilde{\rho}_j)$ *of augmented Schur parameters of* n *is infinite, then*

$$\lim_{\ell\to\infty} \widetilde{r}^{[\ell]} = n, \text{ locally uniformly on } (\mathbb{C} \setminus \mathbb{R}) \cap \operatorname{hol}(n).$$

Proof. With j_0 as in Theorem 4.2, for $j \geq j_0$ we have

$$\begin{aligned} n &= \mathcal{T}_{\widetilde{\Theta}_0 \cdots \widetilde{\Theta}_{j_0-1}}(n_{j_0}), \\ r^{[j]} &= \mathcal{T}_{\widetilde{\Theta}_0 \cdots \widetilde{\Theta}_{j_0-1} \widetilde{\Theta}_{j_0} \cdots \widetilde{\Theta}_{j-1}}(\rho_j) = \mathcal{T}_{\widetilde{\Theta}_0 \cdots \widetilde{\Theta}_{j_0-1}}(t^{[j]}), \end{aligned}$$

where

$$t^{[j]} = \mathcal{T}_{\widetilde{\Theta}_{j_0} \cdots \widetilde{\Theta}_{j-1}}(\rho_j) \to n_{j_0}, \quad j \to \infty, \tag{4.12}$$

uniformly on compact subsets of $\mathbb{C} \setminus \mathbb{R}$. If we write

$$\Theta(z) = \widetilde{\Theta}_0(z) \cdots \widetilde{\Theta}_{j_0}(z) = \frac{1}{(z - z_1^*)^s} \begin{pmatrix} a(z) & b(z) \\ c(z) & d(z) \end{pmatrix},$$

where $s = \deg \Theta$ and a, b, c and d are polynomials, then we have

$$n - r^{[j]} = \frac{(n_{j_0} - t^{[j]}) \det \Theta}{(c\, n_{j_0} + d)(c\, t^{[j]} + d)}. \tag{4.13}$$

We claim that the factor $c\, n_{j_0} + d$ in the denominator of the quotient on the right-hand side does not vanish on $\mathbb{C}^+ \cap \operatorname{hol}(n)$. Let $z_0 \in \mathbb{C}^+ \cap \operatorname{hol}(n)$ be such that

$$c(z_0)\, n_{j_0}(z_0) + d(z_0) = 0.$$

Then, since n is holomorphic at z_0 and

$$n = T_\Theta(n_{j_0}) = \frac{a\,n_{j_0} + b}{c\,n_{j_0} + d},$$

we also have that

$$a(z_0)n_{j_0}(z_0) + b(z_0) = 0.$$

Thus

$$\Theta(z_0)\begin{pmatrix} n_{j_0}(z_0) \\ 1 \end{pmatrix} = 0,$$

which implies $\det\,\Theta(z_0) = 0$, and hence $z_0 = z_1$. From [3, Theorem 7.1] (and the formula for $\alpha(z)$ in the proof of this theorem) we have that

$$a - nc = \begin{pmatrix} 1 & -n \end{pmatrix} \Theta \begin{pmatrix} 1 \\ 0 \end{pmatrix}$$

has a zero at z_1 of order exactly $\deg\Theta$. The formula

$$c\,n_{j_0} + d = \frac{\det\Theta}{a - n\,c}$$

now implies that

$$c(z_1)\,n_{j_0}(z_1) + d(z_1) \neq 0.$$

This completes the proof of the claim. Let K be a compact subset of $\mathbb{C}^+ \cap \mathrm{hol}\,(n)$. Then the claim and the uniform convergence $t^{[j]} \to n_{j_0}$ on K imply that there exists an $\varepsilon > 0$ such that for j large enough

$$\left|c(z)n_{j_0}(z) + d(z)\right| > \varepsilon, \quad \left|c(z)t^{[j]}(z) + d(z)\right| > \varepsilon, \quad z \in K,$$

and the theorem follows from (4.13) and (4.12). □

The Schur algorithm applied to a nonconstant generalized Nevanlinna function n yields the sequence of augmented Schur parameters of n defined by (I)–(III) at the end of Subsection 4.1. Such a sequence is an augmented Schur sequence as defined by (α)–(γ) at the beginning of Subsection 4.2. In fact the following relation holds.

Theorem 4.7. *The Schur algorithm gives a one-to-one correspondence between the class of nonconstant functions $n \in \mathbf{N}$ and the set of augmented Schur sequences $(\widetilde{\rho}_j)$. If the augmented Schur sequence $(\widetilde{\rho}_j)$ is the sequence of augmented Schur parameters of n, then*

$$\kappa_-(n) = \#\{j \geq 0 \,:\, \mathrm{Im}\,\rho_j < 0\} + \sum_{j\geq 0} k_j\,,$$

where $k_j = 0$ if $\widetilde{\rho}_j$ is just a number.

Proof. The one-to-one correspondence between the rational nonconstant functions $n \in \mathbf{N}$ and the finite augmented Schur sequences follows from using the formula

$$n = \widetilde{r}^{[m]} = T_{\widetilde{\Theta}_0\cdots\widetilde{\Theta}_m}(\rho_m) \text{ for some integer } m \geq 1,$$

and applying Proposition 4.4 (ii) and Theorem 4.5. The one-to-one correspondence between the nonrational functions and the infinite augmented Schur sequences can be proved by using Theorem 4.2, Proposition 4.4 (iv), Theorem 4.6, Corollary 3.7, and the relation

$$n = \lim_{j\to\infty} \widetilde{r}^{[j]} = \mathcal{T}_{\widetilde{\Theta}_0 \cdots \widetilde{\Theta}_{j_0-1}}(n_{j_0}),$$

where n_{j_0} is the Nevanlinna function

$$n_{j_0} = \lim_{j\to\infty} \mathcal{T}_{\widetilde{\Theta}_{j_0} \cdots \widetilde{\Theta}_{j-1}}(\rho_j).$$

The formula for $\kappa_-(n)$ follows from Theorem 4.1 □

Remark 4.8. For a generalized Nevanlinna function n, the relation (3.20), which connects the Schur parameters with the Taylor coefficients, holds for $j \geq j_0$ with j_0 as in Theorem 4.2.

References

[1] L.V. Ahlfors, *Complex Analysis*, 3rd ed., McGraw Hill Book Company, New York, 1979.

[2] D. Alpay, T.Ya. Azizov, A. Dijksma, and H. Langer, *The Schur algorithm for generalized Schur functions III: Factorizations of J-unitary matrix polynomials*, Linear Algebra Appl. **369** (2003), 113–144.

[3] D. Alpay, A. Dijksma, and H. Langer, *J_ℓ-unitary factorization and the Schur algorithm for Nevanlinna functions in an indefinite setting*, Linear Algebra Appl. **419** (2006), 675–709.

[4] D. Alpay, A. Dijksma, and H. Langer, *The transformation of Issai Schur and related topics in an indefinite setting*, Operator Theory: Adv. Appl. **176**, Birkhäuser Verlag, Basel, 2007, 1–98.

[5] D. Alpay, A. Dijksma, H. Langer, and Y. Shondin, *The Schur transformation for generalized Nevanlinna functions: interpolation and self-adjoint operator realizations*, Complex Analysis and Operator Theory **1**(2) (2007), 169–210.

[6] D. Alpay and H. Dym, *On applications of reproducing kernel spaces to the Schur algorithm and rational J-unitary factorization*, Operator Theory: Adv. Appl. **18**, Birkhäuser Verlag, Basel, 1986, 89–159.

[7] D. Alpay and H. Dym, *Structure invariant spaces of vector-valued functions, Hermitian matrices and a generalization of the Iohvidov laws*, Linear Algebra Appl. **137/138** (1990), 137–181.

[8] H. Bart, I. Gohberg, and M.A. Kaashoek, *Minimal factorization of matrix and operator functions*, Operator Theory: Adv. Appl. **1**, Birkhäuser Verlag, Basel, 1979.

[9] M.-J. Bertin, A. Decomps-Guilloux, M. Grandet-Hugot, M. Pathiaux-Delefosse, and J.-P. Schreiber, *Pisot and Salem numbers*, Birkhäuser Verlag, Basel, 1992.

[10] A. Dijksma and G. Wanjala, *Generalized Schur functions and augmented Schur parameters*, Operator Theory: Adv. Appl. **162**, Birkhäuser Verlag, Basel, 2005, 135–144.

[11] H. Dym and V. Katsnelson, *Contributions of Issai Schur to analysis*, Studies in memory of Issai Schur (Chevaleret/Rehovot, 2000), Progr. Math., **210**, Birkhäuser Verlag, Boston, 2003, xci–clxxxviii.

[12] M.Kaltenbäck, H.Winkler, and H.Woracek, *Singularities of generalized strings*, Operator Theory: Adv. Appl. **163**, Birkhäuser Verlag, Basel, 2006, 191–248.

D. Alpay
Department of Mathematics
Ben-Gurion University of the Negev
P.O. Box 653
84105 Beer-Sheva, Israel
e-mail: `dany@math.bgu.ac.il`

A. Dijksma
Department of Mathematics
University of Groningen
P.O. Box 407
9700 AK Groningen, The Netherlands
e-mail: `dijksma@math.rug.nl`

H. Langer
Institute for Analysis and Scientific Computing
Vienna University of Technology
Wiedner Hauptstrasse 8–10
1040 Vienna, Austria
e-mail: `hlanger@mail.zserv.tuwien.ac.at`

Operator Theory:
Advances and Applications, Vol. 188, 31–36

On Domains of Powers of Linear Operators and Finite Rank Perturbations

Tomas Ya. Azizov, Jussi Behrndt, Friedrich Philipp and Carsten Trunk

Abstract. Let S and T be linear operators in a linear space such that $S \subset T$. In this note an estimate for the codimension of $\operatorname{dom} S^n$ in $\operatorname{dom} T^n$ in terms of the codimension of $\operatorname{dom} S$ in $\operatorname{dom} T$ is obtained. An immediate consequence is that for any polynomial p the operator $p(S)$ is a finite-dimensional restriction of the operator $p(T)$ whenever S is a finite-dimensional restriction of T. The general results are applied to a perturbation problem of self-adjoint definitizable operators in Krein spaces.

Mathematics Subject Classification (2000). Primary 47A05; Secondary 47B50.

Keywords. Powers of operators, Krein spaces, symmetric and self-adjoint operators, finite rank perturbations, definitizable operators.

1. Introduction

Let S and T be linear operators in a vector space and assume that S is a restriction of T, i.e., the domain $\operatorname{dom} S$ of S is a subset of the domain $\operatorname{dom} T$ of T and $Sx = Tx$ holds for all $x \in \operatorname{dom} S$. Clearly, for any $n \in \mathbb{N}$ also the nth power S^n of S is a restriction of the nth power T^n of T. In this note we verify the useful formula

$$\dim \left(\frac{\operatorname{dom} T^n}{\operatorname{dom} S^n} \right) \leq n \cdot \dim \left(\frac{\operatorname{dom} T}{\operatorname{dom} S} \right), \qquad n \in \mathbb{N}, \tag{1.1}$$

which relates the codimension of $\operatorname{dom} S^n$ in $\operatorname{dom} T^n$ with the codimension of $\operatorname{dom} S$ in $\operatorname{dom} T$. Note, that if S is a finite-dimensional restriction of T and p is a polynomial, then (1.1) immediately implies that also $p(S)$ is a finite-dimensional restriction of $p(T)$. This observation will be used to improve a classical result from P. Jonas and H. Langer on finite rank perturbations of definitizable self-adjoint operators in Krein spaces, cf., [JL, Theorem 1] and [ABT, Theorem 2.2].

The work of T.Ya. Azizov was supported by the Russian Foundation for Basic Research (RFBR), Grant 05-01-00203-a.

2. Domains of powers of linear operators

Let $\mathcal{X}$ be a vector space and let $\mathcal{M}$ and $\mathcal{N}$ be subspaces of $\mathcal{X}$. Then it is well known (cf., e.g., [K, §7.6]) that there exists a subspace $\mathcal{G} \subset \mathcal{X}$ with

$$\mathcal{X} = \mathcal{M} \dotplus \mathcal{G} \quad \text{and} \quad \mathcal{N} = (\mathcal{N} \cap \mathcal{M}) \dotplus (\mathcal{N} \cap \mathcal{G}), \tag{2.1}$$

where $\dotplus$ denotes the direct sum of two subspaces. The subspaces $\mathcal{M}$ and $\mathcal{N}$ will be called isomorphic (in signs: $\mathcal{M} \cong \mathcal{N}$) if there exists a bijective linear mapping $F : \mathcal{M} \to \mathcal{N}$.

Let S and T be linear operators in $\mathcal{X}$ defined on the subspaces $\operatorname{dom} S$ and $\operatorname{dom} T$, respectively, and assume that $\operatorname{dom} S \subset \operatorname{dom} T$ and $Sx = Tx$, $x \in \operatorname{dom} S$, holds, i.e., S is a restriction of T. We shall also write $S \subset T$ in the sense of graphs. If $\mathcal{X}$ is a normed vector space the operators S and T may be unbounded and nonclosed. In the following, a relation between the codimension of $\operatorname{dom} S^n$ in $\operatorname{dom} T^n$, $n \in \mathbb{N}$, and the codimension of $\operatorname{dom} S$ in $\operatorname{dom} T$ will be established, cf. (1.1) and Proposition 2.3. First two auxiliary statements will be proved, the following one of which may be of independent interest.

Proposition 2.1. *Let S and T be linear operators in $\mathcal{X}$ and assume that S is a restriction of T. Then for every $n \in \mathbb{N}$ and $k \in \{0, 1, \ldots, n\}$ we have*

$$\frac{\operatorname{dom} T^n}{\operatorname{dom} S^n} \cong \frac{\operatorname{dom} T^k \cap \operatorname{ran} T^{n-k}}{\operatorname{dom} S^k \cap \operatorname{ran} S^{n-k}} \times \frac{\ker T^{n-k}}{\ker S^{n-k}}.$$

Proof. Let $n \in \mathbb{N}$, $k \in \{0, 1, \ldots, n\}$, set

$$\mathcal{Y} := \frac{\operatorname{dom} T^k \cap \operatorname{ran} T^{n-k}}{\operatorname{dom} S^k \cap \operatorname{ran} S^{n-k}} \quad \text{and} \quad \mathcal{Z} := \frac{\ker T^{n-k}}{\ker S^{n-k}},$$

and denote the cosets in $\mathcal{Y}$ and $\mathcal{Z}$ by $[\,\cdot\,]_{\mathcal{Y}}$ and $[\,\cdot\,]_{\mathcal{Z}}$, respectively. Since

$$\ker S^{n-k} = \ker T^{n-k} \cap \operatorname{dom} S^n$$

by (2.1) there exists a subspace $\mathcal{G} \subset \operatorname{dom} T^n$ with

$$\operatorname{dom} T^n = \operatorname{dom} S^n \dotplus \mathcal{G} \tag{2.2}$$

$$\ker T^{n-k} = \ker S^{n-k} \dotplus (\ker T^{n-k} \cap \mathcal{G}). \tag{2.3}$$

By Q we denote the projection in $\operatorname{dom} T^n$ onto $\operatorname{dom} S^n$ with respect to the decomposition (2.2). Now, we choose projections P_S in $\operatorname{dom} S^n$ onto $\ker S^{n-k}$ and $P_{\mathcal{G}}$ in $\mathcal{G}$ onto $\ker T^{n-k} \cap \mathcal{G}$ and define $P : \operatorname{dom} T^n \to \operatorname{dom} T^n$ by

$$Px := P_S Qx + P_{\mathcal{G}}(x - Qx), \quad x \in \operatorname{dom} T^n.$$

Then P is a projection in $\operatorname{dom} T^n$ onto $\ker T^{n-k}$ with

$$P \operatorname{dom} S^n = \ker S^{n-k}. \tag{2.4}$$

Let the linear mapping $F : \operatorname{dom} T^n \to \mathcal{Y} \times \mathcal{Z}$ be defined by

$$Fx := \bigl\{[T^{n-k}x]_{\mathcal{Y}}, [Px]_{\mathcal{Z}}\bigr\}, \quad x \in \operatorname{dom} T^n.$$

In the following we will show

$$\ker F = \operatorname{dom} S^n \quad \text{and} \quad \operatorname{ran} F = \mathcal{Y} \times \mathcal{Z}. \tag{2.5}$$

Let $x \in \operatorname{dom} T^n$ such that $Fx = 0$. Then we have $T^{n-k}x \in \operatorname{dom} S^k \cap \operatorname{ran} S^{n-k}$ and $Px \in \ker S^{n-k}$. Thus, there exists $u \in \operatorname{dom} S^n$ such that $T^{n-k}x = S^{n-k}u$ which implies $y := x - u \in \ker T^{n-k}$. Hence by (2.4)

$$y = Py = Px - Pu \in \ker S^{n-k}$$

and $x = u + y \in \operatorname{dom} S^n$ follows. Conversely, if $x \in \operatorname{dom} S^n$, then

$$T^{n-k}x = S^{n-k}x \in \operatorname{dom} S^k \cap \operatorname{ran} S^{n-k}$$

and $Px \in \ker S^{n-k}$ (see (2.4)), i.e., $Fx = 0$. In order to see that F is surjective, let $y \in \operatorname{dom} T^k \cap \operatorname{ran} T^{n-k}$ and $z \in \ker T^{n-k}$. Then there exists $x' \in \operatorname{dom} T^n$ with $T^{n-k}x' = y$ and for the vector $x := x' - Px' + z$ we have $T^{n-k}x = y$ and $Px = z$ and thus $Fx = \{[y]_{\mathcal{Y}}, [z]_{\mathcal{Z}}\}$. This establishes (2.5) which gives

$$\frac{\operatorname{dom} T^n}{\operatorname{dom} S^n} = \frac{\operatorname{dom} F}{\ker F} \cong \operatorname{ran} F = \mathcal{Y} \times \mathcal{Z}. \qquad \square$$

Lemma 2.2. *Let $\mathcal{M}_0, \mathcal{M}_1, \mathcal{N}_0, \mathcal{N}_1 \subset \mathcal{X}$ be subspaces of $\mathcal{X}$ such that $\mathcal{M}_0 \subset \mathcal{M}_1$ and $\mathcal{N}_0 \subset \mathcal{N}_1$. Then we have*

$$\dim \frac{\mathcal{M}_1 \cap \mathcal{N}_1}{\mathcal{M}_0 \cap \mathcal{N}_0} \leq \dim \frac{\mathcal{M}_1}{\mathcal{M}_0} + \dim \frac{\mathcal{N}_1}{\mathcal{N}_0}.$$

Proof. Denote the cosets in $(\mathcal{M}_1 \cap \mathcal{N}_1)/(\mathcal{M}_0 \cap \mathcal{N}_0)$ (resp. $\mathcal{M}_1/\mathcal{M}_0$, $\mathcal{N}_1/\mathcal{N}_0$) by $[\cdot]_{\mathcal{M}\cap\mathcal{N}}$ (resp. $[\cdot]_{\mathcal{M}}$, $[\cdot]_{\mathcal{N}}$). Then the mapping

$$F : \frac{\mathcal{M}_1 \cap \mathcal{N}_1}{\mathcal{M}_0 \cap \mathcal{N}_0} \to \frac{\mathcal{M}_1}{\mathcal{M}_0} \times \frac{\mathcal{N}_1}{\mathcal{N}_0}, \quad F[x]_{\mathcal{M}\cap\mathcal{N}} := ([x]_{\mathcal{M}}, [x]_{\mathcal{N}})$$

is a well-defined linear injection. $\square$

The following proposition is the main result of this section.

Proposition 2.3. *Let S and T be linear operators in $\mathcal{X}$ and assume that S is a restriction of T. Then for every $n \in \mathbb{N}$ we have*

$$\dim \left(\frac{\operatorname{dom} T^n}{\operatorname{dom} S^n} \right) \leq n \cdot \dim \left(\frac{\operatorname{dom} T}{\operatorname{dom} S} \right). \tag{2.6}$$

Corollary 2.4. *Let S and T be linear operators in $\mathcal{X}$ and assume that S is a finite-dimensional restriction of T, i.e., $\dim(\operatorname{dom} T/\operatorname{dom} S) < \infty$. Then for any polynomial p the operator $p(S)$ is a finite-dimensional restriction of the operator $p(T)$.*

Proof of Proposition 2.3. The assertion will be proved by induction. Obviously (2.6) is true for $n = 1$. Suppose that (2.6) holds for some $n \in \mathbb{N}$. Then with the help of Proposition 2.1 and Lemma 2.2 we obtain

$$\begin{aligned} \dim \left(\frac{\operatorname{dom} T^{n+1}}{\operatorname{dom} S^{n+1}} \right) &= \dim \left(\frac{\operatorname{dom} T^n \cap \operatorname{ran} T}{\operatorname{dom} S^n \cap \operatorname{ran} S} \right) + \dim \left(\frac{\ker T}{\ker S} \right) \\ &\leq \dim \left(\frac{\operatorname{dom} T^n}{\operatorname{dom} S^n} \right) + \left[\dim \left(\frac{\operatorname{ran} T}{\operatorname{ran} S} \right) + \dim \left(\frac{\ker T}{\ker S} \right) \right]. \end{aligned}$$

Now, an application of Proposition 2.1 for the case $k = 0$ and $n = 1$ implies the assertion of Proposition 2.3. $\square$

We note that relation (2.6) is in general not an equality. As an example, consider the case $T = I$ and $S = T \upharpoonright M$ where $\operatorname{codim} M = 1$.

3. Finite rank perturbations of definitizable operators

In this section we apply the results of the previous section to symmetric operators of finite defect in Krein spaces. For the basic theory of Krein spaces and linear operators acting therein we refer to the monographs [AI] and [B].

A (possibly nondensely defined) operator S in a Krein space $(\mathcal{K}, [\cdot\,,\cdot])$ is called *symmetric* if $[Sx, x]$ is real for all $x \in \operatorname{dom} S$. Recall also that a closed symmetric operator S in $\mathcal{K}$ is said to be of *defect* $m \in \mathbb{N}$ if there exists a self-adjoint extension A of S in $\mathcal{K}$, i.e., $S \subset A = A^+$ (where A^+ denotes the adjoint of A with respect to $[\cdot\,,\cdot]$), such that $\dim(A/S) = m$. Observe that A can be multi-valued if $\operatorname{dom} S$ is not dense in $\mathcal{K}$. However, it is always possible to choose a self-adjoint extension A of S which is an operator; then the defect of S coincides with $\dim(\operatorname{dom} A/\operatorname{dom} S)$.

A point $\lambda \in \mathbb{C}$ is said to be a *point of regular type* of a closed operator T in the Krein space $\mathcal{K}$ if $\ker(T - \lambda) = \{0\}$ and $\operatorname{ran}(T - \lambda)$ is closed. The set of points of regular type of T will be denoted by $r(T)$.

A self-adjoint operator A in $\mathcal{K}$ is said to be *definitizable* if its resolvent set $\rho(A)$ is nonempty and there exists a real polynomial q, $q \neq 0$, such that

$$[q(A)x, x] \geq 0 \quad \text{for all} \quad x \in \operatorname{dom} q(A).$$

We refer to [L] for a detailed study of the spectral properties of definitizable operators in Krein spaces. It was shown by P. Jonas and H. Langer in [JL, Theorem 1] that a definitizable operator remains definitizable under finite rank perturbations in resolvent sense if the perturbed operator is self-adjoint and has a nonempty resolvent set. This result was recently improved in [ABT, Theorem 2.2] where the assumption on the nonemptiness of the resolvent set of the perturbed operator was dropped. In the following we give a new simple proof of [ABT, Theorem 2.2] which makes use of the results in Section 2.

Theorem 3.1. *Let A and B be self-adjoint operators in the Krein space $\mathcal{K}$ and assume that the symmetric operator*

$$S := A \upharpoonright \operatorname{dom} S = B \upharpoonright \operatorname{dom} S, \quad \operatorname{dom} S := \big\{x \in \operatorname{dom} A \cap \operatorname{dom} B : Ax = Bx\big\},$$

is of finite defect. Then A is definitizable if and only if B is definitizable.

Proof. Assume that A is definitizable and let $q \neq 0$ be a real definitizing polynomial for A. We have to verify that $\rho(B)$ is nonempty. Then the assumption that S is of finite defect implies that

$$(A - \lambda)^{-1} - (B - \lambda)^{-1}, \qquad \lambda \in \rho(A) \cap \rho(B),$$

is a finite rank operator and therefore the statement follows from [JL, Theorem 1].

By [L] the set $\mathbb{C}\backslash\mathbb{R}$ with the exception of at most finitely many points belongs to $\rho(A)$ and it follows from $\sigma(q(A)) = q(\sigma(A))$ (cf. [DS, VII.9, Theorem 10]) that

$\rho(q(A)) \cap (\mathbb{C} \setminus \mathbb{R}) \neq \varnothing$. Therefore $q(A)$ is a self-adjoint nonnegative operator in the Krein space $\mathcal{K}$ and hence

$$\mathbb{C} \setminus \mathbb{R} \subset \rho(q(A)) \tag{3.1}$$

holds. As $\rho(A) \subset r(S)$, and thus $r(S) \neq \varnothing$, a slight variation of the proof of [DS, VII.9, Theorem 7] shows that $q(S)$ is a closed operator. By Corollary 2.4 $q(S)$ has finite defect, and therefore also the symmetric extension $q(B)$ of $q(S)$ has finite defect and is closed. Thus, there exists a self-adjoint extension T of $q(B)$ which is an operator.

Next we show in the same way as in the proof of [CL, Proposition 1.1] that T has nonempty resolvent set. Since the symmetric form $[q(S)\cdot,\cdot]$ on $\operatorname{dom} q(S)$ is nonnegative, the form $[T\cdot,\cdot]$ on $\operatorname{dom} T$ has at most $n := \dim(\operatorname{dom} T/\operatorname{dom}(q(S)))$ negative squares. Moreover, (3.1) implies $\mathbb{C} \setminus \mathbb{R} \subset r(q(S))$ and thus $\operatorname{ran}(T-\lambda)$ is closed for all $\lambda \in \mathbb{C} \setminus \mathbb{R}$. Assume now that there are $n+1$ different eigenvalues $\lambda_1, \dots, \lambda_{n+1}$ of T in one of the open half-planes with corresponding eigenvectors $x_1, \dots, x_{n+1}$. As $\operatorname{dom} T$ is dense, [B, Lemma I.10.4] implies that there exist vectors $z_1, \dots, z_{n+1} \in \operatorname{dom} T$ such that $[x_i, z_j] = \delta_{ij}$, $i, j = 1, \dots, n+1$. By setting $y_j := \lambda_j^{-1} z_j \in \operatorname{dom} T$ we have $[Tx_i, y_j] = \delta_{ij}$, $i, j = 1, \dots, n+1$, and it follows that

$$\mathcal{L} := \Big(\operatorname{span}\{x_1, \dots, x_{n+1}, y_1, \dots, y_{n+1}\}, [T\cdot,\cdot]\Big)$$

is a Krein space which contains the neutral subspace $\operatorname{span}\{x_1, \dots, x_{n+1}\}$. Hence $\mathcal{L}$ also contains an $(n+1)$-dimensional negative subspace, which is impossible as $[T\cdot,\cdot]$ has at most n negative squares. This shows that there exists a pair $\mu, \bar{\mu} \in r(T)$ which by the self-adjointness of T implies $\mu, \bar{\mu} \in \rho(T)$. The operator T is thus definitizable, see, e.g., [L, I.3, Example (c)] and, in particular, the set $\mathbb{C} \setminus \mathbb{R}$ with the possible exception of at most finitely many points belongs to $\rho(T)$.

For all $\lambda \in \rho(T) \cap \rho(A)$ we have $\ker(q(B)-\lambda) = \{0\}$ and both $\operatorname{ran}(q(B)-\lambda)$ and $\operatorname{ran}(B-\lambda)$ are closed. This together with $\sigma_p(q(B)) = q(\sigma_p(B))$ implies that there exists a point $\mu \in \mathbb{C}$ such that $\mu, \bar{\mu} \in r(B)$. But then the self-adjointness of B yields $\mu, \bar{\mu} \in \rho(B)$ and, in particular, $\rho(B) \neq \varnothing$. □

References

[AI] T.Ya. Azizov and I.S. Iokhvidov, *Linear Operators in Spaces with an Indefinite Metric*, John Wiley & Sons, 1989.

[ABT] T.Ya. Azizov, J. Behrndt and C. Trunk, *On finite rank perturbations of definitizable operators*, J. Math. Anal. Appl. **339** (2008), 1161–1168.

[B] J. Bognar, *Indefinite Inner Product Spaces*, Springer, 1974.

[CL] B. Curgus and H. Langer, *A Krein space approach to symmetric ordinary differential operators with an indefinite weight function*, J. Differential Equations **79** (1989), 31–61.

[DS] N. Dunford and J. Schwartz, *Linear Operators Part I, General Theory*, Interscience Publishers, 1958.

[JL] P. Jonas and H. Langer, *Compact perturbations of definitizable operators*, J. Operator Theory **2** (1979), 63–77.

[K] G. Köthe, *Topological Vector Spaces I*, Springer, 1969.

[L] H. Langer, *Spectral functions of definitizable operators in Krein spaces*, in: Functional Analysis: Proceedings of a Conference held in Dubrovnik, Yugoslavia, November 2-14, 1981, Lecture Notes in Math. **948**, 1–46, 1982.

Tomas Ya. Azizov
Department of Mathematics
Voronezh State University
Universitetskaya pl. 1
394693 Voronezh, Russia
e-mail: `azizov@math.vsu.ru`

Jussi Behrndt and Friedrich Philipp
Institut für Mathematik, MA 6-4
Technische Universität Berlin
Straße des 17. Juni 136
10623 Berlin, Germany
e-mail: `behrndt@math.tu-berlin.de`
e-mail: `webfritzi@gmx.de`

Carsten Trunk
Institut für Mathematik
Technische Universität Ilmenau
Postfach 10 05 65
98684 Ilmenau, Germany
e-mail: `carsten.trunk@tu-ilmenau.de`

Operator Theory:
Advances and Applications, Vol. 188, 37–48

Finite-dimensional de Branges Subspaces Generated by Majorants

Anton Baranov and Harald Woracek

Abstract. If $\mathcal{H}(E)$ is a de Branges space and ω is a nonnegative function on $\mathbb{R}$, define a de Branges subspace of $\mathcal{H}(E)$ by

$$\mathcal{R}_\omega(E) = \mathrm{Clos}_{\mathcal{H}(E)} \left\{ F \in \mathcal{H}(E) : \ \exists C > 0 : |E^{-1}F| \leq C\omega \text{ on } \mathbb{R} \right\}.$$

It is known that one-dimensional de Branges subspaces generated in this way are related to minimal majorants. We investigate finite-dimensional de Branges subspaces, their representability in terms of majorants, and their relation to minimal majorants.

Mathematics Subject Classification (2000). Primary 46E20; Secondary 30D15, 46E22.

Keywords. de Branges subspace, admissible majorant, Beurling-Malliavin Theorem.

1. Introduction

The theory of Hilbert spaces $\mathcal{H}(E)$ of entire functions founded by L.de Branges is an important branch of analysis. After its foundation in [dB1]–[dB6], it was further developed by many authors. It is an example for a fruitful interplay of function theory and operator theory, and has applications in mathematical physics, see, e.g., [R].

Recently, in the context of model subspaces $H^2(\mathbb{C}^+) \ominus \Theta H^2(\mathbb{C}^+)$ of the Hardy space $H^2(\mathbb{C}^+)$, V. Havin and J. Mashregi introduced the notion of *admissible majorants*, i.e., functions ω on the real line which majorize a nonzero element of the space, cf. [HM1], [HM2]. The interest to this problem was motivated by the famous Beurling–Malliavin Multiplier Theorem, cf. [HJ], [K]. The approach suggested in [HM1], [HM2] has led to a new, essentially simpler, proof of the Beurling–Malliavin Theorem cf. [HMN]. These ideas were further developped in [BH], [BBH], where interesting connections between majorization and other problems of function theory (polynomial approximation, quasianalitycity) were discovered.

A de Branges space $\mathcal{H}(E)$ is isomorphic to the model subspace $H^2(\mathbb{C}^+) \ominus \frac{E^\#}{E} H^2(\mathbb{C}^+)$. Hence the theory of admissible majorants can be applied to de Branges spaces. Due to the rich (analytic) structure of de Branges spaces, however, much more specific results than in the general case can be expected.

In de Branges' theory the notion of *de Branges subspaces*, i.e., subspaces of a space $\mathcal{H}(E)$ which are themselves de Branges spaces, plays an outstanding role. At this point a link with the theory of admissible majorants occurs: Given an admissible majorant ω for a de Branges space $\mathcal{H}(E)$, the space

$$\mathcal{R}_\omega(E) := \operatorname{Clos}_{\mathcal{H}(E)} \left\{ F \in \mathcal{H}(E) : \exists C > 0 : |E^{-1}F| \leq C\omega \text{ on } \mathbb{R} \right\} \tag{1.1}$$

is a de Branges subspace of $\mathcal{H}(E)$. This relationship was investigated in [BW]. There the set of all those de Branges subspaces of a given space $\mathcal{H}(E)$ which can be represented in this way was determined, and it was shown that minimal majorants correspond to one-dimensional de Branges subspaces.

In the present paper we investigate the representability of finite-dimensional de Branges subspaces by means of admissible majorants. These considerations are based on our previous work [BW]. Moreover, the relation of finite-dimensional de Branges subspaces with minimal majorants is made explicit.

Let us briefly describe the content of this paper. In the preliminary Section 2, we set up some notation and recall some basic facts on de Branges spaces which are essential for furher use. In Section 3 we prove our main result, Theorem 3.8. Besides the results of [BW], it is based on a thorough understanding of the family of majorants $\omega^{[k]}(x) := (1+|x|)^k \omega(x)$, $k \in \mathbb{N}_0 := \mathbb{N} \cup \{0\}$. Finally, in Section 4, we turn to the case of infinite-dimensional de Branges subspaces. In this general setting, the situation is more complicated. However, some positive result can be established, cf. Proposition 4.2.

2. Preliminaries on de Branges spaces

An entire function E is said to belong to the *Hermite-Biehler class* $\mathcal{HB}$, if it satisfies

$$|E(\bar{z})| < |E(z)|, \quad z \in \mathbb{C}^+ .$$

In what follows, for any function F, we denote by $F^\#$ the function $F^\#(z) := \overline{F(\bar{z})}$.

2.1. Definition. If $E \in \mathcal{HB}$, the *de Branges space* $\mathcal{H}(E)$ is defined as the set of all entire functions F which have the property that

$$\frac{F}{E}, \frac{F^\#}{E} \in H^2(\mathbb{C}^+) .$$

Moreover, $\mathcal{H}(E)$ will be endowed with the norm

$$\|F\|_E := \left(\int_{\mathbb{R}} \left| \frac{F(t)}{E(t)} \right|^2 dt \right)^{1/2}, \quad F \in \mathcal{H}(E) .$$

It is shown in [dB7, Theorem 21] that $\mathcal{H}(E)$ is a Hilbert space with respect to the norm $\|.\|_E$.

2.2. Definition. A subset $\mathcal{L}$ of a de Branges space $\mathcal{H}(E)$ is called a *de Branges subspace*, if it is itself, with the norm inherited from $\mathcal{H}(E)$, a de Branges space. The set of all de Branges subspaces of a given space $\mathcal{H}(E)$ will be denoted as $\operatorname{Sub}(E)$.

The fact that $\mathcal{L}$ is a de Branges subspace of $\mathcal{H}(E)$ thus means that there exists $E_1 \in \mathcal{H}B$ such that $\mathcal{L} = \mathcal{H}(E_1)$ and $\|F\|_{E_1} = \|F\|_E$, $F \in \mathcal{L}$.

If F is an entire function, denote by $\mathfrak{d}(F)$ its zero-divisor, i.e., the map $\mathfrak{d}(F) : \mathbb{C} \to \mathbb{N}_0$ which assigns to each point w its multiplicity as a zero of F. If $\mathcal{H}(E)$ is a de Branges space, set

$$\mathfrak{d}(\mathcal{H}(E))(w) := \min \left\{ \mathfrak{d}(F)(w) : F \in \mathcal{H}(E) \right\}.$$

It is shown in [dB7] that $\mathfrak{d}(\mathcal{H}(E))(w) = 0$, $w \in \mathbb{C} \setminus \mathbb{R}$, and $\mathfrak{d}(\mathcal{H}(E))(w) = \mathfrak{d}(E)(w)$, $w \in \mathbb{R}$.

If $E \in \mathcal{H}B$ and $\mathfrak{d} : \mathbb{R} \to \mathbb{N}_0$ are given, we denote

$$\operatorname{Sub}^{\mathfrak{d}}(E) := \left\{ \mathcal{L} \in \operatorname{Sub}(E) : \mathfrak{d}(\mathcal{L}) = \mathfrak{d} \right\}.$$

Those subspaces $\mathcal{L} \in \operatorname{Sub}(E)$ with $\mathfrak{d}(\mathcal{L}) = \mathfrak{d}(\mathcal{H}(E))$ are the most interesting ones. To shorten notation we put $\operatorname{Sub}^s(E) := \operatorname{Sub}^{\mathfrak{d}(\mathcal{H}(E))}(E)$.

A milestone in de Branges' theory is the Ordering Theorem for subspaces of a space $\mathcal{H}(E)$, cf. [dB7, Theorem 35] (we state only a somewhat weaker version which suffices for our needs):

2.3. de Branges' Ordering Theorem: *Let $\mathcal{H}(E)$ be a de Branges space and let $\mathfrak{d} : \mathbb{R} \to \mathbb{N}_0$. Then $\operatorname{Sub}^{\mathfrak{d}}(E)$ is totally ordered with respect to set-theoretic inclusion.*

Even more about the structure of the chain $\operatorname{Sub}^{\mathfrak{d}}(E)$ is known. For every $\mathcal{H} \in \operatorname{Sub}^{\mathfrak{d}}(E)$, put

$$\mathcal{H}_- := \operatorname{Clos} \bigcup_{\substack{\mathcal{L} \in \operatorname{Sub}^{\mathfrak{d}}(E) \\ \mathcal{L} \subsetneq \mathcal{H}}} \mathcal{L}, \quad \mathcal{H}_+ := \bigcap_{\substack{\mathcal{L} \in \operatorname{Sub}^{\mathfrak{d}}(E) \\ \mathcal{L} \supsetneq \mathcal{H}}} \mathcal{L}.$$

Then $\mathcal{H}_-, \mathcal{H}_+ \in \operatorname{Sub}^{\mathfrak{d}}(E)$ and

$$\dim(\mathcal{H}/\mathcal{H}_-),\ \dim(\mathcal{H}/\mathcal{H}_+) \in \{0, 1\}.$$

2.4. *Example.* Fundamental examples of de Branges spaces arise from canonical systems of differential equations, see [dB7, Theorems 37,38], [GK], [HSW]. Let H be a 2×2-matrix-valued function defined for $t \in [0, l]$, such that $H(t)$ is real and nonnegative, the entries of $H(t)$ belong to $L^1([0, l])$ and $H(t)$ does not vanish on any nonempty interval. We call an interval $(\alpha, \beta) \subseteq [0, l]$ H-indivisible, if for some $\varphi \in \mathbb{R}$ and some scalar function $h(t)$ we have

$$H(t) = h(t) \left(\cos\varphi, \sin\varphi \right)^T \left(\cos\varphi, \sin\varphi \right), \quad \text{a.e. } t \in (\alpha, \beta).$$

Let $W(t,z)$ be the (unique) solution of the initial value problem

$$\frac{\partial}{\partial t}W(t,z)\begin{pmatrix}0 & -1\\ 1 & 0\end{pmatrix} = zW(t,z)H(t),\ t\in[0,l], \qquad W(0,z)=I\,,$$

and put $E_t(z) := A_t(z) - iB_t(z)$, $t\in[0,l]$, where $(A_t(z),B_t(z)) := (1,0)W(t,z)$. Then

(i) $E_t \in \mathcal{H}B$, $t\in(0,l]$, and $E_0 = 1$.

(ii) If $0 < s \leq t \leq l$, then $\mathcal{H}(E_s) \subseteq \mathcal{H}(E_t)$ and the set-theoretic inclusion map is contractive. If s is not an inner point of an H-indivisible interval, it is actually isometric.

(iii) We have

$$\operatorname{Sub}^s(\mathcal{H}(E_l)) = \big\{\mathcal{H}(E_t) : t \text{ not inner point of an } H\text{-indivisible interval}\big\}\,.$$

2.5. *Example.* The *Paley-Wiener space* $\mathcal{P}W_a$, $a>0$, is defined as the space of all entire functions of exponential type at most a, whose restrictions to the real axis belong to $L^2(\mathbb{R})$. The norm in $\mathcal{P}W_a$ is given by the usual L^2-norm,

$$\|F\| := \Big(\int_{\mathbb{R}} |F(t)|^2\,dt\Big)^{1/2}, \qquad F\in\mathcal{P}W_a\,.$$

By a theorem of Paley and Wiener, the space $\mathcal{P}W_a$ is the image under the Fourier transform of $L^2(-a,a)$. If in Example 2.4 we take $H(t)=I$, $t\in[0,l]$, we obtain $E_t(z) = e^{-itz}$. It is a consequence of a theorem of M.G. Krein, cf. [RR, Examples/Addenda 2, p. 134], that the space $\mathcal{H}(e^{-itz})$ coincides with $\mathcal{P}W_t$.

We see from Example 2.4, (iii), that $\operatorname{Sub}^s(\mathcal{P}W_a) = \{\mathcal{P}W_b : 0<b\leq a\}$.

In the present paper we will mainly deal with finite-dimensional de Branges subspace of a given space $\mathcal{H}(E)$.

2.6. Definition. Let $E\in\mathcal{H}B$ and $\mathfrak{d}:\mathbb{R}\to\mathbb{N}_0$. Define

$$\operatorname{FSub}(E) := \big\{\mathcal{L}\in\operatorname{Sub}(E) : \dim\mathcal{L}<\infty\big\}\,,$$

$$\operatorname{FSub}^{\mathfrak{d}}(E) := \operatorname{FSub}(E)\cap\operatorname{Sub}^{\mathfrak{d}}(E), \quad \operatorname{FSub}^s(E) := \operatorname{FSub}(E)\cap\operatorname{Sub}^s(E)\,.$$

Moreover, put

$$\delta(\mathfrak{d},E) := \sup\big\{\dim\mathcal{L} : \mathcal{L}\in\operatorname{FSub}^{\mathfrak{d}}(E)\big\}\,.$$

The structure of the chain $\operatorname{FSub}^{\mathfrak{d}}(E)$ is very simple. This can be deduced from the following statement which, in particular, applies to a finite-dimensional de Branges subspace $\mathcal{H}(E_1)$ of a given space $\mathcal{H}(E)$.

2.7. *Example.* If $\mathcal{H}(E_1)$ is any finite-dimensional de Branges space, $n{:=}\dim\mathcal{H}(E_1)$, then there exists a function $S\in\mathcal{H}(E_1)$, $S=S^\#$, such that

$$\mathcal{H}(E_1) = S\cdot\operatorname{span}\{1,z,\ldots,z^{n-1}\}\,.$$

The chain $\operatorname{Sub}^s(E_1)$ is given as

$$\operatorname{Sub}^s(E_1) = \Big\{S\cdot\operatorname{span}\{1\}, S\cdot\operatorname{span}\{1,z\},\ldots,S\cdot\operatorname{span}\{1,z,\ldots,z^{n-1}\}\Big\}\,.$$

3. Representation of finite-dimensional subspaces by majorants

3.1. Definition. Let $E \in \mathcal{HB}$. A nonnegative function ω on the real axis $\mathbb{R}$ is called an *admissible majorant* for the space $\mathcal{H}(E)$, if there exists a function $F \in \mathcal{H}(E) \setminus \{0\}$ such that $|E(x)^{-1}F(x)| \leq \omega(x)$, $x \in \mathbb{R}$. The set of all admissible majorants for $\mathcal{H}(E)$ is denoted by $\operatorname{Adm}(E)$.

If $\omega \in \operatorname{Adm}(E)$, the space $\mathcal{R}_\omega(E)$ defined by (1.1) is a de Branges subspace of $\mathcal{H}(E)$, cf. [BW, Proposition 3.2]. Moreover, by [BW, Theorem 3.4], a de Branges subspace $\mathcal{H}(E_1)$ of $\mathcal{H}(E)$ is of the form $\mathcal{R}_\omega(E)$ for some majorant ω if and only if $\operatorname{mt}\frac{E_1}{E} = 0$. Here $\operatorname{mt} f$ is the mean type of a function f in the class $N(\mathbb{C}^+)$ of functions of bounded type in the upper half-plane:

$$\operatorname{mt} f := \limsup_{y\to+\infty} \frac{1}{y} \log|f(iy)|\,,$$

For a function $\omega : \mathbb{R} \to [0,\infty)$, we define $\mathfrak{d}(\omega) : \mathbb{R} \to \mathbb{N}_0 \cup \{\infty\}$ as the function which assigns to a point $v \in \mathbb{R}$ the minimum of all numbers $n \in \mathbb{N}_0$ such that there exists a neighbourhood $U \subseteq \mathbb{R}$ of v with the property

$$\inf_{\substack{z\in U\\ |z-v|^n\neq 0}} \frac{|\omega(z)|}{|z-v|^n} > 0\,.$$

For functions $\omega_1, \omega_2 : \mathbb{R} \to [0,\infty)$ we will write

$$\omega_1 \lesssim \omega_2 \;:\Leftrightarrow\; \exists C>0: \omega_1(x) \leq C\omega_2(x), x\in\mathbb{R}\,,$$

$$\omega_1 \asymp \omega_2 \;:\Leftrightarrow\; \omega_1 \lesssim \omega_2 \text{ and } \omega_2 \lesssim \omega_1\,.$$

3.2. Lemma. *Let $E \in \mathcal{HB}$ and $\omega \in \operatorname{Adm}(E)$. Then $\mathfrak{d}(\mathcal{R}_\omega(E)) = \mathfrak{d}(\mathcal{H}(E)) + \mathfrak{d}(\omega)$.*

Proof. Let $v \in \mathbb{R}$ be fixed. If $F \in \mathcal{H}(E)$, then $E^{-1}F$ is analytic in a neighbourhood of v, and $\mathfrak{d}_{E^{-1}F}(v) = \mathfrak{d}_F(v) - \mathfrak{d}_E(v)$. Hence, for some sufficiently small neighbourhood $U \subseteq \mathbb{R}$ of v, we have

$$\left|\frac{F(x)}{E(x)}\right| \asymp |x-v|^{\mathfrak{d}_F(v)-\mathfrak{d}_E(v)}, \quad x\in U\,.$$

If $F \in R_\omega(E)$, we obtain $|x-v|^{\mathfrak{d}_F(v)-\mathfrak{d}_E(v)} \lesssim \omega(x)$, $x \in U$, and thus $\mathfrak{d}(\omega)(v) \leq \mathfrak{d}_F(v) - \mathfrak{d}_E(v)$. Since $\mathfrak{d}(\mathcal{R}_\omega(E))(v) = \min\{\mathfrak{d}_F(v) : F \in R_\omega(E)\}$, this yields

$$\mathfrak{d}(\mathcal{R}_\omega(E))(v) \geq \mathfrak{d}(\omega)(v) + \mathfrak{d}(\mathcal{H}(E))(v)\,.$$

In order to prove the converse inequality, let $F \in R_\omega(E)$ with $\mathfrak{d}_F(v) > \mathfrak{d}(\omega)(v) + \mathfrak{d}(\mathcal{H}(E))(v)$ be given. Since $\mathfrak{d}_F(v) > \mathfrak{d}(\mathcal{H}(E))(v)$ the function $G(z) := (z-v)^{-1}F(z)$ belongs to the space $\mathcal{H}(E)$. Let $U \subseteq \mathbb{R}$ be a neighbourhood of v such that

$$\inf_{\substack{z\in U\cap D\\ |z-w|^{\mathfrak{d}(\omega)(v)}\neq 0}} \frac{|\omega(z)|}{|z-w|^{\mathfrak{d}(\omega)(v)}} > 0\,.$$

Moreover, let U be chosen so small that $E^{-1}F$ is analytic at every point of U. Since $\mathfrak{d}(E^{-1}F)(v) > \mathfrak{d}(\omega)(v)$, the function $\frac{F(z)}{(z-v)^{\mathfrak{d}(\omega)(v)+1}E(z)}$ is analytic, and hence bounded, on U. It follows that

$$\Big|\frac{F(x)}{(x-v)^{\mathfrak{d}(\omega)(v)+1}E(x)}\Big| \lesssim \frac{\omega(x)}{|x-v|^{\mathfrak{d}(\omega)(v)}}, \quad x \in U,$$

and hence

$$\Big|\frac{G(x)}{E(x)}\Big| = \Big|\frac{F(x)}{(x-v)E(x)}\Big| \lesssim \omega(x), \quad x \in U.$$

For $x \notin U$, we have $\frac{1}{|x-v|} \lesssim 1$, and hence

$$\Big|\frac{G(x)}{E(x)}\Big| = \Big|\frac{F(x)}{(x-v)E(x)}\Big| \lesssim |F(x)| \lesssim \omega(x), \quad x \in \mathbb{R}\setminus U.$$

Altogether we see that $G \in R_\omega(E)$.

Since $\mathfrak{d}(\mathcal{R}_\omega(E))(v) = \min\{\mathfrak{d}_F(v) : F \in R_\omega(E)\}$, we conclude that

$$\mathfrak{d}(\mathcal{R}_\omega(E))(v) \le \mathfrak{d}(\omega)(v) + \mathfrak{d}(\mathcal{H}(E))(v). \qquad \square$$

If $\mathfrak{d} : \mathbb{R} \to \mathbb{N}_0$ is given, we shall denote $\operatorname{Adm}^{\mathfrak{d}}(E) := \{\omega \in \operatorname{Adm}(E) : \mathfrak{d}(\omega) = \mathfrak{d}\}$. The majorants with $\mathfrak{d}(\omega) = 0$ are those of biggest interest. By Lemma 3.2 they generate de Branges subspaces in $\operatorname{Sub}^s(E)$.

We now proceed to the study of a family $\omega^{[k]}$ of majorants, which is vital for our consideration of finite-dimensional de Branges subspaces.

3.3. Definition. Let $\omega : \mathbb{R} \to [0,\infty)$. For $k \in \mathbb{Z}$ define

$$\omega^{[k]}(x) := (1+|x|)^k \omega(x), \ x \in \mathbb{R}.$$

If, additionally, $E \in \mathcal{H}B$ is given, put

$$\alpha(\omega, E) := \inf\big\{k \in \mathbb{Z} : \omega^{[k]} \in \operatorname{Adm}(E)\big\} \in \mathbb{Z} \cup \{\pm\infty\}.$$

Note that, clearly, $(\omega^{[k]})^{[l]} = \omega^{[k+l]}$, $\mathfrak{d}(\omega^{[k]}) = \mathfrak{d}(\omega)$, and $\omega^{[k]} \le \omega^{[l]}$ for $k \le l$.

As the following (trivial) example shows, $\alpha(\omega, E)$ may assume any prescribed value in $\mathbb{Z} \cup \{\pm\infty\}$.

3.4. *Example.* Let $E(z) := z + i$, then $\mathcal{H}(E) = \operatorname{span}\{1\}$. Hence a function ω is an admissible majorant for $\mathcal{H}(E)$ if and only if it is bounded away from zero. It is obvious that for any given $n \in \mathbb{Z} \cup \{\pm\infty\}$ we can find $\omega : \mathbb{R} \to (0,\infty)$ such that $\alpha(\omega, E) = n$.

3.5. Lemma. *Let* $E \in \mathcal{H}B$ *and* $\omega \in \operatorname{Adm}(E)$.

(i) *We have* $\dim\big(\mathcal{R}_{\omega^{[1]}}(E)/\mathcal{R}_\omega(E)\big) \le 1$.

(ii) *Assume that* $\dim \mathcal{R}_\omega(E) < \infty$ *and that*

$$\exists\, \mathcal{L} \in \operatorname{Sub}^{\mathfrak{d}(\omega)}(E) : \dim\big(\mathcal{L}/\mathcal{R}_\omega(E)\big) = 1. \tag{3.1}$$

Then $\dim(\mathcal{R}_{\omega^{[1]}}(E)/\mathcal{R}_\omega(E)) = 1$, *i.e.,* $\mathcal{L} = \mathcal{R}_{\omega^{[1]}}(E)$.

Proof. Let G be entire and not identically zero, let $v \in \mathbb{C} \setminus \mathbb{R}$ such that $G(v) \neq 0$, and consider the difference quotient operator

$$\rho : F(z) \mapsto \frac{F(z) - \frac{F(v)}{G(v)} G(z)}{z - v} .$$

If $\mathcal{H}$ is a de Branges space which contains the function G, then $\rho|_{\mathcal{H}}$ is a bounded linear operator of $\mathcal{H}$ into itself, and

$$\ker\big(\rho|_{\mathcal{H}}\big) = \operatorname{span}\{G\}, \quad \operatorname{ran}\big(\rho|_{\mathcal{H}}\big) = \operatorname{dom} \mathcal{S}(\mathcal{H}) .$$

Here $\mathcal{S}(\mathcal{H})$ denotes the operator of multiplication by z in $\mathcal{H}$. In particular, by [dB7, Theorem 29], we have $\dim(\mathcal{H}/\overline{\operatorname{ran}(\rho|_{\mathcal{H}})}) \in \{0,1\}$.

Proof of (i): Choose $G \in R_\omega(E) \setminus \{0\}$. We have the estimate

$$|(\rho F)(x)| \leq \frac{1}{|x-v|} \cdot |F(x)| + \frac{1}{|x-v|} \left|\frac{F(v)}{G(v)}\right| \cdot |G(x)|, \ x \in \mathbb{R} .$$

Hence $\rho(R_{\omega[1]}(E)) \subseteq R_\omega(E)$ which, by the continuity of $\rho|_{\mathcal{R}_{\omega[1]}(E)}$, implies that $\rho(\mathcal{R}_{\omega[1]}(E)) \subseteq \mathcal{R}_\omega(E)$. Thus also $\overline{\rho(\mathcal{R}_{\omega[1]}(E))} \subseteq \mathcal{R}_\omega(E)$. We conclude that

$$\dim(\mathcal{R}_{\omega[1]}(E)/\mathcal{R}_\omega(E)) \in \{0,1\} .$$

Proof of (ii): Choose G in $R_{\omega[1]}(E) \setminus \{0\}$. By the already proved part (i) of the present lemma, and the fact that $\operatorname{Sub}^{\mathfrak{d}(\omega)}(E)$ is a chain, we have $\mathcal{R}_{\omega[1]}(E) \subseteq \mathcal{L}$. The estimate

$$|F(x)| \leq \left|\frac{F(v)}{G(v)}\right| \cdot |G(x)| + |x-v| \cdot |(\rho F)(x)|, \ x \in \mathbb{R} ,$$

shows that

$$(\rho|_{\mathcal{L}})^{-1}(R_\omega(E)) \subseteq R_{\omega[1]}(E) .$$

However, by finite-dimensionality,

$$\operatorname{ran}(\rho|_{\mathcal{L}}) = \operatorname{dom} \mathcal{S}(\mathcal{L}) = \overline{\operatorname{dom} \mathcal{S}(\mathcal{L})} = \mathcal{R}_\omega(E) = R_\omega(E) .$$

It follows that

$$\mathcal{L} = (\rho|_{\mathcal{L}})^{-1}\big(\operatorname{ran}(\rho|_{\mathcal{L}})\big) \subseteq R_{\omega[1]}(E) = \mathcal{R}_{\omega[1]}(E) .$$

□

An inductive application of this lemma yields the following result.

3.6. Proposition. *Let* $E \in \mathcal{HB}$, $\omega \in \operatorname{Adm}(E)$, *and assume that* $\dim \mathcal{R}_\omega(E) < \infty$. *Then*

$$\dim\big(\mathcal{R}_{\omega[k+1]}(E)/\mathcal{R}_{\omega[k]}(E)\big) = \begin{cases} 1 & , \ 0 \leq k < \delta(\mathfrak{d}(\omega)) - \dim \mathcal{R}_\omega(E) \\ 0 & , \ k \geq \delta(\mathfrak{d}(\omega)) - \dim \mathcal{R}_\omega(E) \end{cases}$$

Proof. We will show by induction on k that

$$\dim \mathcal{R}_{\omega[k]}(E) = \begin{cases} \dim \mathcal{R}_\omega(E) + k, & 0 \leq k \leq \delta(\mathfrak{d}(\omega)) - \dim \mathcal{R}_\omega(E), \\ \delta(\mathfrak{d}(\omega)), & k > \delta(\mathfrak{d}(\omega)) - \dim \mathcal{R}_\omega(E). \end{cases}$$

For $k=0$ this is trivial. Assume that $0<k\leq\delta(\mathfrak{d}(\omega))-\dim\mathcal{R}_\omega(E)$ and that $\dim\mathcal{R}_{\omega^{[k-1]}}(E)=\dim\mathcal{R}_\omega(E)+(k-1)$. Then, by the definition of $\delta(\mathfrak{d}(\omega))$, there exists $\mathcal{L}\in\operatorname{FSub}^{\mathfrak{d}(\omega)}(E)$ with $\mathcal{L}\supsetneq\mathcal{R}_{\omega^{[k-1]}}(E)$. By the structure of the chain $\operatorname{FSub}^{\mathfrak{d}(\omega)}(E)$ we can choose $\mathcal{L}$ such that $\dim(\mathcal{L}/\mathcal{R}_{\omega^{[k-1]}}(E))=1$. Lemma 3.5, (ii), implies that $\dim\mathcal{R}_{\omega^{[k]}}(E)=\dim\mathcal{R}_{\omega^{[k-1]}}(E)+1$.

If $\delta(\mathfrak{d}(\omega))=\infty$, we are done. Otherwise, by the already proved, we have $\dim\mathcal{R}_{\omega^{[k_0]}}(E)=\delta(\mathfrak{d}(\omega))$ for $k_0:=\delta(\mathfrak{d}(\omega))-\dim\mathcal{R}_\omega(E)$. It follows from Lemma 3.5, (i), and the definition of $\delta(\mathfrak{d}(\omega))$ that $\mathcal{R}_{\omega^{[k]}}(E)=\mathcal{R}_{\omega^{[k_0]}}(E)$ for all $k\geq k_0$. □

The relation $\lesssim$ is reflexive and transitive, and hence induces an order on the factor set $\operatorname{Adm}(E)/_{\asymp}$. If we speak of minimal elements we always refer to this order.

3.7. Lemma. *Let $E\in\mathcal{H}B$ and $\mathfrak{d}:\mathbb{R}\to\mathbb{N}_0$.*

(i) *Let $\omega\in\operatorname{Adm}^{\mathfrak{d}}(E)$. Then $\omega/_{\asymp}$ is minimal in $\operatorname{Adm}^{\mathfrak{d}}(E)/_{\asymp}$ if and only if $\omega/_{\asymp}$ is minimal in $\operatorname{Adm}(E)/_{\asymp}$.*

(ii) *The set $\operatorname{Adm}^{\mathfrak{d}}(E)/_{\asymp}$ contains at most one minimal element.*

Proof. The assertion (i) is seen in exactly the same way as [BW, Lemma 4.7]. The second item is then a consequence of [BW, Corollary 4.4]. □

We can now settle the question when, and in which way, finite-dimensional de Branges subspaces can be represented by majorants. Put

$$\mathfrak{R}(E):=\big\{\mathcal{R}_\omega(E):\omega\in\operatorname{Adm}(E)\big\}.$$

3.8. Theorem. *Let $E\in\mathcal{H}B$ and $\mathfrak{d}:\mathbb{R}\to\mathbb{N}_0$. Then the following conditions are equivalent:*

(i) *$\operatorname{Adm}^{\mathfrak{d}}(E)/_{\asymp}$ contains a minimal element;*

(ii) *$\operatorname{FSub}^{\mathfrak{d}}(E)\cap\mathfrak{R}(E)\neq\emptyset$;*

(iii) *$\operatorname{FSub}^{\mathfrak{d}}(E)\neq\emptyset$ and $\operatorname{Sub}^{\mathfrak{d}}(E)\subseteq\mathfrak{R}(E)$.*

In this case, if $\mathcal{L}\in\operatorname{FSub}^{\mathfrak{d}}(E)$, we have

$$\mathcal{L}=\mathcal{R}_{\omega_0^{[k]}}(E),$$

where ω_0 is any representant of the minimal element of $\operatorname{Adm}^{\mathfrak{d}}(E)/_{\asymp}$ and $k=\dim\mathcal{L}-1$.

Proof. (i) ⇒ (ii): Let $\omega/_{\asymp}$ be the minimal element of $\operatorname{Adm}^{\mathfrak{d}}(E)$ (unique by Lemma 3.7). Then, by Lemma 3.7, $\omega/_{\asymp}$ is minimal in $\operatorname{Adm}(E)$. By [BW, Theorem 4.2], $\dim\mathcal{R}_\omega(E)=1$.

(ii) ⇒ (iii): Pick $\mathcal{H}(E_1)\in\operatorname{FSub}^{\mathfrak{d}}(E)\cap\mathfrak{R}(E)$. Then, by [BW, Theorem 3.4], we have $\operatorname{mt}\frac{E_1}{E}=0$. Let $\mathcal{H}(E_2)\in\operatorname{Sub}^{\mathfrak{d}}(E)$ be given. If $\mathcal{H}(E_1)\subseteq\mathcal{H}(E_2)$, we conclude from

$$\operatorname{mt}\frac{E_2}{E}=\max_{F\in\mathcal{H}(E_2)}\operatorname{mt}\frac{F}{E}\leq 0,\quad \operatorname{mt}\frac{E_1}{E_2}=\max_{F\in\mathcal{H}(E_1)}\operatorname{mt}\frac{F}{E_2}\leq 0,$$

that $\operatorname{mt}\frac{E_2}{E}=0$. Otherwise, if $\mathcal{H}(E_2)\subseteq\mathcal{H}(E_1)$, we see from Example 2.7 that also $\operatorname{mt}\frac{E_2}{E}=0$. It follows from [BW, Theorem 3.4] that in either case $\mathcal{H}(E_2)\in\mathfrak{R}(E)$.

(iii) $\Rightarrow$ (i): Since $\operatorname{FSub}^{\mathfrak{d}}(E)\neq\emptyset$, there exists a one-dimensional subspace $\mathcal{L}\in\operatorname{Sub}^{\mathfrak{d}}(E)$. Since $\mathcal{L}\in\mathfrak{R}(E)$, we obtain from [BW, Theorem 4.2] a minimal element $\omega/_{\asymp}$ of $\operatorname{Adm}(E)/_{\asymp}$ such that $\mathcal{L}=\mathcal{R}_\omega(E)$. Since $\mathfrak{d}(\omega)=\mathfrak{d}$, $\omega/_{\asymp}$ minimal in $\operatorname{Adm}^{\mathfrak{d}}(E)$.

Representation of $\mathcal{L}$: Assume that one (and hence all) of (i)–(iii) hold and that $\mathcal{L}\in\operatorname{FSub}^{\mathfrak{d}}(E)$. Let $\omega_0/_{\asymp}$ be the minimal element of $\operatorname{Adm}^{\mathfrak{d}}(E)$. Then $\dim\mathcal{R}_{\omega_0}(E)=1$. Since, clearly, $\delta(\mathfrak{d})\geq\dim\mathcal{L}$, we obtain from Proposition 3.6 that

$$\mathcal{L}=\mathcal{R}_{\omega_0^{[\dim\mathcal{L}-1]}}(E)\,. \qquad \square$$

3.9. *Remark.* Let us note that the (equivalent) conditions of Theorem 3.8 are not always satisfied. This is seen for example by taking $E(z):=(z+i)e^{-iz}$. For this function E, the space $\mathcal{H}(E)$ contains the function 1. Thus $\operatorname{span}\{1\}\in\operatorname{FSub}^{s}(E)$. However, by [BW, Theorem 3.4], we have $\mathfrak{R}(E)=\{\mathcal{H}(E)\}$.

In Proposition 3.6 we have clarified the behaviour of $\mathcal{R}_{\omega^{[k]}}(E)$ for $k\geq0$. It seems that the situation for $k<0$ is more complicated. In this place let us only note the following corollary of Proposition 3.6.

3.10. Corollary. *Let $E\in\mathcal{H}B$ and $\omega\in\operatorname{Adm}(E)$. Assume that $\dim\mathcal{R}_\omega(E)<\infty$ and that* (3.1) *holds. Then $\alpha(\omega,E)\leq1-\dim\mathcal{R}_\omega(E)$ and*

$$\dim\big(\mathcal{R}_{\omega^{[k+1]}}(E)/\mathcal{R}_{\omega^{[k]}}(E)\big)=1,\ 1-\dim\mathcal{R}_\omega(E)\leq k<0\,.$$

Proof. The space $\mathcal{R}_{\omega^{[-1]}}(E)$ is finite dimensional and the assumption (3.1) holds for it since it holds for the bigger space $\mathcal{R}_\omega(E)$. Hence we may apply Proposition 3.6, and obtain that

$$\dim\mathcal{R}_\omega(E)=\dim\mathcal{R}_{\omega^{[-1]}}(E)+1\,.$$

The assertion follows by induction. $\square$

4. The family $\omega^{[k]}$ for $\dim\mathcal{R}_\omega(E)=\infty$

Our treatment of finite-dimensional de Branges subspaces in $\mathfrak{R}(E)$ was based on Lemma 3.5. Let us show by an example that the assumption '$\dim\mathcal{R}_\omega(E)<\infty$' in part (ii) of this lemma cannot be dropped.

4.1. *Example.* We shall construct $E\in\mathcal{H}B$ and $\omega\in\operatorname{Adm}^0(E)$ such that (3.1) holds, but $\mathcal{R}_{\omega^{[k]}}(E)=\mathcal{R}_\omega(E)$ for all $k\in\mathbb{N}\cup\{0\}$.

Consider the canonical system on $[0,2]$ with Hamiltonian

$$H(t):=\begin{cases} I &,\ t\in[0,1),\\ \begin{pmatrix}1&0\\0&0\end{pmatrix}, & t\in[1,2],\end{cases}$$

and put $E := E_2$, cf. Example 2.4. Then $E \in \mathcal{H}B$, $\mathfrak{d}(E) \equiv 0$, and E is explicitly given as

$$E(z) = \cos z + i(\sin z + z\cos z)\,.$$

We have

$$\operatorname{Sub}^s(\mathcal{H}(E)) = \big\{\mathcal{P}W_a : 0 < a \leq 1\big\} \cup \{\mathcal{H}(E)\}\,,$$

and $\operatorname{codim}_{\mathcal{H}(E)} \mathcal{P}W_1 = 1$. In fact,

$$\mathcal{H}(E) = \mathcal{P}W_1 \dot{+} \operatorname{span}\{\cos z\}\,.$$

It follows from [BW, Theorem 3.4] that $\mathcal{R}_\omega(E) \supseteq \mathcal{P}W_1$ for all $\omega \in \operatorname{Adm}^0(E)$, i.e., $\mathcal{R}_\omega(E) \in \{\mathcal{P}W_1, \mathcal{H}(E)\}$.

We give some estimates on E. Let $x \in \mathbb{R}$, $|x| \geq 1$, and assume that $|\cos x| \leq \frac{1}{2|x|}$. Then $|\sin x| \geq \sqrt{3}/2$ and

$$|\sin x + x\cos x| \geq \big||\sin x| - |x\cos x|\big| \geq \frac{\sqrt{3}-1}{2}\,.$$

It follows that

$$|E(x)| \geq \min\Big\{\min_{t\in[-1,1]} |E(t)|, \frac{\sqrt{3}-1}{2}, \frac{1}{2|x|}\Big\} \gtrsim \frac{1}{1+|x|},\quad x \in \mathbb{R}\,. \tag{4.1}$$

Trivially, we have the following estimate from above:

$$|E(x)| \leq 1 + |x|,\; x \in \mathbb{R}\,. \tag{4.2}$$

We show that for all $\beta \in (0,1)$ the function $\omega_\beta(x) := e^{-|x|^\beta}$ belongs to $\operatorname{Adm}^0(E)$. Choose $\beta' \in (\beta, 1)$. It is well known (see [HJ, p. 276] or [K, p. 159]) that there exists $F \in \mathcal{P}W_1 \setminus \{0\}$ such that

$$|F(x)| \lesssim e^{-|x|^{\beta'}},\; x \in \mathbb{R}\,.$$

It follows that $F \in \mathcal{H}(E)$ and, by (4.1),

$$\Big|\frac{F(x)}{E(x)}\Big| \lesssim |F(x)|(1+|x|) \lesssim e^{-|x|^{\beta'}}(1+|x|) \lesssim e^{-|x|^\beta},\; x \in \mathbb{R}\,.$$

We show that $\mathcal{R}_{\omega_\beta}(E) = \mathcal{P}W_1$. We already know that $\mathcal{R}_{\omega_\beta}(E) \supseteq \mathcal{P}W_1$. Assume that $\mathcal{R}_{\omega_\beta}(E) \supsetneq \mathcal{P}W_1$, then also $R_{\omega_\beta}(E) \supsetneq \mathcal{P}W_1$. Hence there exist $\lambda \in \mathbb{C}$, $\lambda \neq 0$, and $F_0 \in \mathcal{P}W_1$, such that $F_0 + \lambda\cos z \in R_{\omega_\beta}(E)$, i.e.,

$$\Big|\frac{F_0(x) + \lambda\cos x}{E(x)}\Big| \lesssim \omega_\beta(x)\,.$$

By (4.2),

$$|\cos x| \lesssim (1+|x|)\,\omega_\beta(x) + |F_0(x)|,\; x \in \mathbb{R}\,.$$

We have reached a contradiction, since both F_0 and $(1+|x|)\omega_\beta(x)$ belong to $L^2(\mathbb{R})$.

For all $\beta \in (0,1)$ and $k \in \mathbb{N} \cup \{0\}$ we have $\mathcal{R}_{\omega^{[k]}}(E) = \mathcal{P}W_1$. Choose $\beta' \in (0,\beta)$, then $(1+|x|)^k e^{-|x|^\beta} \lesssim e^{-|x|^{\beta'}}$, i.e., $\omega_\beta^{[k]} \lesssim \omega_{\beta'}$. Hence $\mathcal{R}_{\omega_\beta^{[k]}}(E) \subseteq \mathcal{R}_{\omega_{\beta'}}(E) = \mathcal{P}W_1$.

Although Example 4.1 shows that the statement of Proposition 3.6 is not true without the assumption that $\dim \mathcal{R}_\omega(E) < \infty$, still there always exist some representing majorants which behave nicely in this respect.

For $\mathcal{L} \in \operatorname{Sub}(E)$ define

$$\delta_+(\mathcal{L}) := \sup\left\{ \dim \mathcal{H}/\mathcal{L} : \mathcal{H} \in \operatorname{Sub}^{\mathfrak{d}(\mathcal{L})}(E), \dim \mathcal{H}/\mathcal{L} < \infty \right\}.$$

4.2. Proposition. *Let $E \in \mathcal{H}B$ and $\mathcal{L} \in \mathfrak{R}(E) \cap \operatorname{Sub}(E)$. Then there exists $\omega \in \operatorname{Adm}(E)$ such that $\mathcal{R}_\omega(E) = \mathcal{L}$ and*

$$\dim\left(\mathcal{R}_{\omega^{[k+1]}}(E)/\mathcal{R}_{\omega^{[k]}}(E)\right) = \begin{cases} 1, & 0 \le k < \delta_+(\mathcal{L}), \\ 0, & k \ge \delta_+(\mathcal{L}). \end{cases}$$

Proof. Write $\mathcal{L} = \mathcal{H}(E_0)$. Assume that $\mathcal{H}(E_1) \in \operatorname{Sub}^{\mathfrak{d}(\mathcal{L})}(E)$ is such that $n := \dim(\mathcal{H}(E_1)/\mathcal{H}(E_0)) < \infty$. Then there exists a 2×2-matrix polynomial $M(z)$ of degree n, such that ($A_j := \frac{1}{2}(E_j + E_j^{\#})$, $B_j := \frac{i}{2}(B_j - B_j^{\#})$ for $j = 0, 1$)

$$\left(A_1(z), B_1(z)\right) = \left(A_0(z), B_0(z)\right) M(z). \tag{4.3}$$

Put

$$\omega(x) := \frac{|E_0(x)|}{(1+|x|)|E(x)|}, \quad \omega_1(x) := \frac{|E_1(x)|}{(1+|x|)|E(x)|}.$$

Then, by the proof of sufficiency in [BW, Theorem 3.4], we have

$$\mathcal{L} = \mathcal{H}(E_0) = \mathcal{R}_\omega(E), \ \mathcal{H}(E_1) = \mathcal{R}_{\omega_1}(E).$$

However, we see from (4.3) that $\omega_1 \lesssim \omega^{[n]}$. This implies

$$\mathcal{H}(E_1) = \mathcal{R}_{\omega_1}(E) \subseteq \mathcal{R}_{\omega^{[n]}}(E).$$

By Lemma 3.5, (i), we have $\dim(\mathcal{R}_{\omega^{[n]}}(E)/\mathcal{R}_\omega(E)) \le n$, and hence it follows that $\mathcal{H}(E_1) = \mathcal{R}_{\omega^{[n]}}(E)$. □

References

[BH] A.D. Baranov, V.P. Havin: *Admissible majorants for model subspaces and arguments of inner functions,* Functional. Anal. Appl. 40(4) (2006), 249–263.

[BBH] A.D. Baranov, A.A. Borichev, V.P. Havin: *Majorants of meromorphic functions with fixed poles,* Indiana Univ. Math. J. 56(4) (2007), 1595–1628..

[BW] A. Baranov, H. Woracek: *Subspaces of de Branges spaces generated by majorants,* Canad. Math. J., to appear.

[dB1] L. de Branges: *Some mean squares of entire functions,* Proc. Amer. Math. Soc. 10 (1959), 833–839.

[dB2] L. de Branges: *Some Hilbert spaces of entire functions,* Proc. Amer. Math. Soc. 10 (1959), 840–846.

[dB3] L. de Branges: *Some Hilbert spaces of entire functions,* Trans. Amer. Math. Soc. 96 (1960), 259–295.

[dB4] L. de Branges: *Some Hilbert spaces of entire functions II,* Trans. Amer. Math. Soc. 99 (1961), 118–152.

[dB5] L. de Branges: *Some Hilbert spaces of entire functions III,* Trans. Amer. Math. Soc. 100 (1961), 73–115.

[dB6] L. de Branges: *Some Hilbert spaces of entire functions IV,* Trans. Amer. Math. Soc. 105 (1962), 43–83.

[dB7] L. de Branges: *Hilbert spaces of entire functions,* Prentice-Hall, London 1968.

[GK] I. Gohberg, M.G. Krein: *Theory and applications of Volterra operators in Hilbert space,* Translations of Mathematical Monographs, AMS. Providence, Rhode Island, 1970.

[HSW] S. Hassi, H.S.V. de Snoo, H. Winkler: *Boundary-value problems for two-dimensional canonical systems,* Integral Equations Operator Theory 36 (4) (2000), 445–479.

[HJ] V. Havin, B. Jöricke: *The uncertainty principle in harmonic analysis,* Springer-Verlag, Berlin 1994.

[HM1] V.P. Havin, J. Mashreghi: *Admissible majorants for model subspaces of H^2. Part I: fast winding of the generating inner function,* Can. J. Math. 55 (6) (2003), 1231–1263.

[HM2] V.P. Havin, J. Mashreghi: *Admissible majorants for model subspaces of H^2. Part I: slow winding of the generating inner function,* Can. J. Math. 55 (6) (2003), 1264-1301.

[HMN] V.P. Havin, J. Mashreghi, F.L. Nazarov: *Beurling–Malliavin multiplier theorem: the 7th proof,* St. Petersburg Math. J. 17 (5) (2006), 699–744.

[K] P. Koosis: *The Logarithmic Integral II,* Cambridge Stud. Adv. Math. **21**, 1992.

[R] C. Remling: *Schrödinger operators and de Branges spaces,* J. Funct. Anal. 196(2) (2002), 323–394.

[RR] M. Rosenblum, J. Rovnyak: *Topics in Hardy classes and univalent functions,* Birkhäuser Verlag, Basel 1994.

Anton Baranov
Department of Mathematics and Mechanics
Saint Petersburg State University
28 Universitetski pr.
198504 St.Petersburg, Russia
e-mail: `abaranov@kth.se`

Harald Woracek
Institut for Analysis and Scientific Computing
Vienna University of Technology
Wiedner Hauptstr. 8–10/101
A-1040 Wien, Austria
e-mail: `harald.woracek@tuwien.ac.at`

Operator Theory:
Advances and Applications, Vol. 188, 49–85

Trace Formulae for Dissipative and Coupled Scattering Systems

Jussi Behrndt, Mark M. Malamud and Hagen Neidhardt

Abstract. For scattering systems consisting of a (family of) maximal dissipative extension(s) and a self-adjoint extension of a symmetric operator with finite deficiency indices, the spectral shift function is expressed in terms of an abstract Titchmarsh-Weyl function and a variant of the Birman-Krein formula is proved.

Mathematics Subject Classification (2000). Primary 47A40; Secondary 47A55, 47B44.

Keywords. Scattering system, scattering matrix, boundary triplet, Titchmarsh-Weyl function, spectral shift function, Birman-Krein formula.

1. Introduction

The main objective of this paper is to apply and to extend results from [9] and [10] on scattering matrices and spectral shift functions for pairs of self-adjoint or maximal dissipative extensions of a symmetric operator A with finite deficiency indices in a Hilbert space $\mathfrak{H}$.

Let us first briefly recall some basic concepts. For a pair of self-adjoint operators H and H_0 in $\mathfrak{H}$ the wave operators $W_\pm(H, H_0)$ of the scattering system $\{H, H_0\}$ are defined by

$$W_\pm(H, H_0) = \text{s-}\lim_{t \to \pm\infty} e^{iHt} e^{-iH_0 t} P_{ac}(H_0),$$

where $P^{ac}(H_0)$ is the projection onto the absolutely continuous subspace of the unperturbed operator H_0. If for instance the resolvent difference

$$(H - z)^{-1} - (H_0 - z)^{-1} \in \mathfrak{S}_1, \qquad z \in \rho(H) \cap \rho(H_0), \tag{1.1}$$

is a trace class operator, then it is well known that the wave operators $W_\pm(H, H_0)$ exist and are isometric in $\mathfrak{H}$, see, e.g., [53]. The scattering operator $S(H, H_0)$ of the scattering system $\{H, H_0\}$ is defined by

$$S(H, H_0) = W_+(H, H_0)^* W_-(H, H_0).$$

$S(H, H_0)$ commutes with H_0 and is unitary on the absolutely continuous subspace of H_0. Therefore $S(H, H_0)$ is unitarily equivalent to a multiplication operator induced by a family $S(H, H_0; \lambda)$ of unitary operators in the spectral representation of H_0. This family is usually called the scattering matrix of the scattering system $\{H, H_0\}$ and is one of the most important quantities in the analysis of scattering processes.

Another important object in scattering and perturbation theory is the so-called spectral shift function introduced by M.G. Krein in [33]. For the case $\mathrm{dom}\,(H) = \mathrm{dom}\,(H_0)$ and $V = H - H_0 \in \mathfrak{S}_1$ a spectral shift function ξ of the pair $\{H, H_0\}$ was defined with the help of the perturbation determinant

$$D_{H/H_0}(z) := \det\left((H - z)(H_0 - z)^{-1}\right). \tag{1.2}$$

Since $\lim_{|\Im\mathrm{m}\,(z)|\to\infty} D_{H/H_0}(z) = 1$ a branch of $z \mapsto \log(D_{H/H_0}(z))$ in the upper half-plane $\mathbb{C}_+$ is fixed by the condition $\log(D_{H/H_0}(z)) \to 0$ as $\Im\mathrm{m}\,(z) \to \infty$ and the spectral shift function is then defined by

$$\xi(\lambda) = \frac{1}{\pi}\Im\mathrm{m}\left(\log\left(D_{H/H_0}(\lambda + i0)\right)\right) = \frac{1}{\pi}\lim_{\varepsilon\to 0}\Im\mathrm{m}\left(\ln\left(D_{H/H_0}(\lambda + i\varepsilon)\right)\right). \tag{1.3}$$

M.G. Krein proved that $\xi \in L_1(\mathbb{R}, d\lambda)$, $\|\xi\|_{L_1} \le \|V\|_1$, and that the trace formula

$$\mathrm{tr}\left((H - z)^{-1} - (H_0 - z)^{-1}\right) = -\int_{\mathbb{R}} \frac{\xi(\lambda)}{(\lambda - z)^2}\,d\lambda \tag{1.4}$$

holds for all $z \in \rho(H) \cap \rho(H_0)$. It turns out that the scattering matrix and the spectral shift function of the pair $\{H, H_0\}$ are related via the Birman-Krein formula:

$$\det\left(S(H, H_0; \lambda)\right) = \exp\left(-2\pi i \xi(\lambda)\right) \qquad \text{for a.e.} \quad \lambda \in \mathbb{R}. \tag{1.5}$$

The trace formula and the Birman-Krein formula can be extended to the case that only the resolvent difference (1.1) of H and H_0 is trace class. Namely, then there exists a real measurable function $\xi \in L^1(\mathbb{R}, (1+\lambda^2)^{-1} d\lambda)$ such that (1.4) and (1.5) hold. However, in this situation it is not immediately clear how to substitute the perturbation determinant in (1.3).

In Section 2 we propose a possible solution of this problem for pairs of self-adjoint extensions A_0 and A_Θ of a densely defined symmetric operator A with finite deficiency indices. Observe that here the resolvent difference is even a finite rank operator. In order to describe the pair $\{A_\Theta, A_0\}$ and a corresponding spectral shift function we use the notion of boundary triplets and associated Weyl functions. More precisely, we choose a boundary triplet $\Pi = \{\mathcal{H}, \Gamma_0, \Gamma_1\}$ for A^* and a self-adjoint parameter Θ in $\mathcal{H}$ such that

$$A_0 = A^* \upharpoonright \ker(\Gamma_0) \quad \text{and} \quad A_\Theta = A^* \upharpoonright \ker(\Gamma_1 - \Theta\Gamma_0)$$

holds. If $M(\cdot)$ is the Weyl function associated with this boundary triplet it is shown in Theorem 2.4 (see also [9] and [34] for special cases) that a spectral shift function $\xi(\cdot)$ of the pair $\{A_\Theta, A_0\}$ can be chosen as

$$\begin{aligned}\xi_\Theta(\lambda) &= \frac{1}{\pi}\Im\mathrm{m}\,\big(\mathrm{tr}\big(\log(M(\lambda+i0)-\Theta)\big)\big)\\ &= \frac{1}{\pi}\Im\mathrm{m}\,\big(\log\big(\det(M(\lambda+i0)-\Theta)\big)\big)+2k, \quad k\in\mathbb{Z}.\end{aligned} \tag{1.6}$$

By comparing (1.6) with (1.3) it is clear that $\det\big(M(z)-\Theta\big)$ plays a similar role as the perturbation determinant (1.2) for additive perturbations. Moreover, a simple proof of the Birman-Krein formula (1.5) in this situation is obtained in Section 2.5 by using the representation

$$S_\Theta(\lambda) = I_{\mathcal{H}_{M(\lambda)}} + 2i\sqrt{\Im\mathrm{m}\,(M(\lambda))}\big(\Theta - M(\lambda)\big)^{-1}\sqrt{\Im\mathrm{m}\,(M(\lambda))} \tag{1.7}$$

of the scattering matrix $S_\Theta(\cdot) = S(A_\Theta, A_0;\cdot)$ of the scattering system $\{A_\Theta, A_0\}$ from [9], cf. also the work [7] by V.M. Adamyan and B.S. Pavlov.

These results are generalized to maximal dissipative extensions in Section 3. Let again A be a symmetric operator in $\mathfrak{H}$ with finite deficiency and let $\Pi = \{\mathcal{H}, \Gamma_0, \Gamma_1\}$ be a boundary triplet for A^*. If D is a dissipative matrix in $\mathcal{H}$, $\Im\mathrm{m}\,(D) \le 0$, then $A_D = A^* \restriction \ker(\Gamma_1 - D\Gamma_0)$ is a maximal dissipative extension of A. For the scattering system $\{A_D, A_0\}$ the wave operators $W_\pm(A_D, A_0)$, the scattering operator $S(A_D, A_0)$ and the scattering matrix $S(A_D, A_0;\lambda)$ can be defined similarly as in the self-adjoint case. It turns out that the representation (1.7) extends to the dissipative case. More precisely, the Hilbert space $L^2(\mathbb{R}, \mathcal{H}_\lambda, d\lambda)$, $\mathcal{H}_\lambda := \mathrm{ran}\,(\Im\mathrm{m}\,(M(\lambda+i0))$, performs a spectral representation of the absolutely continuous part A_0^{ac} of A_0 and the scattering matrix $S_D(\cdot) := S(A_D, A_0;\cdot)$ of the scattering system $\{A_D, A_0\}$ admits the representation

$$S_D(\lambda) = I_{\mathcal{H}_{M(\lambda)}} + 2i\sqrt{\Im\mathrm{m}\,(M(\lambda))}\big(D - M(\lambda)\big)^{-1}\sqrt{\Im\mathrm{m}\,(M(\lambda))},$$

cf. [10, Theorem 3.8]. With the help of a minimal self-adjoint dilation $\widetilde{K}$ of A_D in the Hilbert space $\mathfrak{H} \oplus L^2(\mathbb{R}, \mathcal{H}_D)$, $\mathcal{H}_D := \mathrm{ran}\,(\Im\mathrm{m}\,(D))$, we verify that there is a spectral shift function η_D of the pair $\{A_D, A_0\}$ such that the trace formula

$$\mathrm{tr}\,\big((A_D - z)^{-1} - (A_0 - z)^{-1}\big) = -\int_\mathbb{R} \frac{\eta_D(\lambda)}{(\lambda - z)^2}\,d\lambda, \quad z \in \mathbb{C}_+,$$

holds and $\eta_D(\cdot)$ admits the representation

$$\begin{aligned}\eta_D(\lambda) &= \frac{1}{\pi}\Im\mathrm{m}\,\big(\mathrm{tr}\big(\log(M(\lambda+i0)-D)\big)\big)\\ &= \frac{1}{\pi}\Im\mathrm{m}\,\big(\log\big(\det(M(\lambda+i0)-D)\big)\big)+2k, \quad k\in\mathbb{Z},\end{aligned}$$

cf. Theorem 3.3. For modified trace formulas the reader is referred to [6]. In Section 3.4 we show that the Birman-Krein formula holds in the modified form

$$\det\big(S_D(\lambda)\big) = \det\big(W_{A_D}(\lambda - i0)\big)\exp\big(-2\pi i\eta_D(\lambda)\big)$$

for a.e. $\lambda \in \mathbb{R}$, where $z \mapsto W_{A_D}(z)$, $z \in \mathbb{C}_-$, is the characteristic function of the maximal dissipative operator A_D. Since by [1, 2, 3, 4] the limit $W_{A_D}(\lambda - i0)^*$ can be regarded as the scattering matrix $S^{LP}(\cdot)$ of an appropriate Lax-Phillips scattering system one gets finally the representation

$$\det\bigl(S_D(\lambda)\bigr) = \overline{\det\bigl(S^{LP}(\lambda)\bigr)} \exp\bigl(-2\pi i \eta_D(\lambda)\bigr)$$

for a.e. $\lambda \in \mathbb{R}$. The results correspond to similar results for additive dissipative perturbations, see [5, 42, 43, 44, 49, 50, 51].

In Section 4 so-called open quantum systems with finite rank coupling are investigated. Here we follow [10]. From the mathematical point of view these open quantum systems are closely related to the Krein-Naimark formula for generalized resolvents and the Štraus family of extensions of a symmetric operator. Recall that the Krein-Naimark formula establishes a one-to-one correspondence between the generalized resolvents $z \mapsto P_{\mathfrak{H}}(\widetilde{L} - z)^{-1} \upharpoonright \mathfrak{H}$ of the symmetric operator A, that is, the compressed resolvents of self-adjoint extensions $\widetilde{L}$ of A in bigger Hilbert spaces, and the class of Nevanlinna families $\tau(\cdot)$ via

$$P_{\mathfrak{H}}(\widetilde{L} - z)^{-1} \upharpoonright \mathfrak{H} = (A_0 - z)^{-1} - \gamma(z)\bigl(\tau(z) + M(z)\bigr)^{-1}\gamma(\bar{z})^*.$$

Here $\Pi_A = \{\mathcal{H}, \Gamma_0, \Gamma_1\}$ is a boundary triplet for A^* and γ and M are the corresponding γ-field and Weyl function, respectively. It can be shown that the generalized resolvent coincides pointwise with the resolvent of the Štraus extension

$$A_{-\tau(z)} := A^* \upharpoonright \ker\bigl(\Gamma_1 + \tau(z)\Gamma_0\bigr),$$

i.e., $P_{\mathfrak{H}}(\widetilde{L} - z)^{-1} \upharpoonright \mathfrak{H} = (A_{-\tau(z)} - z)^{-1}$ holds, and that for $z \in \mathbb{C}_+$ each extension $A_{-\tau(z)}$ of A is maximal dissipative in $\mathfrak{H}$.

Under additional assumptions $\tau(\cdot)$ can be realized as the Weyl function corresponding to a densely defined closed simple symmetric operator T with finite deficiency indices in some Hilbert space $\mathfrak{G}$ and a boundary triplet $\Pi_T = \{\mathcal{H}, \Upsilon_0, \Upsilon_1\}$ for T^*. Then the self-adjoint (exit space) extension $\widetilde{L}$ of A can be recovered as a coupling of the operators A and T corresponding to a coupling of the boundary triplets Π_A and Π_T (see [17] and formula (4.4) below). Let $T_0 = T^* \upharpoonright \ker(\Upsilon_0)$ be the self-adjoint extension of T in $\mathfrak{G}$ corresponding to the boundary mapping Υ_0. We prove in Theorem 4.2 that for the system $\{\widetilde{L}, A_0 \oplus T_0\}$ there exists a spectral shift function $\widetilde{\xi}(\cdot)$ given by

$$\widetilde{\xi}(\lambda) = \frac{1}{\pi}\Im\mathrm{m}\,\Bigl(\mathrm{tr}\bigl(\log(M(\lambda + i0) + \tau(\lambda + i0))\bigr)\Bigr)$$

and that the modified trace formula

$$\begin{aligned}\mathrm{tr}\,\bigl((A_{-\tau(z)} - z)^{-1} - (A_0 - z)^{-1}\bigr) &+ \\ \mathrm{tr}\,\bigl((T_{-M(z)} - z)^{-1} - (T_0 - z)^{-1}\bigr) &= -\int_{\mathbb{R}} \frac{1}{(\lambda - z)^2}\,\widetilde{\xi}(\lambda)\,d\lambda\end{aligned}$$

holds for all $z \in \mathbb{C} \setminus \mathbb{R}$.

With the help of the channel wave operators

$$W_\pm(\widetilde{L}, A_0) = \text{s-}\lim_{t\to\pm\infty} e^{it\widetilde{L}} e^{-itA_0} P^{ac}(A_0)$$

$$W_\pm(\widetilde{L}, T_0) = \text{s-}\lim_{t\to\pm\infty} e^{it\widetilde{L}} e^{-itT_0} P^{ac}(T_0)$$

one then defines the channel scattering operators

$$S_{\mathfrak{H}} := W_+(\widetilde{L}, A_0)^* W_-(\widetilde{L}, A_0) \quad \text{and} \quad S_{\mathfrak{G}} := W_+(\widetilde{L}, T_0)^* W_-(\widetilde{L}, T_0).$$

The corresponding channel scattering matrices $S_{\mathfrak{H}}(\lambda)$ and $S_{\mathfrak{G}}(\lambda)$ are studied in Section 4.3. Here we express these scattering matrices in terms of the functions M and τ in the spectral representations $L^2(\mathbb{R}, d\lambda, \mathcal{H}_{M(\lambda)})$ and $L^2(\mathbb{R}, d\lambda, \mathcal{H}_{\tau(\lambda)})$ of A_0^{ac} and T_0^{ac}, respectively, and finally, with the help of these representations the modified Birman-Krein formula

$$\det(S_{\mathfrak{H}}(\lambda)) = \overline{\det(S_{\mathfrak{G}}(\lambda))} \exp\bigl(-2\pi i \widetilde{\xi}(\lambda)\bigr)$$

is proved in Theorem 4.6.

2. Self-adjoint extensions and scattering

In this section we consider scattering systems consisting of two self-adjoint extensions of a densely defined symmetric operator with equal finite deficiency indices in a separable Hilbert space. We generalize a result on the representation of the spectral shift function of such a scattering system from [9] and we give a short proof of the Birman-Krein formula in this setting.

2.1. Boundary triplets and closed extensions

Let A be a densely defined closed symmetric operator with equal deficiency indices $n_\pm(A) = \dim\ker(A^* \mp i) \leq \infty$ in the separable Hilbert space $\mathfrak{H}$. We use the concept of a boundary triplet for A^* in order to describe the closed extensions $A_\Theta \subset A^*$ of A in $\mathfrak{H}$, see [30] and also [20, 22].

Definition 2.1. *A triplet* $\Pi = \{\mathcal{H}, \Gamma_0, \Gamma_1\}$ *is called a* boundary triplet *for the adjoint operator* A^* *if* $\mathcal{H}$ *is a Hilbert space and* $\Gamma_0, \Gamma_1 : \operatorname{dom}(A^*) \to \mathcal{H}$ *are linear mappings such that*

(i) *the abstract second Green's identity,*

$$(A^* f, g) - (f, A^* g) = (\Gamma_1 f, \Gamma_0 g) - (\Gamma_0 f, \Gamma_1 g),$$

holds for all $f, g \in \operatorname{dom}(A^*)$ *and*

(ii) *the mapping* $\Gamma := (\Gamma_0, \Gamma_1)^\top : \operatorname{dom}(A^*) \longrightarrow \mathcal{H} \times \mathcal{H}$ *is surjective.*

We refer to [20] and [22] for a detailed study of boundary triplets and recall only some important facts. First of all a boundary triplet $\Pi = \{\mathcal{H}, \Gamma_0, \Gamma_1\}$ for A^* exists since the deficiency indices $n_\pm(A)$ of A are assumed to be equal. Then $n_\pm(A) = \dim \mathcal{H}$ holds. We note that a boundary triplet for A^* is not unique.

Namely, if $\Pi' = \{\mathcal{G}', \Gamma_0', \Gamma_1'\}$ is another boundary triplet for A^*, then there exists a boundedly invertible operator $W = (W_{ij})_{i,j=1}^2 \in [\mathcal{G} \oplus \mathcal{G}, \mathcal{G}' \oplus \mathcal{G}']$ with the property

$$W^* \begin{pmatrix} 0 & -iI_{\mathcal{G}'} \\ iI_{\mathcal{G}'} & 0 \end{pmatrix} W = \begin{pmatrix} 0 & -iI_{\mathcal{G}} \\ iI_{\mathcal{G}} & 0 \end{pmatrix},$$

such that

$$\begin{pmatrix} \Gamma_0' \\ \Gamma_1' \end{pmatrix} = \begin{pmatrix} W_{11} & W_{12} \\ W_{21} & W_{22} \end{pmatrix} \begin{pmatrix} \Gamma_0 \\ \Gamma_1 \end{pmatrix}$$

holds. Here and in the following we write $[\mathfrak{K}, \mathcal{K}]$ for the set of bounded everywhere defined linear operators acting from a Hilbert space $\mathfrak{K}$ into a Hilbert space $\mathcal{K}$. For brevity we write $[\mathcal{K}]$ if $\mathcal{K} = \mathfrak{K}$.

An operator A' is called a *proper extension* of A if A' is closed and satisfies $A \subseteq A' \subseteq A^*$. In order to describe the set of proper extensions of A with the help of a boundary triplet $\Pi = \{\mathcal{H}, \Gamma_0, \Gamma_1\}$ for A^* we have to consider the set $\widetilde{\mathcal{C}}(\mathcal{H})$ of closed linear relations in $\mathcal{H}$, that is, the set of closed linear subspaces of $\mathcal{H} \times \mathcal{H}$. Linear operators in $\mathcal{H}$ are identified with their graphs, so that the set $\mathcal{C}(\mathcal{H})$ of closed linear operators is viewed as a subset of $\widetilde{\mathcal{C}}(\mathcal{H})$. For the usual definitions of the linear operations with linear relations, the inverse, the resolvent set and the spectrum we refer to [23]. Recall that the adjoint relation $\Theta^* \in \widetilde{\mathcal{C}}(\mathcal{H})$ of a linear relation Θ in $\mathcal{H}$ is defined as

$$\Theta^* := \left\{ \begin{pmatrix} k \\ k' \end{pmatrix} : (h', k) = (h, k') \text{ for all } \begin{pmatrix} h \\ h' \end{pmatrix} \in \Theta \right\} \tag{2.1}$$

and Θ is said to be *symmetric* (*self-adjoint*) if $\Theta \subseteq \Theta^*$ (resp. $\Theta = \Theta^*$). Note that (2.1) extends the definition of the adjoint operator. A linear relation Θ is called *dissipative* if $\Im\mathrm{m}\,(g', g) \leq 0$ holds for all $\binom{g}{g'} \in \Theta$ and Θ is said to be *maximal dissipative* if Θ is dissipative and each dissipative extension of Θ coincides with Θ itself. In this case the upper half-plane $\mathbb{C}_+ = \{\lambda \in \mathbb{C} : \Im\mathrm{m}\,\lambda > 0\}$ belongs to the resolvent set $\rho(\Theta)$. Furthermore, a linear relation Θ is called *accumulative* (*maximal accumulative*) if $-\Theta$ is dissipative (resp. maximal dissipative). For a maximal accumulative relation Θ we have $\mathbb{C}_- = \{\lambda \in \mathbb{C} : \Im\mathrm{m}\,\lambda < 0\} \subset \rho(\Theta)$.

With a boundary triplet $\Pi = \{\mathcal{H}, \Gamma_0, \Gamma_1\}$ for A^* one associates two self-adjoint extensions of A defined by

$$A_0 := A^* \upharpoonright \ker(\Gamma_0) \quad \text{and} \quad A_1 := A^* \upharpoonright \ker(\Gamma_1).$$

A description of all proper extensions of A is given in the next proposition. Note also that the self-adjointness of A_0 and A_1 is a consequence of Proposition 2.2 (ii).

Proposition 2.2. *Let A be a densely defined closed symmetric operator in $\mathfrak{H}$ and let $\Pi = \{\mathcal{H}, \Gamma_0, \Gamma_1\}$ be a boundary triplet for A^*. Then the mapping*

$$\Theta \mapsto A_\Theta := \Gamma^{-1}\Theta = \{f \in \mathrm{dom}\,(A^*) : \Gamma f = (\Gamma_0 f, \Gamma_1 f)^\top \in \Theta\} \tag{2.2}$$

establishes a bijective correspondence between the set $\widetilde{\mathcal{C}}(\mathcal{H})$ and the set of proper extensions of A. Moreover, for $\Theta \in \widetilde{\mathcal{C}}(\mathcal{H})$ the following assertions hold.

(i) $(A_\Theta)^* = A_{\Theta^*}$.
(ii) A_Θ *is symmetric (self-adjoint) if and only if* Θ *is symmetric (resp. self-adjoint).*
(iii) A_Θ *is dissipative (maximal dissipative) if and only if* Θ *is dissipative (resp. maximal dissipative).*
(iv) A_Θ *is accumulative (maximal accumulative) if and only if* Θ *is accumulative (resp. maximal accumulative).*
(v) A_Θ *is disjoint with* A_0, *that is* $\mathrm{dom}\,(A_\Theta) \cap \mathrm{dom}\,(A_0) = \mathrm{dom}\,(A)$, *if and only if* $\Theta \in \mathcal{C}(\mathcal{H})$. *In this case the extension* A_Θ *in* (2.2) *is given by*

$$A_\Theta = A^* \upharpoonright \ker\big(\Gamma_1 - \Theta\Gamma_0\big). \tag{2.3}$$

We note that (2.3) holds also for linear relations Θ if the expression $\Gamma_1 - \Theta\Gamma_0$ is interpreted in the sense of linear relations.

In the following we shall often be concerned with simple symmetric operators. Recall that a symmetric operator is said to be *simple* if there is no nontrivial subspace which reduces it to a self-adjoint operator. By [32] each symmetric operator A in $\mathfrak{H}$ can be written as the direct orthogonal sum $\widehat{A} \oplus A_s$ of a simple symmetric operator $\widehat{A}$ in the Hilbert space

$$\widehat{\mathfrak{H}} = \mathrm{clospan}\big\{\ker(A^* - \lambda) : \lambda \in \mathbb{C}\backslash\mathbb{R}\big\}$$

and a self-adjoint operator A_s in $\mathfrak{H} \ominus \widehat{\mathfrak{H}}$. Here $\mathrm{clospan}\{\cdot\}$ denotes the closed linear span of a set. Obviously A is simple if and only if $\widehat{\mathfrak{H}}$ coincides with $\mathfrak{H}$.

2.2. Weyl functions and resolvents of extensions

Let A be a densely defined closed symmetric operator in $\mathfrak{H}$ with equal deficiency indices as in Section 2.1. If $\lambda \in \mathbb{C}$ is a point of regular type of A, i.e., $(A-\lambda)^{-1}$ is bounded, we denote the *defect subspace* of A by $\mathcal{N}_\lambda = \ker(A^* - \lambda)$. The following definition can be found in [19, 20, 22].

Definition 2.3. *Let* $\Pi = \{\mathcal{H}, \Gamma_0, \Gamma_1\}$ *be a boundary triplet for the operator* A^* *and let* $A_0 = A^* \upharpoonright \ker(\Gamma_0)$. *The operator-valued functions* $\gamma(\cdot) : \rho(A_0) \to [\mathcal{H}, \mathfrak{H}]$ *and* $M(\cdot) : \rho(A_0) \to [\mathcal{H}]$ *defined by*

$$\gamma(\lambda) := \big(\Gamma_0 \upharpoonright \mathcal{N}_\lambda\big)^{-1} \quad \text{and} \quad M(\lambda) := \Gamma_1\gamma(\lambda), \quad \lambda \in \rho(A_0), \tag{2.4}$$

are called the γ-field *and the* Weyl function, *respectively, corresponding to the boundary triplet* $\Pi = \{\mathcal{H}, \Gamma_0, \Gamma_1\}$.

It follows from the identity $\mathrm{dom}\,(A^*) = \ker(\Gamma_0) \dotplus \mathcal{N}_\lambda$, $\lambda \in \rho(A_0)$, where as above $A_0 = A^* \upharpoonright \ker(\Gamma_0)$, that the γ-field $\gamma(\cdot)$ in (2.4) is well-defined. It is easily seen that both $\gamma(\cdot)$ and $M(\cdot)$ are holomorphic on $\rho(A_0)$ and the relations

$$\gamma(\mu) = \big(I + (\mu - \lambda)(A_0 - \mu)^{-1}\big)\gamma(\lambda), \qquad \lambda, \mu \in \rho(A_0),$$

and

$$M(\lambda) - M(\mu)^* = (\lambda - \bar{\mu})\gamma(\mu)^*\gamma(\lambda), \qquad \lambda, \mu \in \rho(A_0), \tag{2.5}$$

are valid (see [20]). The identity (2.5) yields that $M(\cdot)$ is a $[\mathcal{H}]$-valued *Nevanlinna function*, that is, $M(\cdot)$ is holomorphic on $\mathbb{C}\backslash\mathbb{R}$ and takes values in $[\mathcal{H}]$, $M(\lambda) = M(\bar{\lambda})^*$ for all $\lambda \in \mathbb{C}\backslash\mathbb{R}$ and $\Im m\,(M(\lambda))$ is a nonnegative operator for all λ in the upper half-plane $\mathbb{C}_+$. Moreover, it follows that $0 \in \rho(\Im m\,(M(\lambda)))$ holds. It is important to note that if the operator A is simple, then the Weyl function $M(\cdot)$ determines the pair $\{A, A_0\}$ uniquely up to unitary equivalence, cf. [19, 20].

In the case that the deficiency indices $n_+(A) = n_-(A)$ are finite the Weyl function M corresponding to the boundary triplet $\Pi = \{\mathcal{H}, \Gamma_0, \Gamma_1\}$ is a matrix-valued Nevanlinna function in the finite-dimensional space $\mathcal{H}$. From [24, 25] one gets the existence of the (strong) limit

$$M(\lambda + i0) = \lim_{\epsilon \to +0} M(\lambda + i\epsilon)$$

from the upper half-plane for a.e. $\lambda \in \mathbb{R}$.

Let now $\Pi = \{\mathcal{H}, \Gamma_0, \Gamma_1\}$ be a boundary triplet for A^* with γ-field $\gamma(\cdot)$ and Weyl function $M(\cdot)$. The spectrum and the resolvent set of a proper (not necessarily self-adjoint) extension of A can be described with the help of the Weyl function. If $A_\Theta \subseteq A^*$ is the extension corresponding to $\Theta \in \widetilde{\mathcal{C}}(\mathcal{H})$ via (2.2), then a point $\lambda \in \rho(A_0)$ belongs to $\rho(A_\Theta)$ $(\sigma_i(A_0), i = p, c, r)$ if and only if $0 \in \rho(\Theta - M(\lambda))$ (resp. $0 \in \sigma_i(\Theta - M(\lambda))$, $i = p, c, r$). Moreover, for $\lambda \in \rho(A_0) \cap \rho(A_\Theta)$ the well-known resolvent formula

$$(A_\Theta - \lambda)^{-1} = (A_0 - \lambda)^{-1} + \gamma(\lambda)\big(\Theta - M(\lambda)\big)^{-1}\gamma(\bar{\lambda})^* \tag{2.6}$$

holds. Formula (2.6) is a generalization of the known Krein formula for canonical resolvents. We emphasize that it is valid for any proper extension of A with a non-empty resolvent set. It is worth to note that the Weyl function can also be used to investigate the absolutely continuous and singular continuous spectrum of proper extensions of A, cf. [14].

2.3. Spectral shift function and trace formula

Krein's spectral shift function introduced in [33] is an important tool in the spectral and perturbation theory of self-adjoint operators, in particular, scattering theory. A detailed review on the spectral shift function can be found in, e.g., [12, 13]. Furthermore we mention [26, 27, 28] as some recent papers on the spectral shift function and its various applications.

Recall that for any pair of self-adjoint operators H_1, H_0 in a separable Hilbert space $\mathfrak{H}$ such that the resolvents differ by a trace class operator,

$$(H_1 - \lambda)^{-1} - (H_0 - \lambda)^{-1} \in \mathfrak{S}_1(\mathfrak{H}), \tag{2.7}$$

for some (and hence for all) $\lambda \in \rho(H_1) \cap \rho(H_0)$, there exists a real-valued function $\xi(\cdot) \in L^1_{loc}(\mathbb{R})$ satisfying the conditions

$$\operatorname{tr}\big((H_1 - \lambda)^{-1} - (H_0 - \lambda)^{-1}\big) = -\int_{\mathbb{R}} \frac{1}{(t-\lambda)^2}\,\xi(t)\,dt, \tag{2.8}$$

$\lambda \in \rho(H_1) \cap \rho(H_0)$, and

$$\int_{\mathbb{R}} \frac{1}{1+t^2}\, \xi(t)\, dt < \infty, \tag{2.9}$$

cf. [12, 13, 33]. Such a function ξ is called a *spectral shift function* of the pair $\{H_1, H_0\}$. We emphasize that ξ is not unique, since simultaneously with ξ a function $\xi + c$, $c \in \mathbb{R}$, also satisfies both conditions (2.8) and (2.9). Note that the converse also holds, namely, any two spectral shift functions for a pair of self-adjoint operators $\{H_1, H_0\}$ satisfying (2.7) differ by a real constant. We remark that (2.8) is a special case of the general formula

$$\operatorname{tr}\left(\phi(H_1) - \phi(H_0)\right) = \int_{\mathbb{R}} \phi'(t)\, \xi(t)\, dt,$$

which is valid for a wide class of smooth functions, cf. [47] for a large class of such functions $\phi(\cdot)$.

In Theorem 2.4 below we find a representation for the spectral shift function ξ_Θ of a pair of self-adjoint operators A_Θ and A_0 which are both assumed to be extensions of a densely defined closed simple symmetric operator A with equal finite deficiency indices. For that purpose we use the definition

$$\log(T) := -i \int_0^\infty \left((T+it)^{-1} - (1+it)^{-1} I_{\mathcal{H}}\right) dt \tag{2.10}$$

for an operator T in a finite-dimensional Hilbert space $\mathcal{H}$ satisfying $\Im\mathrm{m}\,(T) \geq 0$ and $0 \notin \sigma(T)$, see, e.g., [26, 48]. A straightforward calculation shows that the relation

$$\det(T) = \exp\bigl(\operatorname{tr}\bigl(\log(T)\bigr)\bigr) \tag{2.11}$$

holds. Observe that

$$\operatorname{tr}\bigl(\log(T)\bigr) = \log\bigl(\det(T)\bigr) + 2k\pi i \tag{2.12}$$

holds for some $k \in \mathbb{Z}$. In [9, Theorem 4.1] it was shown that if $\Pi = \{\mathcal{H}, \Gamma_0, \Gamma_1\}$ is a boundary triplet for A^* with $A_0 = A^* \upharpoonright \ker(\Gamma_0)$ and $A_\Theta = A^* \upharpoonright \ker(\Gamma_1 - \Theta\Gamma_0)$ is a self-adjoint extension of A which corresponds to a self-adjoint matrix Θ in $\mathcal{H}$, then the limit $\lim_{\epsilon\to+0} \log\left(M(\lambda + i\epsilon) - \Theta\right)$ exists for a.e. $\lambda \in \mathbb{R}$ and

$$\xi_\Theta(\lambda) := \frac{1}{\pi} \Im\mathrm{m}\left(\operatorname{tr}\bigl(\log(M(\lambda + i0) - \Theta)\bigr)\right) \tag{2.13}$$

defines a spectral shift function for the pair $\{A_\Theta, A_0\}$. We emphasize that Θ was assumed to be a matrix in [9], so that ξ_Θ in (2.13) is a spectral shift function only for special pairs $\{A_\Theta, A_0\}$. Theorem 2.4 below extends the result from [9] to the case of a self-adjoint relation Θ and hence to arbitrary pairs of self-adjoint extensions $\{A_\Theta, A_0\}$ of A.

To this end we first recall that any self-adjoint relation Θ in $\mathcal{H}$ can be written in the form

$$\Theta = \Theta_{\mathrm{op}} \oplus \Theta_\infty \tag{2.14}$$

with respect to the decomposition $\mathcal{H} = \mathcal{H}_{\rm op} \oplus \mathcal{H}_\infty$, where $\Theta_{\rm op}$ is a self-adjoint operator in $\mathcal{H}_{\rm op} := \overline{\mathrm{dom}\,\Theta}$ and Θ_∞ is a pure relation in $\mathcal{H}_\infty := (\mathrm{dom}\,\Theta)^\perp$, that is,

$$\Theta_\infty = \left\{ \begin{pmatrix} 0 \\ h' \end{pmatrix} : h' \in \mathcal{H}_\infty \right\}. \tag{2.15}$$

Since in the following considerations the space $\mathcal{H}$ is finite dimensional we have $\mathcal{H}_{\rm op} = \mathrm{dom}\,\Theta = \mathrm{dom}\,\Theta_{\rm op}$ and $\Theta_{\rm op}$ is a self-adjoint matrix. If $M(\cdot)$ is the Weyl function corresponding to a boundary triplet $\Pi = \{\mathcal{H}, \Gamma_0, \Gamma_1\}$, then

$$M_{\rm op}(\lambda) := P_{\rm op} M(\lambda) \iota_{\rm op}\,, \tag{2.16}$$

is a $[\mathcal{H}_{\rm op}]$-valued Nevanlinna function. Here $P_{\rm op}$ is the orthogonal projection from $\mathcal{H}$ onto $\mathcal{H}_{\rm op}$ and $\iota_{\rm op}$ denotes the canonical embedding of $\mathcal{H}_{\rm op}$ in $\mathcal{H}$. One verifies that

$$\big(\Theta - M(\lambda)\big)^{-1} = \iota_{\rm op} \big(\Theta_{\rm op} - M_{\rm op}(\lambda)\big)^{-1} P_{\rm op} \tag{2.17}$$

holds for all $\lambda \in \mathbb{C}_+$. The following result generalizes [9, Theorem 4.1], see also [34] for a special case.

Theorem 2.4. *Let A be a densely defined closed simple symmetric operator in the separable Hilbert space $\mathfrak{H}$ with equal finite deficiency indices, let $\Pi = \{\mathcal{H}, \Gamma_0, \Gamma_1\}$ be a boundary triplet for A^* and let $M(\cdot)$ be the corresponding Weyl function. Furthermore, let $A_0 = A^* \upharpoonright \ker(\Gamma_0)$ and let $A_\Theta = A^* \upharpoonright \Gamma^{-1}\Theta$, $\Theta \in \widetilde{\mathcal{C}}(\mathcal{H})$, be a self-adjoint extension of A in $\mathfrak{H}$. Then the limit*

$$\lim_{\epsilon \to +0} \log\big(M_{\rm op}(\lambda + i\epsilon) - \Theta_{\rm op}\big)$$

exists for a.e. $\lambda \in \mathbb{R}$ and the function

$$\xi_\Theta(\lambda) := \frac{1}{\pi} \Im\mathrm{m}\,\Big(\mathrm{tr}\big(\log(M_{\rm op}(\lambda + i0) - \Theta_{\rm op})\big)\Big) \tag{2.18}$$

is a spectral shift function for the pair $\{A_\Theta, A_0\}$ with $0 \le \xi_\Theta(\lambda) \le \dim \mathcal{H}_{\rm op}$.

Proof. Since $\lambda \mapsto M_{\rm op}(\lambda) - \Theta_{\rm op}$ is a Nevanlinna function with values in $[\mathcal{H}_{\rm op}]$ and $0 \in \rho(\Im\mathrm{m}\,(M_{\rm op}(\lambda)))$ for all $\lambda \in \mathbb{C}_+$, it follows that $\log(M_{\rm op}(\lambda) - \Theta_{\rm op})$ is well defined for all $\lambda \in \mathbb{C}_+$ by (2.10). According to [26, Lemma 2.8] the function $\lambda \mapsto \log(M_{\rm op}(\lambda) - \Theta_{\rm op})$, $\lambda \in \mathbb{C}_+$, is a $[\mathcal{H}_{\rm op}]$-valued Nevanlinna function such that

$$0 \le \Im\mathrm{m}\,\big(\log(M_{\rm op}(\lambda) - \Theta_{\rm op})\big) \le \pi I_{\mathcal{H}_{\rm op}}$$

holds for all $\lambda \in \mathbb{C}_+$. Hence the limit $\lim_{\epsilon \to +0} \log(M_{\rm op}(\lambda + i\epsilon) - \Theta_{\rm op})$ exists for a.e. $\lambda \in \mathbb{R}$ (see [24, 25] and Section 2.2) and $\lambda \mapsto \mathrm{tr}(\log(M_{\rm op}(\lambda) - \Theta_{\rm op}))$, $\lambda \in \mathbb{C}_+$, is a scalar Nevanlinna function with the property

$$0 \le \Im\mathrm{m}\,\big(\mathrm{tr}(\log(M_{\rm op}(\lambda) - \Theta_{\rm op}))\big) \le \pi \dim \mathcal{H}_{\rm op}\,, \quad \lambda \in \mathbb{C}_+,$$

that is, the function ξ_Θ in (2.18) satisfies $0 \le \xi_\Theta(\lambda) \le \dim \mathcal{H}_{\rm op}$ for a.e. $\lambda \in \mathbb{R}$.

In order to show that (2.8) holds with H_1, H_0 and ξ replaced by A_Θ, A_0 and ξ_Θ, respectively, we note that the relation

$$\frac{d}{d\lambda}\mathrm{tr}\big(\log(M_{\rm op}(\lambda) - \Theta_{\rm op})\big) = \mathrm{tr}\left((M_{\rm op}(\lambda) - \Theta_{\rm op})^{-1} \frac{d}{d\lambda} M_{\rm op}(\lambda) \right) \tag{2.19}$$

is true for all $\lambda \in \mathbb{C}_+$. This can be shown in the same way as in the proof of [9, Theorem 4.1]. From (2.5) we find

$$\gamma(\bar{\mu})^*\gamma(\lambda) = \frac{M(\lambda) - M(\bar{\mu})^*}{\lambda - \mu}, \qquad \lambda, \mu \in \mathbb{C}\backslash\mathbb{R},\ \lambda \neq \mu, \tag{2.20}$$

and passing in (2.20) to the limit $\mu \to \lambda$ one gets

$$\gamma(\bar{\lambda})^*\gamma(\lambda) = \frac{d}{d\lambda}M(\lambda). \tag{2.21}$$

Making use of formula (2.6) for the canonical resolvents this implies

$$\begin{aligned} \operatorname{tr}\left((A_\Theta - \lambda)^{-1} - (A_0 - \lambda)^{-1}\right) &= -\operatorname{tr}\left((M(\lambda) - \Theta)^{-1}\gamma(\bar{\lambda})^*\gamma(\lambda)\right) \\ &= -\operatorname{tr}\left((M(\lambda) - \Theta)^{-1}\frac{d}{d\lambda}M(\lambda)\right) \end{aligned} \tag{2.22}$$

for all $\lambda \in \mathbb{C}_+$. With respect to the decomposition $\mathcal{H} = \mathcal{H}_{\mathrm{op}} \oplus \mathcal{H}_\infty$ the operator

$$\big(M(\lambda) - \Theta\big)^{-1}\frac{d}{d\lambda}M(\lambda) = \iota_{\mathcal{H}_{\mathrm{op}}}\big(M_{\mathrm{op}}(\lambda) - \Theta_{\mathrm{op}}\big)^{-1}P_{\mathrm{op}}\frac{d}{d\lambda}M(\lambda)$$

is a 2×2 block matrix where the entries in the lower row are zero matrices and the upper left corner is given by

$$(M_{\mathrm{op}}(\lambda) - \Theta_{\mathrm{op}})^{-1}\frac{d}{d\lambda}M_{\mathrm{op}}(\lambda).$$

Therefore (2.22) becomes

$$\begin{aligned} \operatorname{tr}\left((A_\Theta - \lambda)^{-1} - (A_0 - \lambda)^{-1}\right) &= -\operatorname{tr}\left((M_{\mathrm{op}}(\lambda) - \Theta_{\mathrm{op}})^{-1}\frac{d}{d\lambda}M_{\mathrm{op}}(\lambda)\right), \\ &= -\frac{d}{d\lambda}\operatorname{tr}\big(\log(M_{\mathrm{op}}(\lambda) - \Theta_{\mathrm{op}})\big), \end{aligned} \tag{2.23}$$

where we have used (2.19).

Further, by [26, Theorem 2.10] there exists a $[\mathcal{H}_{\mathrm{op}}]$-valued measurable function $t \mapsto \Xi_{\Theta_{\mathrm{op}}}(t)$, $t \in \mathbb{R}$, such that $\Xi_{\Theta_{\mathrm{op}}}(t) = \Xi_{\Theta_{\mathrm{op}}}(t)^*$ and $0 \leq \Xi_{\Theta_{\mathrm{op}}}(t) \leq I_{\mathcal{H}_{\mathrm{op}}}$ for a.e. $\lambda \in \mathbb{R}$ and the representation

$$\log(M_{\mathrm{op}}(\lambda) - \Theta_{\mathrm{op}}) = C + \int_{\mathbb{R}} \Xi_{\Theta_{\mathrm{op}}}(t)\left((t-\lambda)^{-1} - t(1+t^2)^{-1}\right)dt, \quad \lambda \in \mathbb{C}_+,$$

holds with some bounded self-adjoint operator C. Hence

$$\operatorname{tr}\big(\log(M_{\mathrm{op}}(\lambda) - \Theta_{\mathrm{op}})\big) = \operatorname{tr}(C) + \int_{\mathbb{R}} \operatorname{tr}\left(\Xi_{\Theta_{\mathrm{op}}}(t)\right)\left((t-\lambda)^{-1} - t(1+t^2)^{-1}\right)dt$$

for $\lambda \in \mathbb{C}_+$ and we conclude from

$$\begin{aligned} \xi_\Theta(\lambda) &= \lim_{\epsilon \to +0}\frac{1}{\pi}\Im\mathrm{m}\left(\operatorname{tr}(\log(M_{\mathrm{op}}(\lambda + i\epsilon) - \Theta_{\mathrm{op}}))\right) \\ &= \lim_{\epsilon \to +0}\frac{1}{\pi}\int_{\mathbb{R}} \operatorname{tr}\left(\Xi_{\Theta_{\mathrm{op}}}(t)\right)\epsilon\big((t-\lambda)^2 + \epsilon^2\big)^{-1}dt \end{aligned}$$

that $\xi_\Theta(\lambda) = \mathrm{tr}(\Xi_{\Theta_{\mathrm{op}}}(\lambda))$ is true for a.e. $\lambda \in \mathbb{R}$. Therefore we have

$$\frac{d}{d\lambda}\mathrm{tr}\big(\log(M_{\mathrm{op}}(\lambda) - \Theta_{\mathrm{op}})\big) = \int_{\mathbb{R}} (t-\lambda)^{-2}\xi_\Theta(t)\,dt$$

and together with (2.23) we immediately get the trace formula

$$\mathrm{tr}\left((A_\Theta - \lambda)^{-1} - (A_0 - \lambda)^{-1}\right) = -\int_{\mathbb{R}} \frac{1}{(t-\lambda)^2}\,\xi_\Theta(t)\,dt.$$

The integrability condition (2.9) holds because of [26, Theorem 2.10]. This completes the proof of Theorem 2.4. □

2.4. A representation of the scattering matrix

Let again A be a densely defined closed simple symmetric operator in the separable Hilbert space $\mathfrak{H}$ with equal finite deficiency indices and let $\Pi = \{\mathcal{H}, \Gamma_0, \Gamma_1\}$ be a boundary triplet for A^* with $A_0 = A^* \upharpoonright \ker(\Gamma_0)$. Let Θ be a self-adjoint relation in $\mathcal{H}$ and let $A_\Theta = A^* \upharpoonright \Gamma^{-1}\Theta$ be the corresponding self-adjoint extension of A in $\mathfrak{H}$. Since $\dim \mathcal{H}$ is finite by (2.6)

$$\dim\Big(\mathrm{ran}\left((A_\Theta - \lambda)^{-1} - (A_0 - \lambda)^{-1}\right)\Big) < \infty, \quad \lambda \in \rho(A_\Theta) \cap \rho(A_0),$$

and therefore the pair $\{A_\Theta, A_0\}$ forms a so-called *complete scattering system*, that is, the *wave operators*

$$W_\pm(A_\Theta, A_0) := \text{s-}\lim_{t\to\pm\infty} e^{itA_\Theta}e^{-itA_0}P^{ac}(A_0),$$

exist and their ranges coincide with the absolutely continuous subspace $\mathfrak{H}^{ac}(A_\Theta)$ of A_Θ, cf. [8, 31, 52, 53]. $P^{ac}(A_0)$ denotes the orthogonal projection onto the absolutely continuous subspace $\mathfrak{H}^{ac}(A_0)$ of A_0. The *scattering operator* S_Θ of the *scattering system* $\{A_\Theta, A_0\}$ is then defined by

$$S_\Theta := W_+(A_\Theta, A_0)^* W_-(A_\Theta, A_0).$$

If we regard the scattering operator as an operator in $\mathfrak{H}^{ac}(A_0)$, then S_Θ is unitary, commutes with the absolutely continuous part

$$A_0^{ac} := A_0 \upharpoonright \mathrm{dom}\,(A_0) \cap \mathfrak{H}^{ac}(A_0)$$

of A_0 and it follows that S_Θ is unitarily equivalent to a multiplication operator induced by a family $\{S_\Theta(\lambda)\}_{\lambda\in\mathbb{R}}$ of unitary operators in a spectral representation of A_0^{ac}, see, e.g., [8, Proposition 9.57]. This family is called the *scattering matrix* of the scattering system $\{A_\Theta, A_0\}$.

We recall a representation theorem for the scattering matrix $\{S_\Theta(\lambda)\}_{\lambda\in\mathbb{R}}$ in terms of the Weyl function $M(\cdot)$ of the boundary triplet $\Pi = \{\mathcal{H}, \Gamma_0, \Gamma_1\}$ from [9]. For this we consider the Hilbert space $L^2(\mathbb{R}, d\lambda, \mathcal{H})$, where $d\lambda$ is the Lebesgue measure on $\mathbb{R}$. Further, we set

$$\mathcal{H}_{M(\lambda)} := \mathrm{ran}\,\big(\Im\mathrm{m}\,(M(\lambda))\big), \quad M(\lambda) := M(\lambda + i0), \tag{2.24}$$

which defines subspaces of $\mathcal{H}$ for a.e. $\lambda \in \mathbb{R}$. By $P_{M(\lambda)}$ we denote the orthogonal projection from $\mathcal{H}$ onto $\mathcal{H}_{M(\lambda)}$. The family $\{P_{M(\lambda)}\}_{\lambda\in\mathbb{R}}$ is measurable. Hence $\{P_{M(\lambda)}\}_{\lambda\in\mathbb{R}}$ induces a multiplication operator P_M on $L^2(\mathbb{R}, d\lambda, \mathcal{H})$ defined by

$$(P_M f)(\lambda) = P_{M(\lambda)} f(\lambda), \qquad f \in L^2(\mathbb{R}, d\lambda, \mathcal{H}),$$

which is an orthogonal projection.

The subspace $\operatorname{ran}(P_M)$ is denoted by $L^2(\mathbb{R}, d\lambda, \mathcal{H}_{M(\lambda)})$ in the following. We remark that $L^2(\mathbb{R}, d\lambda, \mathcal{H}_{M(\lambda)})$ can be regarded as the direct integral of the Hilbert spaces $\mathcal{H}_{M(\lambda)}$, that is,

$$L^2(\mathbb{R}, d\lambda, \mathcal{H}_{M(\lambda)}) = \int^{\oplus} \mathcal{H}_{M(\lambda)}\, d\lambda.$$

The following theorem was proved in [9].

Theorem 2.5. *Let A be a densely defined closed simple symmetric operator with equal finite deficiency indices in the separable Hilbert space $\mathfrak{H}$ and let $\Pi=\{\mathcal{H},\Gamma_0,\Gamma_1\}$ be a boundary triplet for A^* with corresponding Weyl function $M(\cdot)$. Furthermore, let $A_0 = A^* \upharpoonright \ker(\Gamma_0)$ and let $A_\Theta = A^* \upharpoonright \Gamma^{-1}\Theta$, $\Theta \in \widetilde{\mathcal{C}}(\mathcal{H})$, be a self-adjoint extension of A in $\mathfrak{H}$. Then the following holds:*

(i) *A_0^{ac} is unitarily equivalent to the multiplication operator with the free variable in the Hilbert space $L^2(\mathbb{R}, d\lambda, \mathcal{H}_{M(\lambda)})$.*

(ii) *In the spectral representation $L^2(\mathbb{R}, d\lambda, \mathcal{H}_{M(\lambda)})$ of A_0^{ac} the scattering matrix $\{S_\Theta(\lambda)\}_{\lambda\in\mathbb{R}}$ of the scattering system $\{A_\Theta, A_0\}$ admits the representation*

$$S_\Theta(\lambda) = I_{\mathcal{H}_{M(\lambda)}} + 2i\sqrt{\Im\mathrm{m}\,(M(\lambda))}\big(\Theta - M(\lambda)\big)^{-1}\sqrt{\Im\mathrm{m}\,(M(\lambda))} \tag{2.25}$$

for a.e. $\lambda \in \mathbb{R}$, where $M(\lambda) = M(\lambda + i0)$.

In the next corollary we find a slightly more convenient representation of the scattering matrix $\{S_\Theta(\lambda)\}_{\lambda\in\mathbb{R}}$ of the scattering system $\{A_\Theta, A_0\}$ for the case that Θ is a self-adjoint relation which is decomposed in the form $\Theta = \Theta_{\mathrm{op}} \oplus \Theta_\infty$ with respect to $\mathcal{H} = \mathcal{H}_{\mathrm{op}} \oplus \mathcal{H}_\infty$, cf. (2.14) and (2.15). If $M(\cdot)$ is the Weyl function corresponding to the boundary triplet $\Pi = \{\mathcal{H}, \Gamma_0, \Gamma_1\}$, then the function

$$\lambda \mapsto M_{\mathrm{op}}(\lambda) = P_{\mathrm{op}} M(\lambda)\iota_{\mathrm{op}}$$

from (2.16) is a $[\mathcal{H}_{\mathrm{op}}]$-valued Nevanlinna function, and the subspaces

$$\mathcal{H}_{M_{\mathrm{op}}(\lambda)} := \operatorname{ran}\big(\Im\mathrm{m}\,(M_{\mathrm{op}}(\lambda + i0))\big)$$

of $\mathcal{H}_{M(\lambda)}$ are defined as in (2.24).

Corollary 2.6. *Let the assumptions be as in Theorem 2.5, let $\mathcal{H}_{M_{\mathrm{op}}(\lambda)}$ be as above and $\mathcal{H}^\infty_{M(\lambda)} := \mathcal{H}_{M(\lambda)} \ominus \mathcal{H}_{M_{\mathrm{op}}(\lambda)}$. Then there exists a family $V(\lambda) : \mathcal{H}_{M(\lambda)} \to \mathcal{H}_{M(\lambda)}$ of unitary operators such that the representation*

$$S_\Theta(\lambda) = V(\lambda)\left\{I_{\mathcal{H}^\infty_{M(\lambda)}} \oplus S_{\Theta_{\mathrm{op}}}(\lambda)\right\} V(\lambda)^* \tag{2.26}$$

holds with

$$S_{\Theta_{\rm op}}(\lambda) = I_{\mathcal{H}_{M_{\rm op}(\lambda)}} + 2i\sqrt{\Im{\rm m}\,(M_{\rm op}(\lambda))}\big(\Theta_{\rm op} - M_{\rm op}(\lambda)\big)^{-1}\sqrt{\Im{\rm m}\,(M_{\rm op}(\lambda))}$$

for a.e. $\lambda \in \mathbb{R}$.

Proof. Using (2.25) and (2.17) we find the representation

$$S_\Theta(\lambda) = I_{\mathcal{H}_{M(\lambda)}} + 2i\sqrt{\Im{\rm m}\,(M(\lambda))}\,\iota_{\rm op}\,\big(\Theta_{\rm op} - M_{\rm op}(\lambda)\big)^{-1}P_{\rm op}\sqrt{\Im{\rm m}\,(M(\lambda))}$$

for a.e. $\lambda \in \mathbb{R}$. From the polar decomposition of $\sqrt{\Im{\rm m}\,(M(\lambda))}\,\iota_{\rm op}$ we obtain a family of isometric mappings $V_{\rm op}(\lambda)$ from $\mathcal{H}_{M_{\rm op}(\lambda)}$ onto

$${\rm ran}\,\Big(\sqrt{\Im{\rm m}\,(M(\lambda))}\,\iota_{\rm op}\Big) \subset \mathcal{H}_{M(\lambda)}$$

defined by

$$V_{\rm op}(\lambda)\sqrt{\Im{\rm m}\,(M_{\rm op}(\lambda))} := \sqrt{\Im{\rm m}\,(M(\lambda))}\,\iota_{\rm op}\,.$$

Hence we find

$$\begin{aligned} S_\Theta(\lambda) = I_{\mathcal{H}_{M(\lambda)}} &+ 2iV_{\rm op}(\lambda)\sqrt{\Im{\rm m}\,(M_{\rm op}(\lambda))} \\ &\times \big(\Theta_{\rm op} - M_{\rm op}(\lambda)\big)^{-1}\sqrt{\Im{\rm m}\,(M_{\rm op}(\lambda))}V_{\rm op}(\lambda)^* \end{aligned}$$

for a.e. $\lambda \in \mathbb{R}$. Since the Hilbert space $\mathcal{H}_{M(\lambda)}$ is finite dimensional there is an isometry $V_\infty(\lambda)$ acting from $\mathcal{H}^\infty_{M(\lambda)} = \mathcal{H}_{M(\lambda)} \ominus \mathcal{H}_{M_{\rm op}(\lambda)}$ into $\mathcal{H}_{M(\lambda)}$ such that $V(\lambda) := V_\infty(\lambda) \oplus V_{\rm op}(\lambda)$ defines a unitary operator on $\mathcal{H}_{M(\lambda)}$. This immediately yields (2.26). □

2.5. Birman-Krein formula

An important relation between the spectral shift function and the scattering matrix for a pair of self-adjoint operators for the case of a trace class perturbation was found in [11] by M.S. Birman and M.G. Krein. Subsequently, this relation was called the Birman-Krein formula. Under the assumption that A_Θ and A_0 are self-adjoint extensions of a densely defined symmetric operator A with finite deficiency indices and A_Θ corresponds to a self-adjoint matrix Θ via a boundary triplet $\Pi = \{\mathcal{H}, \Gamma_0, \Gamma_1\}$ for A^* a simple proof for the Birman-Krein formula

$$\det(S_\Theta(\lambda)) = \exp\big(-2\pi i \xi_\Theta(\lambda)\big)$$

was given in [9]. Here $\xi_\Theta(\cdot)$ is the spectral shift function of the pair $\{A_\Theta, A_0\}$ defined by (2.13) and the scattering matrix $\{S_\Theta(\lambda)\}_{\lambda\in\mathbb{R}}$ is given by (2.25).

The following theorem generalizes [9, Theorem 4.1] to the case of a self-adjoint relation Θ (instead of a matrix), so that the Birman-Krein formula is verified for all pairs of self-adjoint extensions of the underlying symmetric operator.

Theorem 2.7. *Let A be a densely defined closed simple symmetric operator in the separable Hilbert space $\mathfrak{H}$ with equal finite deficiency indices, let $\Pi = \{\mathcal{H}, \Gamma_0, \Gamma_1\}$ be a boundary triplet for A^* and let $M(\cdot)$ be the corresponding Weyl function. Furthermore, let $A_0 = A^* \upharpoonright \ker(\Gamma_0)$ and let $A_\Theta = A^* \upharpoonright \Gamma^{-1}\Theta$, $\Theta \in \widetilde{\mathcal{C}}(\mathcal{H})$, be a* *self-adjoint extension of A in $\mathfrak{H}$. Then the spectral shift function $\xi_\Theta(\cdot)$ in* (2.18) *and the scattering matrix $\{S_\Theta(\lambda)\}_{\lambda\in\mathbb{R}}$ of the pair $\{A_\Theta, A_0\}$ are related via*

$$\det\big(S_\Theta(\lambda)\big) = \exp\big(-2\pi i \xi_\Theta(\lambda)\big) \tag{2.27}$$

for a.e. $\lambda \in \mathbb{R}$.

Proof. To verify the Birman-Krein formula we note that by (2.11)

$$\begin{aligned}
&\exp\big(-2i\Im\mathrm{m}\,(\mathrm{tr}(\log(M_{\mathrm{op}}(\lambda) - \Theta_{\mathrm{op}})))\big)\\
&\quad = \exp\big(-\mathrm{tr}(\log(M_{\mathrm{op}}(\lambda) - \Theta_{\mathrm{op}}))\big)\exp\big(\overline{\mathrm{tr}(\log(M_{\mathrm{op}}(\lambda) - \Theta_{\mathrm{op}}))}\big)\\
&\quad = \frac{\overline{\det(M_{\mathrm{op}}(\lambda) - \Theta_{\mathrm{op}})}}{\det(M_{\mathrm{op}}(\lambda) - \Theta_{\mathrm{op}})} = \frac{\det(M_{\mathrm{op}}(\lambda)^* - \Theta_{\mathrm{op}})}{\det(M_{\mathrm{op}}(\lambda) - \Theta_{\mathrm{op}})}
\end{aligned}$$

holds for all $\lambda \in \mathbb{C}_+$. Hence we find

$$\exp\big(-2\pi i \xi_\Theta(\lambda)\big) = \frac{\det\big(M_{\mathrm{op}}(\lambda + i0)^* - \Theta_{\mathrm{op}}\big)}{\det\big(M_{\mathrm{op}}(\lambda + i0) - \Theta_{\mathrm{op}}\big)} \tag{2.28}$$

for a.e. $\lambda \in \mathbb{R}$, where $M_{\mathrm{op}}(\lambda + i0) := \lim_{\epsilon\to+0} M_{\mathrm{op}}(\lambda + i\epsilon)$ exists for a.e. $\lambda \in \mathbb{R}$. It follows from the representation of the scattering matrix in Corollary 2.6 and the identity $\det(I + AB) = \det(I + BA)$ that

$$\begin{aligned}
&\det S_\Theta(\lambda)\\
&\quad = \det\Big(I_{\mathcal{H}_{\mathrm{op}}} + 2i\big(\Im\mathrm{m}\,(M_{\mathrm{op}}(\lambda + i0))\big)\big(\Theta_{\mathrm{op}} - M_{\mathrm{op}}(\lambda + i0)\big)^{-1}\Big)\\
&\quad = \det\Big(I_{\mathcal{H}_{\mathrm{op}}} + \big(M_{\mathrm{op}}(\lambda + i0) - M_{\mathrm{op}}(\lambda + i0)^*\big)\big(\Theta_{\mathrm{op}} - M_{\mathrm{op}}(\lambda + i0)\big)^{-1}\Big)\\
&\quad = \det\Big(\big(M_{\mathrm{op}}(\lambda + i0)^* - \Theta_{\mathrm{op}}\big)\cdot\big(M_{\mathrm{op}}(\lambda + i0) - \Theta_{\mathrm{op}}\big)^{-1}\Big)\\
&\quad = \frac{\det\big(M_{\mathrm{op}}(\lambda + i0)^* - \Theta_{\mathrm{op}}\big)}{\det\big(M_{\mathrm{op}}(\lambda + i0) - \Theta_{\mathrm{op}}\big)}
\end{aligned}$$

holds for a.e. $\lambda \in \mathbb{R}$. Comparing this with (2.28) we obtain (2.27). □

3. Dissipative scattering systems

In this section we investigate scattering systems consisting of a maximal dissipative and a self-adjoint operator, which are both extensions of a common symmetric operator with equal finite deficiency indices. We shall explicitly construct a so-called dilation of the maximal dissipative operator and we calculate the spectral shift function of the dissipative scattering system with the help of this dilation. It will be shown that the scattering matrix of the dissipative scattering system and this spectral shift function are connected via a modified Birman-Krein formula.

3.1. Self-adjoint dilations of maximal dissipative operators

Let A be a densely defined closed simple symmetric operator in the separable Hilbert space $\mathfrak{H}$ with equal finite deficiency indices $n_+(A) = n_-(A) = n < \infty$, let $\Pi = \{\mathcal{H}, \Gamma_0, \Gamma_1\}$, $A_0 = A^* \upharpoonright \ker(\Gamma_0)$, be a boundary triplet for A^* and let $D \in [\mathcal{H}]$ be a dissipative $n \times n$-matrix, i.e., $\Im\mathrm{m}\,(D) \leq 0$. Then by Proposition 2.2 (iii) the closed extension

$$A_D = A^* \upharpoonright \ker(\Gamma_1 - D\Gamma_0)$$

of A corresponding to $\Theta = D$ via (2.2) is maximal dissipative, that is, A_D is dissipative and maximal in the sense that each dissipative extension of A_D in $\mathfrak{H}$ coincides with A_D. Observe that $\mathbb{C}_+$ belongs to $\rho(A_D)$. For $\lambda \in \rho(A_D) \cap \rho(A_0)$ the resolvent of the extension A_D is given by

$$(A_D - \lambda)^{-1} = (A_0 - \lambda)^{-1} + \gamma(\lambda)\big(D - M(\lambda)\big)^{-1}\gamma(\bar{\lambda})^*, \tag{3.1}$$

cf. (2.6). With respect to the decomposition

$$D = \Re\mathrm{e}\,(D) + i\Im\mathrm{m}\,(D)$$

we decompose $\mathcal{H}$ into the orthogonal sum of the finite-dimensional subspaces $\ker(\Im\mathrm{m}\,(D))$ and $\mathcal{H}_D := \mathrm{ran}\,(\Im\mathrm{m}\,(D))$,

$$\mathcal{H} = \ker(\Im\mathrm{m}\,(D)) \oplus \mathcal{H}_D, \tag{3.2}$$

and denote by P_D the orthogonal projection from $\mathcal{H}$ onto $\mathcal{H}_D$ and by ι_D the canonical embedding of $\mathcal{H}_D$ into $\mathcal{H}$. Since $\Im\mathrm{m}\,(D) \leq 0$ the self-adjoint matrix

$$-P_D \Im\mathrm{m}\,(D)\,\iota_D \in [\mathcal{H}_D]$$

is strictly positive and therefore (see, e.g., [18, 22]) the function

$$\lambda \mapsto \begin{cases} -iP_D\Im\mathrm{m}\,(D)\iota_D, & \lambda \in \mathbb{C}_+, \\ iP_D\Im\mathrm{m}\,(D)\,\iota_D, & \lambda \in \mathbb{C}_-, \end{cases}$$

can be realized as the Weyl function corresponding to a boundary triplet of a symmetric operator.

Here the symmetric operator and boundary triplet can be made more explicit, cf. [10, Lemma 3.1]. In fact, let G be the symmetric first-order differential operator in the Hilbert space $L^2(\mathbb{R}, \mathcal{H}_D)$ defined by

$$(Gg)(x) = -ig'(x), \qquad \mathrm{dom}\,(G) = \big\{g \in W_2^1(\mathbb{R}, \mathcal{H}_D) : g(0) = 0\big\}. \tag{3.3}$$

Then G is simple, $n_\pm(G) = \dim \mathcal{H}_D$ and the adjoint operator $G^*g = -ig'$ is defined on

$$\mathrm{dom}\,(G^*) = W_2^1(\mathbb{R}_-, \mathcal{H}_D) \oplus W_2^1(\mathbb{R}_+, \mathcal{H}_D).$$

Moreover, the triplet $\Pi_G = \{\mathcal{H}_D, \Upsilon_0, \Upsilon_1\}$, where

$$\begin{aligned} \Upsilon_0 g &:= \frac{1}{\sqrt{2}}\big(-P_D\Im\mathrm{m}\,(D)\,\iota_D\big)^{-\frac{1}{2}}\big(g(0+) - g(0-)\big), \\ \Upsilon_1 g &:= \frac{i}{\sqrt{2}}\big(-P_D\Im\mathrm{m}\,(D)\,\iota_D\big)^{\frac{1}{2}}\big(g(0+) + g(0-)\big), \end{aligned} \tag{3.4}$$

$g \in \operatorname{dom}(G^*)$, is a boundary triplet for G^* and the extension $G_0 := G^* \upharpoonright \ker(\Upsilon_0)$ of G is the usual self-adjoint first-order differential operator in $L^2(\mathbb{R}, \mathcal{H}_D)$ with domain $\operatorname{dom}(G_0) = W_2^1(\mathbb{R}, \mathcal{H}_D)$ and $\sigma(G_0) = \mathbb{R}$. It is not difficult to see that the defect subspaces of G are given by

$$\ker(G^* - \lambda) = \begin{cases} \operatorname{span}\{x \mapsto e^{i\lambda x}\chi_{\mathbb{R}_+}(x)\xi \,:\, \xi \in \mathcal{H}_D\}, & \lambda \in \mathbb{C}_+, \\ \operatorname{span}\{x \mapsto e^{i\lambda x}\chi_{\mathbb{R}_-}(x)\xi \,:\, \xi \in \mathcal{H}_D\}, & \lambda \in \mathbb{C}_-, \end{cases}$$

and therefore it follows that the Weyl function $\tau(\cdot)$ corresponding to the boundary triplet $\Pi_G = \{\mathcal{H}_D, \Upsilon_0, \Upsilon_1\}$ is given by

$$\tau(\lambda) = \begin{cases} -iP_D \Im\mathrm{m}\,(D)\,\iota_D, & \lambda \in \mathbb{C}_+, \\ iP_D \Im\mathrm{m}\,(D)\,\iota_D, & \lambda \in \mathbb{C}_-. \end{cases}$$

Let A be the densely defined closed simple symmetric operator in $\mathfrak{H}$ from above and let G be the first-order differential operator in (3.3). Clearly,

$$K := \begin{pmatrix} A & 0 \\ 0 & G \end{pmatrix}$$

is a densely defined closed simple symmetric operator in the separable Hilbert space

$$\mathfrak{K} := \mathfrak{H} \oplus L^2(\mathbb{R}, \mathcal{H}_D)$$

with equal finite deficiency indices $n_\pm(K) = n_\pm(A) + n_\pm(G) = n + \dim \mathcal{H}_D < \infty$ and the adjoint is

$$K^* = \begin{pmatrix} A^* & 0 \\ 0 & G^* \end{pmatrix}.$$

The elements in $\operatorname{dom}(K^*) = \operatorname{dom}(A^*) \oplus \operatorname{dom}(G^*)$ will be written in the form $f \oplus g$, $f \in \operatorname{dom}(A^*)$, $g \in \operatorname{dom}(G^*)$. It is straightforward to check that $\widetilde{\Pi} = \{\widetilde{\mathcal{H}}, \widetilde{\Gamma}_0, \widetilde{\Gamma}_1\}$, where $\widetilde{\mathcal{H}} := \mathcal{H} \oplus \mathcal{H}_D$,

$$\widetilde{\Gamma}_0(f \oplus g) := \begin{pmatrix} \Gamma_0 f \\ \Upsilon_0 g \end{pmatrix} \quad \text{and} \quad \widetilde{\Gamma}_1(f \oplus g) := \begin{pmatrix} \Gamma_1 f - \Re\mathrm{e}\,(D)\Gamma_0 f \\ \Upsilon_1 g \end{pmatrix}, \tag{3.5}$$

$f \oplus g \in \operatorname{dom}(K^*)$, is a boundary triplet for K^*. If $\gamma(\cdot), \nu(\cdot)$ and $M(\cdot), \tau(\cdot)$ are the γ-fields and Weyl functions of the boundary triplets $\Pi = \{\mathcal{H}, \Gamma_0, \Gamma_1\}$ and $\Pi_G = \{\mathcal{H}_D, \Upsilon_0, \Upsilon_1\}$, respectively, then one easily verifies that the Weyl function $\widetilde{M}(\cdot)$ and γ-field $\widetilde{\gamma}(\cdot)$ corresponding to the boundary triplet $\widetilde{\Pi} = \{\widetilde{\mathcal{H}}, \widetilde{\Gamma}_0, \widetilde{\Gamma}_1\}$ are given by

$$\widetilde{M}(\lambda) = \begin{pmatrix} M(\lambda) - \Re\mathrm{e}\,(D) & 0 \\ 0 & \tau(\lambda) \end{pmatrix}, \qquad \lambda \in \mathbb{C}\backslash\mathbb{R}, \tag{3.6}$$

and

$$\widetilde{\gamma}(\lambda) = \begin{pmatrix} \gamma(\lambda) & 0 \\ 0 & \nu(\lambda) \end{pmatrix}, \qquad \lambda \in \mathbb{C}\backslash\mathbb{R}, \tag{3.7}$$

respectively. Observe that

$$K_0 := K^* \upharpoonright \ker(\widetilde{\Gamma}_0) = \begin{pmatrix} A_0 & 0 \\ 0 & G_0 \end{pmatrix} \tag{3.8}$$

holds. With respect to the decomposition

$$\widetilde{\mathcal{H}} = \ker(\Im\mathrm{m}\,(D)) \oplus \mathcal{H}_D \oplus \mathcal{H}_D$$

of $\widetilde{\mathcal{H}}$ (cf. (3.2)) we define the linear relation $\widetilde{\Theta}$ in $\widetilde{\mathcal{H}}$ by

$$\widetilde{\Theta} := \left\{ \begin{pmatrix} (u,v,v)^\top \\ (0,-w,w)^\top \end{pmatrix} : u \in \ker(\Im\mathrm{m}\,(D),\ v,w \in \mathcal{H}_D \right\}. \tag{3.9}$$

We leave it to the reader to check that $\widetilde{\Theta}$ is self-adjoint. Hence by Proposition 2.2 the operator

$$\begin{aligned} \widetilde{K} :&= K_{\widetilde{\Theta}} = K^* \upharpoonright \widetilde{\Gamma}^{-1}\widetilde{\Theta} \\ &= \left\{ f \oplus g \in \mathrm{dom}\,(A^*) \oplus \mathrm{dom}\,(G^*) : \left(\widetilde{\Gamma}_0(f\oplus g), \widetilde{\Gamma}_1(f \oplus g)\right)^\top \in \widetilde{\Theta} \right\} \end{aligned}$$

is a self-adjoint extension of the symmetric operator K in $\mathfrak{K} = \mathfrak{H} \oplus L^2(\mathbb{R}, \mathcal{H}_D)$. The following theorem was proved in [10], see also [45, 46] for a special case involving Sturm-Liouville operators with dissipative boundary conditions.

Theorem 3.1. *Let A, $\Pi = \{\mathcal{H}, \Gamma_0, \Gamma_1\}$ and $A_D = A^* \upharpoonright \ker(\Gamma_1 - D\Gamma_0)$ be as above. Furthermore, let G and $\Pi_G = \{\mathcal{H}_D, \Upsilon_0, \Upsilon_1\}$ be given by (3.3) and (3.4), respectively, and let $K = A \oplus G$. Then the self-adjoint extension $\widetilde{K}$ of K has the form*

$$\widetilde{K} = K^* \upharpoonright \left\{ f \oplus g \in \mathrm{dom}\,(K^*) : \begin{array}{r} P_D\Gamma_0 f - \Upsilon_0 g = 0, \\ (I - P_D)(\Gamma_1 - \Re\mathrm{e}\,(D)\Gamma_0)f = 0, \\ P_D(\Gamma_1 - \Re\mathrm{e}\,(D)\Gamma_0)f + \Upsilon_1 g = 0 \end{array} \right\} \tag{3.10}$$

and $\widetilde{K}$ is a minimal self-adjoint dilation of the maximal dissipative operator A_D, that is, for all $\lambda \in \mathbb{C}_+$

$$P_{\mathfrak{H}}\left(\widetilde{K} - \lambda\right)^{-1} \upharpoonright_{\mathfrak{H}} = (A_D - \lambda)^{-1}$$

holds and the minimality condition $\mathfrak{K} = \mathrm{clospan}\{(\widetilde{K} - \lambda)^{-1}\mathfrak{H} : \lambda \in \mathbb{C}\backslash\mathbb{R}\}$ is satisfied. Moreover $\sigma(\widetilde{K}) = \mathbb{R}$.

We note that also in the case where the parameter D is not a dissipative matrix but a maximal dissipative relation in $\mathcal{H}$ a minimal self-adjoint dilation of A_D can be constructed in a similar way as in Theorem 3.1, see [10, Remark 3.3]

3.2. Spectral shift function and trace formula

In order to calculate the spectral shift function of the pair $\{\widetilde{K}, K_0\}$ from (3.8) and (3.10) we write the self-adjoint relation $\widetilde{\Theta}$ from (3.9) in the form $\widetilde{\Theta} = \widetilde{\Theta}_{\mathrm{op}} \oplus \widetilde{\Theta}_\infty$, where

$$\widetilde{\Theta}_{\mathrm{op}} := \left\{ \begin{pmatrix} (u,v,v)^\top \\ (0,0,0)^\top \end{pmatrix} : u \in \ker(\Im\mathrm{m}\,(D)),\ v \in \mathcal{H}_D \right\} \tag{3.11}$$

is the zero operator in the space

$$\widetilde{\mathcal{H}}_{\mathrm{op}} := \left\{ \begin{pmatrix} u \\ v \\ v \end{pmatrix} : u \in \ker(\Im\mathrm{m}\,(D)),\ v \in \mathcal{H}_D \right\}$$

and

$$\widetilde{\Theta}_\infty := \left\{ \begin{pmatrix} (0,0,0)^\top \\ (0,-w,w)^\top \end{pmatrix} : w \in \mathcal{H}_D \right\}$$

is the purely multi-valued relation in the space

$$\widetilde{\mathcal{H}}_\infty = \widetilde{\mathcal{H}} \ominus \widetilde{\mathcal{H}}_{\rm op} = \left\{ \begin{pmatrix} 0 \\ -w \\ w \end{pmatrix} : w \in \mathcal{H}_D \right\}.$$

The orthogonal projection from $\widetilde{\mathcal{H}}$ onto $\widetilde{\mathcal{H}}_{\rm op}$ will be denoted by $\widetilde{P}_{\rm op}$ and the canonical embedding of $\widetilde{\mathcal{H}}_{\rm op}$ in $\widetilde{\mathcal{H}}$ is denoted by $\widetilde{\iota}_{\rm op}$. As an immediate consequence of Theorem 2.4 we find the following representation of a spectral shift function for the pair $\{\widetilde{K}, K_0\}$.

Corollary 3.2. *Let A and G be the symmetric operators from Section 3.1 and let $K = A \oplus G$. Furthermore, let $\widetilde{\Pi} = \{\widetilde{\mathcal{H}}, \widetilde{\Gamma}_0, \widetilde{\Gamma}_1\}$ be the boundary triplet for K^* from (3.5) with Weyl function $\widetilde{M}(\cdot)$ given by (3.6) and define the $[\widetilde{\mathcal{H}}_{\rm op}]$-valued Nevanlinna function by*

$$\widetilde{M}_{\rm op}(\lambda) := \widetilde{P}_{\rm op}\, \widetilde{M}(\lambda)\, \widetilde{\iota}_{\rm op}\,.$$

Then the limit $\lim_{\epsilon\to+0} \widetilde{M}_{\rm op}(\lambda + i\epsilon)$ exists for a.e. $\lambda \in \mathbb{R}$ and the function

$$\xi_{\widetilde{\Theta}}(\lambda) := \frac{1}{\pi} \Im{\rm m}\left({\rm tr}\big(\log(\widetilde{M}_{\rm op}(\lambda + i0))\big)\right) \tag{3.12}$$

is a spectral shift function for the pair $\{\widetilde{K}, K_0\}$ with $0 \le \xi_{\widetilde{\Theta}}(\lambda) \le \dim \widetilde{\mathcal{H}}_{\rm op} = n$.

Observe that the spectral shift function in (3.12) satisfies the trace formula

$${\rm tr}\big((\widetilde{K} - \lambda)^{-1} - (K_0 - \lambda)^{-1}\big) = -\int_{\mathbb{R}} \frac{1}{(t-\lambda)^2}\, \xi_{\widetilde{\Theta}}(t)\, dt \tag{3.13}$$

for $\lambda \in \mathbb{C}\backslash\mathbb{R}$. In the following theorem we calculate the spectral shift function of $\{\widetilde{K}, K_0\}$ in a more explicit form up to a constant $2k$, $k \in \mathbb{Z}$. We mention that the spectral shift function in (3.14) below can be regarded as the spectral shift function of the dissipative scattering system $\{A_D, A_0\}$, cf. [40, 41, 42].

Theorem 3.3. *Let A and G be the symmetric operators from Section 3.1 and let $\Pi = \{\mathcal{H}, \Gamma_0, \Gamma_1\}$ be a boundary triplet for A^* with corresponding Weyl function $M(\cdot)$. Let $D \in [\mathcal{H}]$ be a dissipative $n \times n$-matrix and let $A_D = A^* \upharpoonright \ker(\Gamma_1 - D\Gamma_0)$ be the corresponding maximal dissipative extension of A. Furthermore, let K_0 be as in (3.8) and let $\widetilde{K}$ be the minimal self-adjoint dilation of A_D from (3.10).*

Then the spectral shift function $\xi_{\widetilde{\Theta}}(\cdot)$ of the pair $\{\widetilde{K}, K_0\}$ admits the representation $\xi_{\widetilde{\Theta}}(\cdot) = \eta_D(\cdot) + 2k$ for some $k \in \mathbb{Z}$, where

$$\eta_D(\lambda) := \frac{1}{\pi} \Im{\rm m}\left({\rm tr}\big(\log(M(\lambda + i0) - D)\big)\right) \tag{3.14}$$

for a.e. $\lambda \in \mathbb{R}$, and the modified trace formulas

$${\rm tr}\big((A_D - \lambda)^{-1} - (A_0 - \lambda)^{-1}\big) = -\int_{\mathbb{R}} \frac{1}{(t-\lambda)^2}\, \eta_D(t)\, dt, \quad \lambda \in \mathbb{C}_+, \tag{3.15}$$

and

$$\operatorname{tr}\big((A_D^* - \lambda)^{-1} - (A_0 - \lambda)^{-1}\big) = -\int_{\mathbb{R}} \frac{1}{(t-\lambda)^2}\,\eta_D(t)\,dt, \quad \lambda \in \mathbb{C}_-, \tag{3.16}$$

are valid.

Proof. With the help of the operator

$$V : \mathcal{H} \longrightarrow \mathcal{H}, \quad x \mapsto \begin{pmatrix} (I-P_D)x \\ \frac{1}{\sqrt{2}} P_D x \end{pmatrix}$$

and the unitary operator

$$\widetilde{V} : \mathcal{H} \longrightarrow \widetilde{\mathcal{H}}_{\mathrm{op}}, \quad x \mapsto \begin{pmatrix} (I-P_D)x \\ \frac{1}{\sqrt{2}} P_D x \\ \frac{1}{\sqrt{2}} P_D x \end{pmatrix}$$

one easily verifies that

$$\begin{aligned} \widetilde{V}^* \widetilde{M}_{\mathrm{op}}(\lambda) \widetilde{V} &= V\left(M(\lambda) - \Re\mathrm{e}\,(D) + \begin{pmatrix} 0 & 0 \\ 0 & \tau(\lambda) \end{pmatrix} \right) V \\ &= V(M(\lambda) - D)V \end{aligned} \tag{3.17}$$

holds for all $\lambda \in \mathbb{C}_+$. Using this relation and the definition of $\log(\cdot)$ in (2.10) we get

$$\operatorname{tr}\big(\log\big(\widetilde{M}_{\mathrm{op}}(\lambda)\big)\big) = \operatorname{tr}\big(\log\big(\widetilde{V}^* \widetilde{M}_{\mathrm{op}}(\lambda)\widetilde{V}\big)\big) = \operatorname{tr}\big(\log\big(V(M(\lambda) - D)V\big)\big)$$

and therefore (2.12) (see also [29]) implies

$$\begin{aligned} \frac{d}{d\lambda}\operatorname{tr}\big(\log\big(\widetilde{M}_{\mathrm{op}}(\lambda)\big)\big) &= \frac{d}{d\lambda}\log\big(\det\big(V(M(\lambda)-D)V\big)\big) \\ = \frac{d}{d\lambda}\log\big(\det(M(\lambda)-D)\big) &+ \frac{d}{d\lambda}\log\big(\det V^2\big) = \frac{d}{d\lambda}\operatorname{tr}\big(\log(M(\lambda)-D)\big). \end{aligned}$$

Hence $\operatorname{tr}(\log(\widetilde{M}_{\mathrm{op}}(\cdot)))$ and $\operatorname{tr}(\log(M(\cdot) - D))$ differ by a constant. From

$$\exp\big(\operatorname{tr}\big(\log\big(\widetilde{M}_{\mathrm{op}}(\lambda)\big)\big)\big) = \exp\big(\operatorname{tr}\big(\log(M(\lambda)-D)\big)\big)\det V^2$$

we conclude that there exists $k \in \mathbb{Z}$ such that

$$\Im\mathrm{m}\,\big(\operatorname{tr}\big(\log\big(\widetilde{M}_{\mathrm{op}}(\lambda)\big)\big)\big) = \Im\mathrm{m}\,\big(\operatorname{tr}\big(\log(M(\lambda)-D)\big)\big) + 2k\pi$$

holds. Hence it follows that the spectral shift function $\xi_{\widetilde{\Theta}}$ of the pair $\{\widetilde{K}, K_0\}$ in (3.12) and the function $\eta_D(\cdot)$ in (3.14) differ by $2k$ for some $k \in \mathbb{Z}$.

Next we verify that the trace formulas (3.15) and (3.16) hold. From (2.6) we obtain

$$\begin{aligned} \operatorname{tr}\big((\widetilde{K} - \lambda)^{-1} - (K_0 - \lambda)^{-1}\big) &= \operatorname{tr}\big(\widetilde{\gamma}(\lambda)\big(\widetilde{\Theta} - \widetilde{M}(\lambda)\big)^{-1}\widetilde{\gamma}(\bar{\lambda})^*\big) \\ &= \operatorname{tr}\big(\big(\widetilde{\Theta} - \widetilde{M}(\lambda)\big)^{-1}\widetilde{\gamma}(\bar{\lambda})^*\widetilde{\gamma}(\lambda)\big) \end{aligned}$$

for $\lambda \in \mathbb{C}\backslash\mathbb{R}$. As in (2.21) and (2.22) we find

$$\operatorname{tr}\big((\widetilde{K}-\lambda)^{-1}-(K_0-\lambda)^{-1}\big)=\operatorname{tr}\left(\big(\widetilde{\Theta}-\widetilde{M}(\lambda)\big)^{-1}\frac{d}{d\lambda}\widetilde{M}(\lambda)\right).$$

With the same argument as in the proof of Theorem 2.4 we then conclude

$$\operatorname{tr}\big((\widetilde{K}-\lambda)^{-1}-(K_0-\lambda)^{-1}\big)=\operatorname{tr}\left(\big(\widetilde{\Theta}_{\mathrm{op}}-\widetilde{M}_{\mathrm{op}}(\lambda)\big)^{-1}\frac{d}{d\lambda}\widetilde{M}_{\mathrm{op}}(\lambda)\right). \tag{3.18}$$

Since $\widetilde{\Theta}_{\mathrm{op}}=0$ and $\widetilde{V}$ is unitary it follows from (3.17) that

$$\big(\widetilde{\Theta}_{\mathrm{op}}-\widetilde{M}_{\mathrm{op}}(\lambda)\big)^{-1}=-\widetilde{M}_{\mathrm{op}}(\lambda)^{-1}=-\widetilde{V}V^{-1}\big(M(\lambda)-D\big)^{-1}V^{-1}\widetilde{V}^{*}$$

and

$$\frac{d}{d\lambda}\widetilde{M}_{\mathrm{op}}(\lambda)=\widetilde{V}V\frac{d}{d\lambda}M(\lambda)V\widetilde{V}^{*}$$

holds. This together with (3.18) implies

$$\operatorname{tr}\big((\widetilde{K}-\lambda)^{-1}-(K_0-\lambda)^{-1}\big)=\operatorname{tr}\left(-(M(\lambda)-D)^{-1}\frac{d}{d\lambda}M(\lambda)\right)$$

for all $\lambda \in \mathbb{C}_+$ and with (2.21) we get

$$\operatorname{tr}\big((\widetilde{K}-\lambda)^{-1}-(K_0-\lambda)^{-1}\big)=\operatorname{tr}\big(\gamma(\lambda)(D-M(\lambda))^{-1}\gamma(\bar{\lambda})^{*}\big)$$

as in (2.22). Using (3.1) we obtain

$$\operatorname{tr}\big((\widetilde{K}-\lambda)^{-1}-(K_0-\lambda)^{-1}\big)=\operatorname{tr}\big((A_D-\lambda)^{-1}-(A_0-\lambda)^{-1}\big)$$

for $\lambda \in \mathbb{C}_+$. Taking into account (3.13) we prove (3.15) and (3.16) follows by taking adjoints. □

3.3. Scattering matrices of dissipative and Lax-Phillips scattering systems

In this section we recall some results from [10] on the interpretation of the diagonal entries of the scattering matrix of $\{\widetilde{K},K_0\}$ as scattering matrices of a dissipative and a Lax-Phillips scattering system. For this, let again A and G be the symmetric operators from Section 3.1 and let $\Pi=\{\mathcal{H},\Gamma_0,\Gamma_1\}$ be a boundary triplet for A^* with Weyl function $M(\cdot)$. Let $D\in[\mathcal{H}]$ be a dissipative $n\times n$-matrix and let $A_D=A^*\upharpoonright\ker(\Gamma_1-D\Gamma_0)$ be the corresponding maximal dissipative extension of A. Furthermore, let $\widetilde{\Pi}=\{\widetilde{\mathcal{H}},\widetilde{\Gamma}_0,\widetilde{\Gamma}_1\}$ be the boundary triplet for $K^*=A^*\oplus G^*$ from (3.5) with Weyl function $\widetilde{M}(\cdot)$ given by (3.6), let $\widetilde{\Theta}$ be as in (3.9) and let $\widetilde{K}$ be the minimal self-adjoint dilation of A_D given by (3.10). It follows immediately from Theorem 2.5 that the scattering matrix $\{\widetilde{S}(\lambda)\}_{\lambda\in\mathbb{R}}$ of the complete scattering system $\{\widetilde{K},K_0\}$ is given by

$$\widetilde{S}(\lambda)=I_{\widetilde{\mathcal{H}}_{\widetilde{M}(\lambda)}}+2i\sqrt{\Im\mathrm{m}\,(\widetilde{M}(\lambda))}\big(\widetilde{\Theta}-\widetilde{M}(\lambda)\big)^{-1}\sqrt{\Im\mathrm{m}\,(\widetilde{M}(\lambda))}$$

in the spectral representation $L^2(\mathbb{R},d\lambda,\mathcal{H}_{\widetilde{M}(\lambda)})$ of K_0^{ac}. Here the spaces

$$\widetilde{\mathcal{H}}_{\widetilde{M}(\lambda)}:=\operatorname{ran}\big(\Im\mathrm{m}\,(\widetilde{M}(\lambda+i0))\big)$$

for a.e. $\lambda \in \mathbb{R}$ are defined in analogy to (2.24). This representation can be made more explicit, cf. [10, Theorem 3.6].

Theorem 3.4. *Let A, $\Pi = \{\mathcal{H}, \Gamma_0, \Gamma_1\}$, $M(\cdot)$ and A_D be as above, let $K_0 = A_0 \oplus G_0$ and let $\widetilde{K}$ be the minimal self-adjoint dilation of A_D from Theorem 3.1. Then the following holds:*

(i) *$K_0^{ac} = A_0^{ac} \oplus G_0$ is unitarily equivalent to the multiplication operator with the free variable in $L^2(\mathbb{R}, d\lambda, \mathcal{H}_{M(\lambda)} \oplus \mathcal{H}_D)$.*

(ii) *In $L^2(\mathbb{R}, d\lambda, \mathcal{H}_{M(\lambda)} \oplus \mathcal{H}_D)$ the scattering matrix $\{\widetilde{S}(\lambda)\}_{\lambda \in \mathbb{R}}$ of the complete scattering system $\{\widetilde{K}, K_0\}$ is given by*

$$\widetilde{S}(\lambda) = \begin{pmatrix} I_{\mathcal{H}_{M(\lambda)}} & 0 \\ 0 & I_{\mathcal{H}_D} \end{pmatrix} + 2i \begin{pmatrix} \widetilde{T}_{11}(\lambda) & \widetilde{T}_{12}(\lambda) \\ \widetilde{T}_{21}(\lambda) & \widetilde{T}_{22}(\lambda) \end{pmatrix} \in [\mathcal{H}_{M(\lambda)} \oplus \mathcal{H}_D],$$

for a.e. $\lambda \in \mathbb{R}$, where

$$\begin{aligned}
\widetilde{T}_{11}(\lambda) &= \sqrt{\Im\mathrm{m}\,(M(\lambda))}\big(D - M(\lambda)\big)^{-1}\sqrt{\Im\mathrm{m}\,(M(\lambda))},\\
\widetilde{T}_{12}(\lambda) &= \sqrt{\Im\mathrm{m}\,(M(\lambda))}\big(D - M(\lambda)\big)^{-1}\sqrt{-\Im\mathrm{m}\,(D)},\\
\widetilde{T}_{21}(\lambda) &= \sqrt{-\Im\mathrm{m}\,(D)}\big(D - M(\lambda)\big)^{-1}\sqrt{\Im\mathrm{m}\,(M(\lambda))},\\
\widetilde{T}_{22}(\lambda) &= \sqrt{-\Im\mathrm{m}\,(D)}\big(D - M(\lambda)\big)^{-1}\sqrt{-\Im\mathrm{m}\,(D)}
\end{aligned}$$

and $M(\lambda) = M(\lambda + i0)$.

Observe that the scattering matrix $\{\widetilde{S}(\lambda)\}_{\lambda \in \mathbb{R}}$ of the scattering system $\{\widetilde{K}, K_0\}$ depends only on the dissipative matrix D and the Weyl function $M(\cdot)$ of the boundary triplet $\Pi = \{\mathcal{H}, \Gamma_0, \Gamma_1\}$ for A^*, i.e., $\{\widetilde{S}(\lambda)\}_{\lambda \in \mathbb{R}}$ is completely determined by objects corresponding to the operators A, A_0 and A_D in $\mathfrak{H}$.

In the following we will focus on the so-called *dissipative scattering system* $\{A_D, A_0\}$ and we refer the reader to [15, 16, 36, 37, 38, 39, 40, 41, 42] for a detailed investigation of such scattering systems. We recall that the wave operators $W_\pm(A_D, A_0)$ of the dissipative scattering system $\{A_D, A_0\}$ are defined by

$$W_+(A_D, A_0) = \text{s-}\lim_{t \to +\infty} e^{itA_D^*} e^{-itA_0} P^{ac}(A_0)$$

and

$$W_-(A_D, A_0) = \text{s-}\lim_{t \to +\infty} e^{-itA_D} e^{itA_0} P^{ac}(A_0).$$

The scattering operator

$$S_D := W_+(A_D, A_0)^* W_-(A_D, A_0)$$

of the dissipative scattering system $\{A_D, A_0\}$ will be regarded as an operator in $\mathfrak{H}^{ac}(A_0)$. Then S_D is a contraction which in general is not unitary. Since S_D and A_0^{ac} commute it follows that S_D is unitarily equivalent to a multiplication operator induced by a family $\{S_D(\lambda)\}_{\lambda \in \mathbb{R}}$ of contractive operators in a spectral representation of A_0^{ac}.

With the help of Theorem 3.4 we obtain a representation of the scattering matrix of the dissipative scattering system $\{A_D, A_0\}$ in terms of the Weyl function $M(\cdot)$ of $\Pi = \{\mathcal{H}, \Gamma_0, \Gamma_1\}$ in the following corollary, cf. [10, Corollary 3.8].

Corollary 3.5. *Let A, $\Pi = \{\mathcal{H}, \Gamma_0, \Gamma_1\}$, $A_0 = A^* \restriction \ker(\Gamma_0)$ and $M(\cdot)$ be as above and let $A_D = A^* \restriction \ker(\Gamma_1 - D\Gamma_0)$, $D \in [\mathcal{H}]$, be maximal dissipative. Then the following holds:*

(i) *A_0^{ac} is unitarily equivalent to the multiplication operator with the free variable in $L^2(\mathbb{R}, d\lambda, \mathcal{H}_{M(\lambda)})$.*

(ii) *The scattering matrix $\{S_D(\lambda)\}$ of the dissipative scattering system $\{A_D, A_0\}$ is given by the left upper corner of the scattering matrix $\{\widetilde{S}(\lambda)\}$ in Theorem 3.4, i.e.,*

$$S_D(\lambda) = I_{\mathcal{H}_{M(\lambda)}} + 2i\sqrt{\Im\mathrm{m}\,(M(\lambda))}\big(D - M(\lambda)\big)^{-1}\sqrt{\Im\mathrm{m}\,(M(\lambda))}$$

for all a.e. $\lambda \in \mathbb{R}$, where $M(\lambda) = M(\lambda + i0)$.

In the following we are going to interpret the right lower corner of the scattering matrix $\{\widetilde{S}(\lambda)\}$ of $\{\widetilde{K}, K_0\}$ as the scattering matrix corresponding to a Lax-Phillips scattering system, see, e.g., [8, 35] for further details. To this end we decompose the space $L^2(\mathbb{R}, \mathcal{H}_D)$ into the orthogonal sum of the subspaces

$$\mathcal{D}_- := L^2(\mathbb{R}_-, \mathcal{H}_D) \quad \text{and} \quad \mathcal{D}_+ := L^2(\mathbb{R}_+, \mathcal{H}_D).$$

Then clearly

$$\mathfrak{K} = \mathfrak{H} \oplus L^2(\mathbb{R}, \mathcal{H}_D) = \mathfrak{H} \oplus \mathcal{D}_- \oplus \mathcal{D}_+$$

and we agree to denote the elements in $\mathfrak{K}$ in the form $f \oplus g_- \oplus g_+$, $f \in \mathfrak{H}$, $g_\pm \in \mathcal{D}_\pm$ and $g = g_- \oplus g_+ \in L^2(\mathbb{R}, \mathcal{H}_D)$. By J_+ and J_- we denote the operators

$$J_+ : L^2(\mathbb{R}, \mathcal{H}_D) \to \mathfrak{K}, \quad g \mapsto 0 \oplus 0 \oplus g_+,$$

and

$$J_- : L^2(\mathbb{R}, \mathcal{H}_D) \to \mathfrak{K}, \quad g \mapsto 0 \oplus g_- \oplus 0,$$

respectively. Observe that $J_+ + J_-$ is the embedding of $L^2(\mathbb{R}, \mathcal{H}_D)$ into $\mathfrak{K}$. The subspaces $\mathcal{D}_+$ and $\mathcal{D}_-$ are so-called *outgoing* and *incoming subspaces* for the self-adjoint dilation $\widetilde{K}$ in $\mathfrak{K}$, that is, one has

$$e^{-it\widetilde{K}}\mathcal{D}_\pm \subseteq \mathcal{D}_\pm, \quad t \in \mathbb{R}_\pm, \quad \text{and} \quad \bigcap_{t\in\mathbb{R}} e^{-it\widetilde{K}}\mathcal{D}_\pm = \{0\}.$$

If, in addition, $\sigma(A_0)$ is singular, then

$$\overline{\bigcup_{t\in\mathbb{R}} e^{-it\widetilde{K}}\mathcal{D}_+} = \overline{\bigcup_{t\in\mathbb{R}} e^{-it\widetilde{K}}\mathcal{D}_-} = \mathfrak{K}^{ac}(\widetilde{K})$$

holds. Hence $\{\widetilde{K}, \mathcal{D}_-, \mathcal{D}_+\}$ is a Lax-Phillips scattering system and, in particular, the *Lax-Phillips wave operators*

$$\Omega_\pm := \text{s-}\lim_{t\to\pm\infty} e^{it\widetilde{K}} J_\pm e^{-itG_0} : L^2(\mathbb{R}, \mathcal{H}_D) \to \mathfrak{K}$$

exist, cf. [8]. Since s-$\lim_{t\to\pm\infty} J_{\mp} e^{-itG_0} = 0$ the restrictions of the wave operators $W_\pm(\widetilde{K}, K_0)$ of the scattering system $\{\widetilde{K}, K_0\}$ onto $L^2(\mathbb{R}, \mathcal{H}_D)$ coincide with the Lax-Phillips wave operators $\Omega_\pm$,

$$W_\pm(\widetilde{K}, K_0)\iota_{L^2} = \text{s-}\lim_{t\to\pm\infty} e^{it\widetilde{K}}(J_+ + J_-)e^{-itG_0} = \Omega_\pm.$$

Here ι_{L^2} is the canonical embedding of $L^2(\mathbb{R}, \mathcal{H}_D)$ into $\mathfrak{K}$. Hence the *Lax-Phillips scattering operator* $S^{LP} := \Omega_+^* \Omega_-$ admits the representation

$$S^{LP} = P_{L^2} S(\widetilde{K}, K_0)\,\iota_{L^2}$$

where $S(\widetilde{K}, K_0) = W_+(\widetilde{K}, K_0)^* W_-(\widetilde{K}, K_0)$ is the scattering operator of the scattering system $\{\widetilde{K}, K_0\}$ and P_{L^2} is the orthogonal projection from $\mathfrak{K}$ onto $L^2(\mathbb{R}, \mathcal{H}_D)$. Hence the Lax-Phillips scattering operator S^{LP} is a contraction in $L^2(\mathbb{R}, \mathcal{H}_D)$ and commutes with the self-adjoint differential operator G_0. Therefore S^{LP} is unitarily equivalent to a multiplication operator induced by a family $\{S^{LP}(\lambda)\}_{\lambda\in\mathbb{R}}$ of contractive operators in $L^2(\mathbb{R}, \mathcal{H}_D)$; this family is called the *Lax-Phillips scattering matrix*.

The above considerations together with Theorem 3.4 immediately imply the following corollary on the representation of the Lax-Phillips scattering matrix, cf. [10, Corollary 3.10].

Corollary 3.6. *Let $\{\widetilde{K}, \mathcal{D}_-, \mathcal{D}_+\}$ be the Lax-Phillips scattering system considered above and let A, $\Pi = \{\mathcal{H}, \Gamma_0, \Gamma_1\}$, A_D, $M(\cdot)$ and G_0 be as in the beginning of this section. Then the following holds:*

(i) *$G_0 = G_0^{ac}$ is unitarily equivalent to the multiplication operator with the free variable in $L^2(\mathbb{R}, \mathcal{H}_D) = L^2(\mathbb{R}, d\lambda, \mathcal{H}_D)$.*

(ii) *In $L^2(\mathbb{R}, d\lambda, \mathcal{H}_D)$ the Lax-Phillips scattering matrix $\{S^{LP}(\lambda)\}_{\lambda\in\mathbb{R}}$ admits the representation*

$$S^{LP}(\lambda) = I_{\mathcal{H}_D} + 2i\sqrt{-\Im\mathrm{m}\,(D)}\big(D - M(\lambda)\big)^{-1}\sqrt{-\Im\mathrm{m}\,(D)} \tag{3.19}$$

for a.e. $\lambda \in \mathbb{R}$, where $M(\lambda) = M(\lambda + i0)$.

Let again A_D be the maximal dissipative extension of A corresponding to the maximal dissipative matrix $D \in [\mathcal{H}]$ and let $\mathcal{H}_D = \mathrm{ran}\,(\Im\mathrm{m}\,(D))$. By [21] the characteristic function $W_{A_D}(\cdot)$ of the operator A_D is given by

$$\begin{aligned} &W_{A_D} : \mathbb{C}_- \to [\mathcal{H}_D] \\ &\mu \mapsto I_{\mathcal{H}_D} - 2i\sqrt{-\Im\mathrm{m}\,(D)}\big(D^* - M(\mu)\big)^{-1}\sqrt{-\Im\mathrm{m}\,(D)}. \end{aligned} \tag{3.20}$$

The function W_{A_D} determines the completely non-self-adjoint part of A_D uniquely up to unitary equivalence.

Comparing (3.19) and (3.20) we obtain the famous relation between the Lax-Phillips scattering matrix and the characteristic function discovered originally by Adamyan and Arov in [1, 2, 3, 4], cf. [10, Corollary 3.11] for another proof and further development.

Corollary 3.7. *Let the assumption be as in Corollary* 3.6. *Then the Lax-Phillips scattering matrix* $\{S^{LP}(\lambda)\}_{\lambda\in\mathbb{R}}$ *and the characteristic function* $W_{A_D}(\cdot)$ *of the maximal dissipative operator* A_D *are related by*

$$S^{LP}(\lambda) = W_{A_D}(\lambda - i0)^*$$

for a.e. $\lambda \in \mathbb{R}$.

3.4. A modified Birman-Krein formula for dissipative scattering systems

Let $\{\widetilde{K}, K_0\}$ be the complete scattering system from the previous subsections and let $\{\widetilde{S}(\lambda)\}_{\lambda\in\mathbb{R}}$ be the corresponding scattering matrix. If $\xi_{\widetilde{\Theta}}(\cdot)$ is the spectral shift function in (3.12), then the Birman-Krein formula

$$\det(S_{\widetilde{\Theta}}(\lambda)) = \exp\bigl(-2\pi i\xi_{\widetilde{\Theta}}(\lambda)\bigr)$$

holds for a.e. $\lambda \in \mathbb{R}$, see Theorem 2.7. In the next theorem we prove a variant of the Birman-Krein formula for dissipative scattering systems.

Theorem 3.8. *Let* A *and* G *be the symmetric operators from Section* 3.1 *and let* $\Pi = \{\mathcal{H}, \Gamma_0, \Gamma_1\}$ *be a boundary triplet for* A^* *with Weyl function* $M(\cdot)$. *Let* $D \in [\mathcal{H}]$ *be dissipative and let* $A_D = A^* \upharpoonright \ker(\Gamma_1 - D\Gamma_0)$ *be the corresponding maximal dissipative extension of* A. *Then the spectral shift function* $\eta_D(\cdot)$ *of the pair* $\{A_D, A_0\}$ *given by* (3.14) *and the scattering matrices* $\{S_D(\lambda)\}_{\lambda\in\mathbb{R}}$ *and* $\{S^{LP}(\lambda)\}_{\lambda\in\mathbb{R}}$ *from Corollary* 3.5 *and Corollary* 3.6 *are related via*

$$\det(S_D(\lambda)) = \overline{\det(S^{LP}(\lambda))}\,\exp\bigl(-2\pi i\eta_D(\lambda)\bigr) \tag{3.21}$$

and

$$\det(S^{LP}(\lambda)) = \overline{\det(S_D(\lambda))}\,\exp\bigl(-2\pi i\eta_D(\lambda)\bigr) \tag{3.22}$$

for a.e. $\lambda \in \mathbb{R}$.

Proof. Let $\widetilde{K}$ be the minimal self-adjoint dilation of A_D from (3.10) corresponding to the self-adjoint parameter $\widetilde{\Theta}$ in (3.9) via the boundary triplet $\widetilde{\Pi} = \{\widetilde{\mathcal{H}}, \widetilde{\Gamma}_0, \widetilde{\Gamma}_1\}$. Taking into account Corollary 2.6 it follows that the scattering matrix $\{\widetilde{S}(\lambda)\}_{\lambda\in\mathbb{R}}$ of the scattering system $\{\widetilde{K}, K_0\}$ satisfies

$$\det(\widetilde{S}(\lambda)) = \det(\widetilde{S}_{\widetilde{\Theta}_{\mathrm{op}}}(\lambda)), \tag{3.23}$$

where $\widetilde{\Theta}_{\mathrm{op}}$ is the operator part of $\widetilde{\Theta}$ from (3.11) and

$$\widetilde{S}_{\widetilde{\Theta}_{\mathrm{op}}}(\lambda) = I_{\widetilde{\mathcal{H}}_{\widetilde{M}_{\mathrm{op}}(\lambda)}} + 2i\sqrt{\Im\mathrm{m}\,(\widetilde{M}_{\mathrm{op}}(\lambda))}\bigl(\widetilde{\Theta}_{\mathrm{op}} - \widetilde{M}_{\mathrm{op}}(\lambda)\bigr)^{-1}\sqrt{\Im\mathrm{m}\,(\widetilde{M}_{\mathrm{op}}(\lambda))}$$

for a.e. $\lambda \in \mathbb{R}$. Making use of $\widetilde{\Theta}_{\mathrm{op}} = 0$ (see (3.11)) and formula (3.17) we obtain

$$\begin{aligned}
\det\bigl(\widetilde{S}_{\widetilde{\Theta}_{\mathrm{op}}}(\lambda)\bigr) &= \det\Bigl(I_{\widetilde{\mathcal{H}}_{\widetilde{M}_{\mathrm{op}}(\lambda)}} + 2i\Im\mathrm{m}\,(\widetilde{M}_{\mathrm{op}}(\lambda))\,\bigl(\widetilde{\Theta}_{\mathrm{op}} - \widetilde{M}_{\mathrm{op}}(\lambda)\bigr)^{-1}\Bigr) \\
&= \det\bigl(I_{\mathcal{H}} - 2i\Im\mathrm{m}\,(M(\lambda) - D)(M(\lambda) - D)^{-1}\bigr) \\
&= \frac{\det(M(\lambda)^* - D^*)}{\det(M(\lambda) - D)}.
\end{aligned}$$

Hence

$$\frac{\det(M(\lambda)^* - D)}{\det(M(\lambda)^* - D^*)}\det\big(\widetilde{S}_{\widetilde{\Theta}_{\rm op}}(\lambda)\big) = \frac{\det(M(\lambda)^* - D)}{\det(M(\lambda) - D)}.$$

Obviously, we have

$$\frac{\det(M(\lambda)^* - D)}{\det(M(\lambda)^* - D^*)} = \det\big(I_{\mathcal{H}} - 2i\Im\mathrm{m}\,(D)(M(\lambda)^* - D^*)^{-1}\big)$$

and since

$$\begin{aligned}\det&\big(I_{\mathcal{H}} - 2i\Im\mathrm{m}\,(D)(M(\lambda)^* - D^*)^{-1}\big)\\ &= \det\big(I_{\mathcal{H}} - 2i\sqrt{-\Im\mathrm{m}\,(D)}(D^* - M(\lambda)^*)^{-1}\sqrt{-\Im\mathrm{m}\,(D)}\big)\\ &= \overline{\det(S_{LP}(\lambda))}\end{aligned}$$

we get

$$\frac{\det(M(\lambda)^* - D)}{\det(M(\lambda)^* - D^*)}\det\big(\widetilde{S}_{\widetilde{\Theta}_{\rm op}}(\lambda)\big) = \overline{\det(S_{LP}(\lambda))}\det\big(\widetilde{S}_{\widetilde{\Theta}_{\rm op}}(\lambda)\big).$$

Similarly, we find

$$\begin{aligned}&\frac{\det(M(\lambda)^* - D)}{\det(M(\lambda) - D)}\\ &\qquad= \det\big(I_{\mathcal{H}} + 2i\sqrt{\Im\mathrm{m}\,(M(\lambda)}(D - M(\lambda))^{-1}\sqrt{\Im\mathrm{m}\,(M(\lambda)}\,\big)\\ &\qquad= \det(S_D(\lambda)),\end{aligned}$$

so that the relation

$$\overline{\det(S_{LP}(\lambda))}\det\big(\widetilde{S}_{\widetilde{\Theta}_{\rm op}}(\lambda)\big) = \det(S_D(\lambda))$$

holds for a.e. $\lambda \in \mathbb{R}$. Hence the Birman-Krein formula

$$\det\big(\widetilde{S}(\lambda)\big) = \exp\big(-2\pi i\xi_{\widetilde{\Theta}}(\lambda)\big),$$

which connects the scattering matrix of $\{\widetilde{K}, K_0\}$ and the spectral shift function $\xi_{\widetilde{\Theta}}(\cdot)$ in (3.12), Theorem 3.3 and (3.23) immediately imply (3.21) and (3.22) for a.e. $\lambda \in \mathbb{R}$. □

4. Coupled scattering systems

In the following we investigate so-called coupled scattering systems in a similar form as in [10], where, roughly speaking, the fixed dissipative scattering system in the previous section is replaced by a family of dissipative scattering systems which can be regarded as an open quantum system. These maximal dissipative operators form a Štraus family of extensions of a symmetric operator and their resolvents coincide pointwise with the resolvent of a certain self-adjoint operator in a bigger Hilbert space. The spectral shift functions of the dissipative scattering systems are explored and a variant of the Birman-Krein formula is proved.

4.1. Štraus family and coupling of symmetric operators

Let A be a densely defined closed simple symmetric operator with equal finite deficiency indices $n_\pm(A)$ in the separable Hilbert space $\mathfrak{H}$ and let $\Pi_A = \{\mathcal{H}, \Gamma_0, \Gamma_1\}$ be a boundary triplet for A^* with γ-field $\gamma(\cdot)$ and Weyl function $M(\cdot)$. Furthermore, let T be a densely defined closed simple symmetric operator with equal finite deficiency indices $n_\pm(T) = n_\pm(A)$ in the separable Hilbert space $\mathfrak{G}$ and let $\Pi_T = \{\mathcal{H}, \Upsilon_0, \Upsilon_1\}$ be a boundary triplet of T^* with γ-field $\nu(\cdot)$ and Weyl function $\tau(\cdot)$.

Observe that $-\tau(\lambda) \in [\mathcal{H}]$ is a dissipative matrix for each $\lambda \in \mathbb{C}_+$ and therefore by Proposition 2.2

$$A_{-\tau(\lambda)} := A^* \upharpoonright \ker\big(\Gamma_1 + \tau(\lambda)\Gamma_0\big), \quad \lambda \in \mathbb{C}_+, \tag{4.1}$$

is a family of maximal dissipative extensions of A in $\mathfrak{H}$. This family is called the *Štraus family of A associated with τ*. Since the limit $\tau(\lambda) := \tau(\lambda + i0)$ exists for a.e. $\lambda \in \mathbb{R}$ the Štraus family admits an extension to the real axis for a.e. $\lambda \in \mathbb{R}$. Analogously the Štraus family

$$T_{-M(\lambda)} := T^* \upharpoonright \ker\big(\Upsilon_1 + M(\lambda)\Upsilon_0\big), \quad \lambda \in \mathbb{C}_+,$$

of T associated with M consists of maximal dissipative extensions of T in $\mathfrak{G}$ and admits an extension to the real axis for a.e. $\lambda \in \mathbb{R}$. Sometimes it is convenient to define the Štraus family also on $\mathbb{C}_-$, in this case the extensions $A_{-\tau(\lambda)}$ and $T_{-M(\lambda)}$ are maximal accumulative for $\lambda \in \mathbb{C}_-$, cf. Proposition 2.2.

In a similar way as in Section 3.1 we consider the densely defined closed simple symmetric operator

$$L := \begin{pmatrix} A & 0 \\ 0 & T \end{pmatrix}$$

with equal finite deficiency indices $n_\pm(L) = 2n_\pm(A) = 2n_\pm(T)$ in the separable Hilbert space $\mathfrak{L} = \mathfrak{H} \oplus \mathfrak{G}$. Then obviously $\Pi_L = \{\widetilde{\mathcal{H}}, \widetilde{\Gamma}_0, \widetilde{\Gamma}_1\}$, where $\widetilde{\mathcal{H}} := \mathcal{H} \oplus \mathcal{H}$

$$\widetilde{\Gamma}_0(f \oplus g) := \begin{pmatrix} \Gamma_0 f \\ \Upsilon_0 g \end{pmatrix} \quad \text{and} \quad \widetilde{\Gamma}_1(f \oplus g) := \begin{pmatrix} \Gamma_1 f \\ \Upsilon_1 g \end{pmatrix}, \tag{4.2}$$

$f \in \mathrm{dom}\,(A^*)$, $g \in \mathrm{dom}\,(T^*)$, is a boundary triplet for the adjoint

$$L^* = \begin{pmatrix} A^* & 0 \\ 0 & T^* \end{pmatrix}.$$

The γ-field $\widetilde{\gamma}(\cdot)$ and Weyl function $\widetilde{M}(\cdot)$ corresponding to the boundary triplet $\Pi_L = \{\widetilde{\mathcal{H}}, \widetilde{\Gamma}_0, \widetilde{\Gamma}_1\}$ are given by

$$\widetilde{\gamma}(\lambda) = \begin{pmatrix} \gamma(\lambda) & 0 \\ 0 & \nu(\lambda) \end{pmatrix} \quad \text{and} \quad \widetilde{M}(\lambda) = \begin{pmatrix} M(\lambda) & 0 \\ 0 & \tau(\lambda) \end{pmatrix}, \quad \lambda \in \mathbb{C}\backslash\mathbb{R},$$

cf. (3.6) and (3.7). In the sequel we investigate the scattering system consisting of the self-adjoint operator

$$L_0 := L^* \upharpoonright \ker(\widetilde{\Gamma}_0) = \begin{pmatrix} A_0 & 0 \\ 0 & T_0 \end{pmatrix}, \tag{4.3}$$

where $A_0 = A^* \upharpoonright \ker(\Gamma_0)$ and $T_0 = T^* \upharpoonright \ker(\Upsilon_0)$, and the self-adjoint operator $\widetilde{L} = L^* \upharpoonright \widetilde{\Gamma}^{-1}\Theta$ which corresponds to the self-adjoint relation

$$\Theta := \left\{ \begin{pmatrix} (v,v)^\top \\ (w,-w)^\top \end{pmatrix} : v, w \in \mathcal{H} \right\}$$

in $\widetilde{\mathcal{H}}$. The self-adjoint extension $\widetilde{L}$ of L is sometimes called a coupling of the subsystems $\{\mathfrak{H}, A\}$ and $\{\mathfrak{G}, T\}$, cf. [17]. In the following theorem $\widetilde{L}$ and its connection to the Štraus family in (4.1) is made explicit, cf. [10, 17].

Theorem 4.1. *Let A, $\Pi_A = \{\mathcal{H}, \Gamma_0, \Gamma_1\}$, $M(\cdot)$, T, $\Pi_T = \{\mathcal{H}, \Upsilon_0, \Upsilon_1\}$, $\tau(\cdot)$ and L be as above. Then the self-adjoint extension $\widetilde{L}$ of L in $\mathfrak{L}$ is given by*

$$\widetilde{L} = L^* \upharpoonright \left\{ f \oplus g \in \operatorname{dom}(L^*) : \begin{array}{c} \Gamma_0 f - \Upsilon_0 g = 0 \\ \Gamma_1 f + \Upsilon_1 g = 0 \end{array} \right\} \tag{4.4}$$

and satisfies

$$P_{\mathfrak{H}}\big(\widetilde{L} - \lambda\big)^{-1} \upharpoonright_{\mathfrak{H}} = \big(A_{-\tau(\lambda)} - \lambda\big)^{-1} \quad \textit{and} \quad P_{\mathfrak{G}}\big(\widetilde{L} - \lambda\big)^{-1} \upharpoonright_{\mathfrak{G}} = \big(T_{-M(\lambda)} - \lambda\big)^{-1}$$

for all $\lambda \in \mathbb{C}\backslash\mathbb{R}$. Moreover, the following minimality conditions hold:

$$\mathfrak{L} = \operatorname{clospan}\big\{ \big(\widetilde{L} - \lambda\big)^{-1}\mathfrak{H} : \lambda \in \mathbb{C}\backslash\mathbb{R} \big\} = \operatorname{clospan}\big\{ \big(\widetilde{L} - \lambda\big)^{-1}\mathfrak{K} : \lambda \in \mathbb{C}\backslash\mathbb{R} \big\}.$$

4.2. Spectral shift function and trace formula for a coupled scattering system

Next we calculate the spectral shift function of the complete scattering system $\{\widetilde{L}, L_0\}$. By Theorem 2.4 a spectral shift function $\widetilde{\xi}_\Theta(\cdot)$ is given by

$$\widetilde{\xi}_\Theta(\lambda) = \frac{1}{\pi} \Im\mathrm{m}\left(\operatorname{tr}\big(\log(\widetilde{M}_{\mathrm{op}}(\lambda + i0) - \Theta_{\mathrm{op}})\big)\right) \tag{4.5}$$

for a.e. $\lambda \in \mathbb{R}$, where

$$\Theta_{\mathrm{op}} := \left\{ \begin{pmatrix} (v,v)^\top \\ (0,0)^\top \end{pmatrix} : v \in \mathcal{H} \right\} \tag{4.6}$$

is the operator part of Θ in the space

$$\widetilde{\mathcal{H}}_{\mathrm{op}} := \left\{ \begin{pmatrix} v \\ v \end{pmatrix} : v \in \mathcal{H} \right\} \subset \widetilde{\mathcal{H}} \tag{4.7}$$

and $\widetilde{M}_{\mathrm{op}}(\cdot) = \widetilde{P}_{\mathrm{op}} \widetilde{M}(\cdot)\widetilde{\iota}_{\mathrm{op}}$ denotes compression of the Weyl function $\widetilde{M}(\cdot)$ in $\widetilde{\mathcal{H}}$ onto $\widetilde{\mathcal{H}}_{\mathrm{op}}$. Observe that $\Theta_{\mathrm{op}} = 0$ so that the spectral shift function $\widetilde{\xi}_\Theta(\cdot)$ in (4.5) has the form

$$\widetilde{\xi}_\Theta(\lambda) = \frac{1}{\pi} \Im\mathrm{m}\left(\operatorname{tr}\big(\log(\widetilde{M}_{\mathrm{op}}(\lambda + i0))\big)\right) \tag{4.8}$$

for a.e. $\lambda \in \mathbb{R}$. Furthermore, the trace formula

$$\operatorname{tr}\big((\widetilde{L} - \lambda)^{-1} - (L_0 - \lambda)^{-1}\big) = -\int_{\mathbb{R}} \frac{1}{(t-\lambda)^2}\, \widetilde{\xi}_\Theta(t)\, dt \tag{4.9}$$

holds for all $\lambda \in \mathbb{C}\backslash\mathbb{R}$.

Theorem 4.2. *Let A, $\Pi_A = \{\mathcal{H}, \Gamma_0, \Gamma_1\}$, $M(\cdot)$ and T, $\Pi_T = \{\mathcal{H}, \Upsilon_0, \Upsilon_1\}$, $\tau(\cdot)$ be as in the beginning of Section 4.1. Then the spectral shift function $\widetilde{\xi}_\Theta(\cdot)$ of the pair $\{\widetilde{L}, L_0\}$ admits the representation*

$$\widetilde{\xi}_\Theta(\lambda) = \frac{1}{\pi}\Im\mathrm{m}\left(\mathrm{tr}\big(\log(M(\lambda + i0) + \tau(\lambda + i0))\big)\right) + 2k \tag{4.10}$$

for some $k \in \mathbb{Z}$ and a.e. $\lambda \in \mathbb{R}$. Moreover, the modified trace formula

$$\begin{aligned}\mathrm{tr}\left((A_{-\tau(\lambda)} - \lambda)^{-1} - (A_0 - \lambda)^{-1}\right) +& \\ \mathrm{tr}\left((T_{-M(\lambda)} - \lambda)^{-1} - (T_0 - \lambda)^{-1}\right) &= -\int_{\mathbb{R}} \frac{1}{(t-\lambda)^2}\,\widetilde{\xi}_\Theta(t)\,dt\end{aligned} \tag{4.11}$$

holds for all $\lambda \in \mathbb{C}\backslash\mathbb{R}$.

Proof. With the help of the unitary operator

$$\widetilde{V} : \mathcal{H} \longrightarrow \widetilde{\mathcal{H}}_{\mathrm{op}}, \qquad x \mapsto \frac{1}{\sqrt{2}}\begin{pmatrix} x \\ x \end{pmatrix}, \tag{4.12}$$

we obtain

$$\widetilde{V}^*\widetilde{M}_{\mathrm{op}}(\lambda)\widetilde{V} = \frac{1}{2}\left(M(\lambda) + \tau(\lambda)\right). \tag{4.13}$$

We conclude in the same way as in the proof of Theorem 3.3 that the functions $\mathrm{tr}(\log(\widetilde{M}_{\mathrm{op}}(\cdot)))$ and $\mathrm{tr}(\log(M(\cdot) + \tau(\cdot)))$ differ by a constant and

$$\exp\big(\mathrm{tr}\big(\log(\widetilde{M}_{\mathrm{op}}(\lambda))\big)\big) = \exp\big(\mathrm{tr}\big(\log(M(\lambda) + \tau(\lambda))\big)\big)\det \tfrac{1}{2} I_{\mathcal{H}}$$

implies that there exists $k \in \mathbb{Z}$ such that

$$\Im\mathrm{m}\left(\mathrm{tr}\big(\log(\widetilde{M}_{\mathrm{op}}(\lambda))\big)\right) = \Im\mathrm{m}\left(\mathrm{tr}\big(\log(M(\lambda) + \tau(\lambda))\big)\right) + 2k\pi$$

holds. This together with (4.8) implies (4.10).

In order to verify the trace formula (4.11) note that by (2.6) we have

$$(\widetilde{L} - \lambda)^{-1} - (L_0 - \lambda)^{-1} = \widetilde{\gamma}(\lambda)\big(\widetilde{\Theta} - \widetilde{M}(\lambda)\big)^{-1}\widetilde{\gamma}(\bar{\lambda})^*$$

for all $\lambda \in \rho(\widetilde{L}) \cap \rho(L_0)$. Taking into account (2.17) we get

$$(\widetilde{L} - \lambda)^{-1} - (L_0 - \lambda)^{-1} = -\widetilde{\gamma}(\lambda)\widetilde{\iota}_{\mathrm{op}}\left(\widetilde{M}_{\mathrm{op}}(\lambda)\right)^{-1}\widetilde{P}_{\mathrm{op}}\widetilde{\gamma}(\bar{\lambda})^*$$

and by using

$$\left(\widetilde{M}_{\mathrm{op}}(\lambda)\right)^{-1} = 2\widetilde{V}\big((M(\lambda) + \tau(\lambda)\big)^{-1}\widetilde{V}^*,$$

cf. (4.13), we obtain

$$(\widetilde{L} - \lambda)^{-1} - (L_0 - \lambda)^{-1} = -2\widetilde{\gamma}(\lambda)\widetilde{\iota}_{\mathrm{op}}\widetilde{V}\big(M(\lambda) + \tau(\lambda)\big)^{-1}\widetilde{V}^*\widetilde{P}_{\mathrm{op}}\widetilde{\gamma}(\bar{\lambda})^*$$

which yields

$$\mathrm{tr}\big((\widetilde{L} - \lambda)^{-1} - (L_0 - \lambda)^{-1}\big) = -2\mathrm{tr}\big(\big(M(\lambda) + \tau(\lambda)\big)^{-1}\widetilde{V}^*\widetilde{P}_{\mathrm{op}}\widetilde{\gamma}(\bar{\lambda})^*\widetilde{\gamma}(\lambda)\widetilde{\iota}_{\mathrm{op}}\widetilde{V}\big)$$

for all $\lambda \in \rho(\widetilde{L}) \cap \rho(L_0)$. As in (2.5) we find

$$\widetilde{P}_{\mathrm{op}}\widetilde{\gamma}(\bar{\lambda})^*\widetilde{\gamma}(\lambda)\widetilde{\iota}_{\mathrm{op}} = \widetilde{P}_{\mathrm{op}}\frac{d}{d\lambda}\widetilde{M}(\lambda)\widetilde{\iota}_{\mathrm{op}} = \frac{d}{d\lambda}\widetilde{M}_{\mathrm{op}}(\lambda)$$

and with the help of (4.13) we conclude

$$\widetilde{V}^*\widetilde{P}_{\rm op}\,\widetilde{\gamma}(\bar\lambda)^*\widetilde{\gamma}(\lambda)\widetilde{\iota}_{\rm op}\,\widetilde{V} = \frac{1}{2}\left(\frac{d}{d\lambda}M(\lambda) + \frac{d}{d\lambda}\tau(\lambda)\right).$$

Hence

$$\begin{aligned}&\operatorname{tr}\big((\widetilde{L}-\lambda)^{-1} - (L_0-\lambda)^{-1}\big)\\ &\quad = -\operatorname{tr}\left(\big(M(\lambda)+\tau(\lambda)\big)^{-1}\left(\frac{d}{d\lambda}M(\lambda) + \frac{d}{d\lambda}\tau(\lambda)\right)\right).\end{aligned}$$

Using again (2.5) we find

$$\begin{aligned}&\operatorname{tr}\big((\widetilde{L}-\lambda)^{-1} - (L_0-\lambda)^{-1}\big)\\ &\quad = -\operatorname{tr}\left(\gamma(\lambda)(M(\lambda)+\tau(\lambda))^{-1}\gamma(\bar\lambda)^*\right) - \operatorname{tr}\left(\nu(\lambda)(M(\lambda)+\tau(\lambda))^{-1}\nu(\bar\lambda)^*\right).\end{aligned}$$

By (2.6) the resolvents of the Štraus family of A associated with τ and the Štraus family of T associated with M are given by

$$\big(A_{-\tau(\lambda)}-\lambda\big)^{-1} - (A_0-\lambda)^{-1} = -\gamma(\lambda)\big(M(\lambda)+\tau(\lambda)\big)^{-1}\gamma(\bar\lambda)^* \tag{4.14}$$

and

$$\big(T_{-M(\lambda)}-\lambda\big)^{-1} - (T_0-\lambda)^{-1} = -\nu(\lambda)\big(M(\lambda)+\tau(\lambda)\big)^{-1}\nu(\bar\lambda)^*, \tag{4.15}$$

respectively. Taking into account (4.14), (4.15) and (4.9) we prove (4.11). □

Let us consider the the spectral shift function $\eta_{-\tau(\mu)}(\cdot)$ of the dissipative scattering system $\{A_{-\tau(\mu)}, A_0\}$ for those $\mu\in\mathbb{R}$ for which the limit $\tau(\mu) := \tau(\mu+i0)$ exists. By Theorem 3.3 the function $\eta_{-\tau(\mu)}(\cdot)$ admits the representation

$$\eta_{-\tau(\mu)}(\lambda) = \frac{1}{\pi}\Im m\left(\operatorname{tr}\big(\log(M(\lambda+i0)+\tau(\mu))\big)\right) \tag{4.16}$$

for a.e. $\lambda\in\mathbb{R}$. Moreover, we have

$$\operatorname{tr}\left((A_{-\tau(\mu)}-\lambda)^{-1} - (A_0-\lambda)^{-1}\right) = -\int_{\mathbb{R}}\frac{1}{(t-\lambda)^2}\,\eta_{-\tau(\mu)}(t)\,dt$$

for all $\lambda\in\mathbb{C}_+$, cf. Theorem 3.3. Similarly, we introduce the spectral shift function $\eta_{-M(\mu)}(\cdot)$ of the dissipative scattering system $\{T_{-M(\mu)}, T_0\}$ for those $\mu\in\mathbb{R}$ for which the limit $M(\mu) = M(\mu+i0)$ exists. It follows that

$$\eta_{-M(\mu)}(\lambda) = \frac{1}{\pi}\Im m\left(\operatorname{tr}\big(\log(M(\mu)+\tau(\lambda+i0))\big)\right) \tag{4.17}$$

holds for a.e. $\lambda\in\mathbb{R}$ and

$$\operatorname{tr}\left((T_{-M(\mu)}-\lambda)^{-1} - (T_0-\lambda)^{-1}\right) = -\int_{\mathbb{R}}\frac{1}{(t-\lambda)^2}\,\eta_{-M(\mu)}(t)\,dt$$

is valid for $\lambda\in\mathbb{C}_+$. Hence we get the following corollary.

Corollary 4.3. *Let the assumptions be as in Theorem 4.2, let L_0, $\widetilde{L}$ be as in (4.3), (4.4) and let $\eta_{-\tau(\mu)}(\cdot)$ and $\eta_{-M(\mu)}(\cdot)$ be the spectral shift functions in (4.16) and (4.17), respectively. Then the spectral shift function $\widetilde{\xi}_\Theta(\cdot)$ of the pair $\{\widetilde{L}, L_0\}$ admits the representation*

$$\widetilde{\xi}_\Theta(\lambda) = \eta_{-\tau(\lambda)}(\lambda) + 2k = \eta_{-M(\lambda)}(\lambda) + 2l$$

for a.e. $\lambda \in \mathbb{R}$ and some $k, l \in \mathbb{Z}$.

4.3. Scattering matrices of coupled systems

We investigate the scattering matrix of the scattering system $\{\widetilde{L}, L_0\}$, where $\widetilde{L}$ and L_0 are the self-adjoint operators in $\mathfrak{L} = \mathfrak{H} \oplus \mathfrak{G}$ from (4.4) and (4.3), respectively. By Theorem 2.5 the scattering matrix $\{\widetilde{S}_\Theta(\lambda)\}_{\lambda\in\mathbb{R}}$ of $\{\widetilde{L}, L_0\}$ admits the representation

$$\widetilde{S}_\Theta(\lambda) = I_{\widetilde{\mathcal{H}}_{\widetilde{M}(\lambda)}} + 2i\sqrt{\Im\mathrm{m}\,(\widetilde{M}(\lambda))}\big(\Theta - \widetilde{M}(\lambda)\big)^{-1}\sqrt{\Im\mathrm{m}\,(\widetilde{M}(\lambda))}. \tag{4.18}$$

Here $\widetilde{M}(\cdot)$ is the Weyl function of the boundary triplet $\Pi_L = \{\widetilde{\mathcal{H}}, \widetilde{\Gamma}_0, \widetilde{\Gamma}_1\}$ from (4.2) and

$$\widetilde{\mathcal{H}}_{\widetilde{M}(\lambda)} := \mathrm{ran}\,\big(\Im\mathrm{m}\,\big(\widetilde{M}(\lambda + i0)\big)\big)$$

for a.e. $\lambda \in \mathbb{R}$. In [10] the scattering matrix of $\{\widetilde{L}, L_0\}$ was expressed in terms of the Weyl functions $M(\cdot)$ and $\tau(\cdot)$ of the boundary triplets $\Pi_A = \{\mathcal{H}, \Gamma_0, \Gamma_1\}$ and $\Pi_T = \{\mathcal{H}, \Upsilon_0, \Upsilon_1\}$, respectively. The following representation for $\{\widetilde{S}_\Theta(\lambda)\}_{\lambda\in\mathbb{R}}$ can be deduced from Corollary 2.6.

Theorem 4.4. *Let A, $\Pi_A = \{\mathcal{H}, \Gamma_0, \Gamma_1\}$, $M(\cdot)$ and T, $\Pi_T = \{\mathcal{H}, \Upsilon_0, \Upsilon_1\}$, $\tau(\cdot)$ be as above. Then the following holds:*

(i) *$L_0^{ac} = A_0^{ac} \oplus T_0^{ac}$ is unitarily equivalent to the multiplication operator with the free variable in $L^2(\mathbb{R}, d\lambda, \mathcal{H}_{M(\lambda)} \oplus \mathcal{H}_{\tau(\lambda)})$.*

(ii) *In $L^2(\mathbb{R}, d\lambda, \mathcal{H}_{M(\lambda)} \oplus \mathcal{H}_{\tau(\lambda)})$ the scattering matrix $\{\widetilde{S}_\Theta(\lambda)\}_{\lambda\in\mathbb{R}}$ of the complete scattering system $\{\widetilde{L}, L_0\}$ is given by*

$$\widetilde{S}_\Theta(\lambda) = I_{\mathcal{H}_{M(\lambda)}\oplus\mathcal{H}_{\tau(\lambda)}} - 2i\begin{pmatrix}\widetilde{T}_{11}(\lambda) & \widetilde{T}_{12}(\lambda)\\ \widetilde{T}_{21}(\lambda) & \widetilde{T}_{22}(\lambda)\end{pmatrix} \in [\mathcal{H}_{M(\lambda)} \oplus \mathcal{H}_{\tau(\lambda)}],$$

for a.e. $\lambda \in \mathbb{R}$ where

$$\widetilde{T}_{11}(\lambda) = \sqrt{\Im\mathrm{m}\,(M(\lambda))}\big(M(\lambda) + \tau(\lambda)\big)^{-1}\sqrt{\Im\mathrm{m}\,(M(\lambda))},$$
$$\widetilde{T}_{12}(\lambda) = \sqrt{\Im\mathrm{m}\,(M(\lambda))}\big(M(\lambda) + \tau(\lambda)\big)^{-1}\sqrt{\Im\mathrm{m}\,(\tau(\lambda))},$$
$$\widetilde{T}_{21}(\lambda) = \sqrt{\Im\mathrm{m}\,(\tau(\lambda))}\big(M(\lambda) + \tau(\lambda)\big)^{-1}\sqrt{\Im\mathrm{m}\,(M(\lambda))},$$
$$\widetilde{T}_{22}(\lambda) = \sqrt{\Im\mathrm{m}\,(\tau(\lambda))}\big(M(\lambda) + \tau(\lambda)\big)^{-1}\sqrt{\Im\mathrm{m}\,(\tau(\lambda))}$$

and $M(\lambda) = M(\lambda + i0)$, $\tau(\lambda) = \tau(\lambda + i0)$.

Let $J_{\mathfrak{H}} : \mathfrak{H} \longrightarrow \mathfrak{L}$ and $J_{\mathfrak{G}} : \mathfrak{G} \longrightarrow \mathfrak{L}$ be the natural embedding operators of the subspaces $\mathfrak{H}$ and $\mathfrak{G}$ into $\mathfrak{L}$, respectively. The wave operators

$$W_\pm(\widetilde{L}, A_0) := \text{s-}\lim_{t\to\pm\infty} e^{it\widetilde{L}} J_{\mathfrak{H}} e^{-itA_0} P^{ac}(A_0)$$

and

$$W_\pm(\widetilde{L}, T_0) := \text{s-}\lim_{t\to\pm\infty} e^{it\widetilde{L}} J_{\mathfrak{G}} e^{-itT_0} P^{ac}(T_0)$$

are called the *channel wave operators* or *partial wave operators*. The *channel scattering operators* $S_{\mathfrak{H}}$ and $S_{\mathfrak{G}}$ are defined by

$$S_{\mathfrak{H}} := W_+(\widetilde{L}, A_0)^* W_-(\widetilde{L}, A_0) \quad \text{and} \quad S_{\mathfrak{G}} := W_+(\widetilde{L}, T_0)^* W_-(\widetilde{L}, T_0).$$

The channel scattering operators $S_{\mathfrak{H}}$ and $S_{\mathfrak{G}}$ are contractions in $\mathfrak{H}^{ac}(A_0)$ and $\mathfrak{G}^{ac}(T_0)$ and commute with A_0 and T_0, respectively. Hence, there are measurable families of contractions

$$\{S_{\mathfrak{H}}(\lambda)\}_{\lambda\in\mathbb{R}} \quad \text{and} \quad \{S_{\mathfrak{G}}(\lambda)\}_{\lambda\in\mathbb{R}} \tag{4.19}$$

such that the multiplication operators induced by these families in the spectral representations $L^2(\mathbb{R}, d\lambda, \mathcal{H}_{M(\lambda)})$ and $L^2(\mathbb{R}, d\lambda, \mathcal{H}_{\tau(\lambda)})$ of A_0^{ac} and T_0^{ac}, respectively, are unitarily equivalent to the channel scattering operators $S_{\mathfrak{H}}$ and $S_{\mathfrak{G}}$. The multiplication operators in (4.19) are called *channel scattering matrices*.

Corollary 4.5. *Let A, $\Pi_A = \{\mathcal{H}, \Gamma_0, \Gamma_1\}$, $M(\cdot)$ and T, $\Pi_T = \{\mathcal{H}, \Upsilon_0, \Upsilon_1\}$, $\tau(\cdot)$ be as above. Then the following holds:*

(i) *A_0^{ac} and T_0^{ac} are unitarily equivalent to the multiplication operators with the free variable in $L^2(\mathbb{R}, d\lambda, \mathcal{H}_{M(\lambda)})$ and $L^2(\mathbb{R}, d\lambda, \mathcal{H}_{\tau(\lambda)})$, respectively.*

(ii) *In $L^2(\mathbb{R}, d\lambda, \mathcal{H}_{M(\lambda)})$ and $L^2(\mathbb{R}, d\lambda, \mathcal{H}_{\tau(\lambda)})$ the channel scattering matrices $\{S_{\mathfrak{H}}(\lambda)\}_{\lambda\in\mathbb{R}}$ and $\{S_{\mathfrak{G}}(\lambda)\}_{\lambda\in\mathbb{R}}$ are given by*

$$S_{\mathfrak{H}}(\lambda) = I_{\mathcal{H}_{M(\lambda)}} - 2i\sqrt{\Im\mathrm{m}\,(M(\lambda))}\big(M(\lambda) + \tau(\lambda)\big)^{-1}\sqrt{\Im\mathrm{m}\,(M(\lambda))}$$

and

$$S_{\mathfrak{G}}(\lambda) = I_{\mathcal{H}_{\tau(\lambda)}} - 2i\sqrt{\Im\mathrm{m}\,(\tau(\lambda))}\big(M(\lambda) + \tau(\lambda)\big)^{-1}\sqrt{\Im\mathrm{m}\,(\tau(\lambda))}$$

for a.e. $\lambda \in \mathbb{R}$.

4.4. A modified Birman-Krein formula for coupled scattering systems

In a similar way as in Section 3.4 we prove a variant of the Birman-Krein formula for the coupled scattering system $\{\widetilde{L}, L_0\}$, where $\widetilde{L}$ and L_0 are as in (4.4) and (4.3), respectively. First of all it is clear that the scattering matrix $\{\widetilde{S}_\Theta(\lambda)\}_{\lambda\in\mathbb{R}}$ of $\{\widetilde{L}, L_0\}$ and the spectral shift function $\widetilde{\xi}_\Theta(\cdot)$ from (4.8) are connected via the usual Birman-Krein formula

$$\det\big(\widetilde{S}_\Theta(\lambda)\big) = \exp\big(-2\pi i\widetilde{\xi}_\Theta(\lambda)\big) \tag{4.20}$$

for a.e. $\lambda \in \mathbb{R}$, cf. Theorem 2.7. With the help of the channel scattering matrices from (4.19) and Corollary 4.5 we find the following modified Birman-Krein formula.

Theorem 4.6. *Let A and T be as in Section 4.1 and let $\{\widetilde{L}, L_0\}$ be the complete scattering system from above. Then the spectral shift function $\widetilde{\xi}_\Theta(\cdot)$ of the pair $\{\widetilde{L}, L_0\}$ in (4.8) is related with the channel scattering matrices $\{S_{\mathfrak{H}}(\lambda)\}_{\lambda\in\mathbb{R}}$ and $\{S_{\mathfrak{G}}(\lambda)\}_{\lambda\in\mathbb{R}}$ in (4.19) via*

$$\det(S_{\mathfrak{H}}(\lambda)) = \overline{\det(S_{\mathfrak{G}}(\lambda))}\exp\bigl(-2\pi i\widetilde{\xi}_\Theta(\lambda)\bigr) \tag{4.21}$$

and

$$\det(S_{\mathfrak{G}}(\lambda)) = \overline{\det(S_{\mathfrak{H}}(\lambda))}\exp\bigl(-2\pi i\widetilde{\xi}_\Theta(\lambda)\bigr) \tag{4.22}$$

for a.e. $\lambda\in\mathbb{R}$.

Proof. Let $\{\widetilde{S}_\Theta(\lambda)\}_{\lambda\in\mathbb{R}}$ be the scattering matrix of $\{\widetilde{L}, L_0\}$ from (4.18). Making use of Corollary 2.6 we obtain

$$\det\bigl(\widetilde{S}_\Theta(\lambda)\bigr) = \det\bigl(\widetilde{S}_{\Theta_{\rm op}}(\lambda)\bigr) \tag{4.23}$$

where $\Theta_{\rm op} = 0 \in [\widetilde{\mathcal{H}}_{\rm op}]$ is the operator part of Θ in $\widetilde{\mathcal{H}}_{\rm op}$, cf. (4.6), (4.7), and $\widetilde{S}_{\Theta_{\rm op}}(\lambda)$ is given by

$$\widetilde{S}_{\Theta_{\rm op}}(\lambda) = I_{\mathcal{H}_{\widetilde{M}_{\rm op}(\lambda)}} - 2i\sqrt{\Im\mathrm{m}\,(\widetilde{M}_{\rm op}(\lambda))}\bigl(\widetilde{M}_{\rm op}(\lambda)\bigr)^{-1}\sqrt{\Im\mathrm{m}\,(\widetilde{M}_{\rm op}(\lambda))}$$

for a.e. $\lambda\in\mathbb{R}$. Here $\widetilde{M}_{\rm op}(\cdot) = \widetilde{P}_{\rm op}\widetilde{M}(\cdot)\widetilde{\iota}_{\rm op}$ is the compression of the Weyl function corresponding to the boundary triplet $\Pi_L = \{\widetilde{\mathcal{H}}, \widetilde{\Gamma}_0, \widetilde{\Gamma}_1\}$ onto the space $\widetilde{\mathcal{H}}_{\rm op}$. Let $\widetilde{V}$ be as in (4.12). Then we have

$$\widetilde{M}_{\rm op}(\lambda) = \frac{1}{2}\widetilde{V}(M(\lambda)+\tau(\lambda))\widetilde{V}^* \quad\text{and}\quad \widetilde{M}_{\rm op}(\lambda)^{-1} = 2\widetilde{V}(M(\lambda)+\tau(\lambda))^{-1}\widetilde{V}^*,$$

cf. (4.13), and therefore we get

$$\det\bigl(\widetilde{S}_{\Theta_{\rm op}}(\lambda)\bigr) = \det\bigl(I_{\mathcal{H}} - 2i\Im\mathrm{m}\,\bigl(M(\lambda)+\tau(\lambda)\bigr)\bigl(M(\lambda)+\tau(\lambda)\bigr)^{-1}\bigr).$$

This yields

$$\det\bigl(\widetilde{S}_{\Theta_{\rm op}}(\lambda)\bigr) = \frac{\overline{\det(M(\lambda)+\tau(\lambda))}}{\det(M(\lambda)+\tau(\lambda))}$$

for a.e. $\lambda\in\mathbb{R}$ and hence

$$\frac{\overline{\det(M(\lambda)+\tau(\lambda)^*)}}{\overline{\det(M(\lambda)+\tau(\lambda))}}\det\bigl(\widetilde{S}_{\Theta_{\rm op}}(\lambda)\bigr) = \frac{\det(M(\lambda)^*+\tau(\lambda))}{\det(M(\lambda)+\tau(\lambda))}$$

for a.e. $\lambda\in\mathbb{R}$. On the other hand, as a consequence of Corollary 4.5 we obtain

$$\det(S_{\mathfrak{H}}(\lambda)) = \frac{\det(M(\lambda)^*+\tau(\lambda))}{\det(M(\lambda)+\tau(\lambda))}$$

and

$$\det(S_{\mathfrak{G}}(\lambda)) = \frac{\det(M(\lambda)+\tau(\lambda)^*)}{\det(M(\lambda)+\tau(\lambda))}$$

for a.e. $\lambda\in\mathbb{R}$ and therefore we find

$$\overline{\det(S_{\mathfrak{G}}(\lambda))}\det\bigl(\widetilde{S}_{\Theta_{\rm op}}(\lambda)\bigr) = \det(S_{\mathfrak{H}}(\lambda))$$

for a.e. $\lambda \in \mathbb{R}$. Taking into account (4.20) and (4.23) we obtain (4.21). The relation (4.22) follows from (4.21). □

Making use of Corollary 4.3 we obtain the following form for the relations (4.21) and (4.22).

Corollary 4.7. *Let the assumptions be as in Theorem* 4.6 *and let* $\eta_{-\tau(\mu)}(\cdot)$ *and* $\eta_{-M(\mu)}(\cdot)$ *be as in* (4.16) *and* (4.17), *respectively. Then the channel scattering matrices* $\{S_{\mathfrak{H}}(\lambda)\}_{\lambda\in\mathbb{R}}$ *and* $\{S_{\mathfrak{G}}(\lambda)\}_{\lambda\in\mathbb{R}}$ *are connected with the functions* $\lambda \mapsto \eta_{-\tau(\lambda)}(\lambda)$ *and* $\lambda \mapsto \eta_{-M(\lambda)}(\lambda)$ *via*

$$\det(S_{\mathfrak{H}}(\lambda)) = \overline{\det(S_{\mathfrak{G}}(\lambda))}\exp\bigl(-2\pi i\eta_{-\tau(\lambda)}(\lambda)\bigr)$$

and

$$\det(S_{\mathfrak{G}}(\lambda)) = \overline{\det(S_{\mathfrak{H}}(\lambda))}\exp\bigl(-2\pi i\eta_{-M(\lambda)}(\lambda)\bigr)$$

for a.e. $\lambda \in \mathbb{R}$.

References

[1] Adamjan, V.M.; Arov, D.Z.: *On a class of scattering operators and characteristic operator-functions of contractions*, Dokl. Akad. Nauk SSSR 160 (1965), 9–12.

[2] Adamjan, V.M.; Arov, D.Z.: *On scattering operators and contraction semigroups in Hilbert space*, Dokl. Akad. Nauk SSSR 165 (1965), 9–12.

[3] Adamjan, V.M.; Arov, D.Z.: *Unitary couplings of semi-unitary operators*, Mat. Issled. 1 (1966) vyp. 2, 3–64.

[4] Adamjan, V.M.; Arov, D.Z.: *Unitary couplings of semi-unitary operators*, Akad. Nauk Armjan. SSR Dokl. 43 (1966) no. 5, 257–263.

[5] Adamjan, V.M.; Neidhardt, H.: *On the summability of the spectral shift function for pair of contractions and dissipative operators*, J. Operator Theory 24 (1990), no. 1, 187–205.

[6] Adamjan, V.M.; Pavlov, B.S.: *Trace formula for dissipative operators*, Vestnik Leningrad. Univ. Mat. Mekh. Astronom. 1979, no. 2, 5–9, 118.

[7] Adamyan, V.M.; Pavlov, B.S.: *Null-range potentials and M.G. Krein's formula for generalized resolvents*, Zap. Nauchn. Semin. Leningr. Otd. Mat. Inst. Steklova 149 (1986) 7–23 (Russian); translation in J. Sov. Math. 42 no. 2 (1988) 1537–1550.

[8] Baumgärtel, H.; Wollenberg, M.: *Mathematical Scattering Theory*, Akademie-Verlag, Berlin, 1983.

[9] Behrndt, J.; Malamud, M.M.; Neidhardt, H.: *Scattering matrices and Weyl function*, to appear in Proc. London. Math. Soc.

[10] Behrndt, J.; Malamud, M.M.; Neidhardt, H.: *Scattering theory for open quantum systems with finite rank coupling*, Math. Phys. Anal. Geom. 10 (2007), 313–358.

[11] Birman, M.S.; Krein, M.G.: *On the theory of wave operators and scattering operators*, Dokl. Akad. Nauk SSSR 144 (1962), 475–478.

[12] Birman, M.S.; Yafaev, D.R.: *Spectral properties of the scattering matrix*, Algebra i Analiz 4 (1992), no. 6, 1–27; translation in St. Petersburg Math. J. 4 (1993), no. 6, 1055–1079.

[13] Birman, M.S.; Yafaev, D.R.: *The spectral shift function. The papers of M.G. Kreĭn and their further development*, Algebra i Analiz 4 (1992), no. 5, 1–44; translation in St. Petersburg Math. J. 4 (1993), no. 5, 833–870

[14] Brasche, J.F.; Malamud, M.M.; Neidhardt, H.: *Weyl function and spectral properties of selfadjoint extensions*, Integral Equations Operator Theory 43 (2002), no. 3, 264–289.

[15] Davies, E.B.: *Two-channel Hamiltonians and the optical model of nuclear scattering*, Ann. Inst. H. Poincaré Sect. A (N.S.) 29 (1978), no. 4, 395–413.

[16] Davies, E.B.: *Nonunitary scattering and capture. I. Hilbert space theory*, Comm. Math. Phys. 71 (1980), no. 3, 277–288.

[17] Derkach, V.A.; Hassi, S.; Malamud, M.M.; de Snoo, H.: *Generalized resolvents of symmetric operators and admissibility*, Methods Funct. Anal. Topology 6 (2000), 24–53.

[18] Derkach, V.A.; Hassi, S.; Malamud, M.M.; de Snoo, H.: *Boundary relations and their Weyl families*, Trans. Amer. Math. Soc., 358 (2006), 5351–5400.

[19] Derkach, V.A.; Malamud, M.M.: *On the Weyl function and Hermite operators with gaps*, Dokl. Akad. Nauk SSSR 293 (1987), no. 5, 1041–1046.

[20] Derkach, V.A.; Malamud, M.M.: *Generalized resolvents and the boundary value problems for Hermitian operators with gaps*, J. Funct. Anal. 95 (1991), 1–95.

[21] Derkach, V.A.; Malamud, M.M.: *Characteristic functions of linear operators*, Russian Acad. Sci. Dokl. Math. 45 (1992), 417–424.

[22] Derkach, V.A.; Malamud, M.M.: *The extension theory of hermitian operators and the moment problem*, J. Math. Sci. (New York) 73 (1995), 141–242.

[23] Dijksma, A.; de Snoo, H.: *Symmetric and selfadjoint relations in Krein spaces I*, Operator Theory: Advances and Applications 24, Birkhäuser, Basel (1987), 145–166.

[24] Donoghue, W.F.: *Monotone Matrix Functions and Analytic Continuation*, Springer, Berlin-New York, 1974.

[25] Garnett, J.B.: *Bounded Analytic Functions*, Academic Press, New York-London, 1981.

[26] Gesztesy, F.; Makarov, K.A.; Naboko, S.N.: *The spectral shift operator*, in *Mathematica results in quantum mechanics*, J. Dittrich, P. Exner, M. Tater (eds.), Operator Theory: Advances and Applications 108, Birkhäuser, Basel, 1999, 59–90.

[27] Gesztesy, F.; Makarov, K.A.: *The Ξ operator and its relation to Krein's spectral shift function*, J. Anal. Math. 81 (2000), 139–183.

[28] Gesztesy, F.; Makarov, K.A.: *Some applications of the spectral shift operator*, in *Operator theory and its applications*, A.G. Ramm, P.N. Shivakumar and A.V. Strauss (eds.), Fields Institute Communication Series 25, Amer. Math. Soc., Providence, RI, 2000, 267–292.

[29] Gohberg, I.; Krein, M.G.: *Introduction to the theory of linear nonselfadjoint operators*, Translations of Mathematical Monographs, Vol. 18 American Mathematical Society, Providence, R.I. 1969.

[30] Gorbachuk, V.I.; Gorbachuk, M.L.: *Boundary Value Problems for Operator Differential Equations*, Mathematics and its Applications (Soviet Series) 48, Kluwer Academic Publishers Group, Dordrecht, 1991.

[31] Kato, T.: *Perturbation Theory for Linear Operators*, Die Grundlehren der mathematischen Wissenschaften, Band 132, 2nd edition Springer, Berlin-New York, 1976.

[32] Krein, M.G.: *Basic propositions of the theory of representations of hermitian operators with deficiency index* (m, m), Ukrain. Mat. Z. 1 (1949), 3–66.

[33] Krein, M.G.: *On perturbation determinants and a trace formula for unitary and selfadjoint operators*, Dokl. Akad. Nauk SSSR 144 (1962), 268–271.

[34] Langer, H.; de Snoo, H.; Yavrian, V.A.: *A relation for the spectral shift function of two self-adjoint extensions*, Operator Theory: Advances and Applications 127 (Birkhäuser, Basel, 2001), 437–445.

[35] Lax, P.D.; Phillips, R.S.: *Scattering Theory*, Academic Press, New York-London 1967.

[36] Martin, Ph.A.: *Scattering theory with dissipative interactions and time delay*, Nuovo Cimento B (11) 30 (1975), no. 2, 217–238.

[37] Naboko, S.N.: *Wave operators for nonselfadjoint operators and a functional model*, Zap. Naučn. Sem. Leningrad. Otdel. Mat. Inst. Steklov. (LOMI) 69 (1977), 129–135.

[38] Naboko, S.N.: *Functional model of perturbation theory and its applications to scattering theory*, Trudy Mat. Inst. Steklov. 147 (1980), 86–114, 203.

[39] Neidhardt, H.: *Scattering theory of contraction semigroups*. Report MATH 1981, 5. Akademie der Wissenschaften der DDR, Institut für Mathematik, Berlin, 1981.

[40] Neidhardt, H.: *A dissipative scattering theory*. Operator Theory: Advances and Applications 14 (1984), Birkhäuser Verlag Basel, 1984, 197–212.

[41] Neidhardt, H.: *A nuclear dissipative scattering theory*, J. Operator Theory 14 (1985), 57–66.

[42] Neidhardt, H.: *Eine mathematische Streutheorie für maximal dissipative Operatoren*. Report MATH, 86-3. Akademie der Wissenschaften der DDR, Institut für Mathematik, Berlin, 1986.

[43] Neidhardt, H.: *Scattering matrix and spectral shift of the nuclear dissipative scattering theory*, Operator Theory: Advances and Applications 24, Birkhäuser, Basel, 1987, 237–250.

[44] Neidhardt, H.: *Scattering matrix and spectral shift of the nuclear dissipative scattering theory. II*, J. Operator Theory 19 (1988), no. 1, 43–62.

[45] Pavlov, B.S.: *Dilation theory and spectral analysis of nonselfadjoint differential operators*, Mathematical programming and related questions (Proc. Seventh Winter School, Drogobych, 1974), Theory of operators in linear spaces (Russian), pp. 3–69, Central. Ekonom. Mat. Inst. Akad. Nauk SSSR, Moscow, 1976.

[46] Pavlov, B.S.: *Spectral analysis of a dissipative singular Schrödinger operator in terms of a functional model*, Partial differential equations, VIII, 87–153, Encyclopaedia Math. Sci., 65, Springer, Berlin, 1996.

[47] Peller, V.V.: *Hankel operators in the theory of perturbations of unitary and selfadjoint operators*, Funktsional. Anal. i Prilozhen. 19 (1985), no. 2, 37–51.

[48] Potapov, V.P.: *The multiplicative structure of J-contractive matrix functions* (Russian), Trudy Moskov. Mat. Obshch. 4 (1955), 125–236.

[49] Rybkin, A.V.: *Trace formulas for resonances*, Teoret. Mat. Fiz. 56 (1983), no. 3, 439–447.

[50] Rybkin, A.V.: *The spectral shift function for a dissipative and a selfadjoint operator, and trace formulas for resonances*, Mat. Sb. (N.S.) 125(167) (1984), no. 3, 420–430.

[51] Rybkin, A.V.: *The discrete and the singular spectrum in the trace formula for a contractive and a unitary operator*, Funktsional. Anal. i Prilozhen. 23 (1989), no. 3, 84–85.

[52] J. Weidmann, Lineare Operatoren in Hilberträumen. Teil II: Anwendungen, B.G. Teubner, Stuttgart, 2003.

[53] Yafaev, D.R.: *Mathematical Scattering Theory: General Theory*, Translations of Mathematical Monographs, Vol. 105, American Mathematical Society, Providence, RI, 1992.

Jussi Behrndt
Technische Universität Berlin
Institut für Mathematik, MA 6–4
Straße des 17. Juni 136
D–10623 Berlin, Germany
e-mail: behrndt@math.tu-berlin.de

Mark M. Malamud
Donetsk National University
Department of Mathematics
Universitetskaya 24
83055 Donetsk, Ukraine
e-mail: mmm@telenet.dn.ua

Hagen Neidhardt
Weierstraß-Institut für
Angewandte Analysis und Stochastik
Mohrenstr. 39
D-10117 Berlin, Germany
e-mail: neidhardt@wias-berlin.de

Operator Theory:
Advances and Applications, Vol. 188, 87–112

Approximation of $\mathcal{N}_\kappa^\infty$-functions I: Models and Regularization

Aad Dijksma, Annemarie Luger and Yuri Shondin

Abstract. The class $\mathcal{N}_\kappa^\infty$ consists of all generalized Nevanlinna functions N with κ negative squares for which the root space at ∞ of the self-adjoint relation in the minimal model (short for self-adjoint operator realization) of N contains a κ-dimensional non-positive subspace. In this paper we discuss two specific models for the function $N \in \mathcal{N}_\kappa^\infty$: one associated with the irreducible representation of N and one associated with a regularized version of this representation which need not be irreducible. The state space in each of these models is a reproducing kernel Pontryagin space whose reproducing kernel is a matrix function constructed from the data in the representation.

Mathematics Subject Classification (2000). Primary 47B25, 47B50, 47B32; Secondary 47A06.

Keywords. Generalized Nevanlinna function, realization, model, reproducing kernel space, Pontryagin space, self-adjoint operator, symmetric operator, linear relation.

1. Introduction

The class $\mathcal{N}_\kappa^\infty$ consists of all generalized Nevanlinna functions with κ negative squares for which ∞ is the only (generalized) pole of non-positive type: Recall that a generalized Nevanlinna function N with κ negative squares is a meromorphic function N on $\mathbb{C} \setminus \mathbb{R}$ such that $N(z) = N(z^*)^*$ and the Nevanlinna kernel

$$K_N(z,w) = \frac{N(z) - N(w)^*}{z - w^*} \tag{1.1}$$

has κ negative squares on $\operatorname{hol}(N)$, the domain of holomorphy of N. The class of such functions is denoted by $\mathcal{N}_\kappa$. The set $\mathcal{N}_0$ coincides with the set of Nevanlinna functions: these are the functions n with $\mathbb{C} \setminus \mathbb{R} \subset \operatorname{hol}(n)$ satisfying $n(z) = n(z^*)^*$

The second author gratefully acknowledges support from the "Fond zur Förderung der wissenschaftlichen Forschung" (FWF, Austria), grant numbers P15540-N05 and J2540-N13.

and $\operatorname{Im} n(z)/\operatorname{Im} z \geq 0$ on $\mathbb{C} \setminus \mathbb{R}$. A function N belongs to $\mathcal{N}_\kappa^\infty$ if and only if it belongs to $\mathcal{N}_\kappa$ and can be written as

$$N(z) = (z - z_0^*)^m n(z)(z - z_0)^m + p(z), \tag{1.2}$$

where $z_0 \in \operatorname{hol}(N)$, m is an integer ≥ 0, p is a real polynomial, and n belongs to $\mathcal{N}_0$ such that $\operatorname{hol}(n) = \operatorname{hol}(N)$. The representation (1.2) can be chosen such that it is irreducible, which means that

$$m \text{ is minimal}, \qquad \lim_{y \to \infty} y^{-1} n(iy) = 0, \tag{1.3}$$

and the normalization condition $\operatorname{Re} n(i) = 0$ holds, and then this representation is unique. The Nevanlinna function in (1.2) has the integral representation

$$n(z) = \int_{\mathbb{R}} \left(\frac{1}{t - z} - \frac{t}{t^2 + 1} \right) d\sigma(t), \tag{1.4}$$

where σ is a nondecreasing function on $\mathbb{R}$ such that

$$\int_{\mathbb{R}} \frac{1}{t^2 + 1} \, d\sigma(t) < \infty.$$

The class $\mathcal{N}_\kappa^\infty$ was first considered in [14], see also [4]. We refer to Definition 2.5 below for equivalent formulations.

This analytic characterization of the class $\mathcal{N}_\kappa^\infty$ can be expressed geometrically as follows. It is well known that a generalized Nevanlinna function admits a realization (or representation)

$$N(z) = N(z_0)^* + (z - z_0^*) \left\langle \left(I_{\mathcal{P}} + (z - z_0)(A - z)^{-1} \right) u, u \right\rangle_{\mathcal{P}} \tag{1.5}$$

in terms of a self-adjoint operator or relation (or "multi-valued operator") A with non-empty resolvent set $\rho(A)$ in a Pontryagin space $\mathcal{P}$ with inner product $\langle \cdot, \cdot \rangle_{\mathcal{P}}$, an element $u \in \mathcal{P}$, and a point $z_0 \in \rho(A)$. A self-adjoint operator realization will be called a model, see Definition 2.1. Models need not be unique, models which are minimal, that is, satisfy

$$\mathcal{P} = \overline{\operatorname{span}} \left\{ \left(I_{\mathcal{P}} + (z - z_0)(A - z)^{-1} \right) u \mid z \in \rho(A) \right\}, \tag{1.6}$$

and which always exist, are unique up to isomorphisms. The class $\mathcal{N}_\kappa^\infty$ consists of all generalized Nevanlinna functions with κ negative squares for which the root space at ∞ of the self-adjoint relation in the minimal model of N contains a κ-dimensional non-positive subspace. The number m in the irreducible representation (1.2) of N is the dimension of the isotropic subspace of the root space at ∞, hence $m \leq \kappa$. The higher m, the more singular we consider the function N. Singular functions from the class $\mathcal{N}_\kappa^\infty$ with $m = \kappa$ appear naturally in the theory of strongly singular perturbations, see for example [5]; regular functions from this class with $m = 0$ appear in boundary eigenvalue problems with boundary conditions which depend polynomially on the eigenvalue parameter.

The purpose of the paper, Parts I and II together, is to study the approximation of a singular function $N \in \mathcal{N}_\kappa^\infty$ by a sequence of regular functions $N^{\{\ell\}} \in \mathcal{N}_\kappa^\infty$, as $\ell \to \infty$, from the point of view of associated models. This problem was inspired by the recent papers [20] and [19], where approximations of strongly singular perturbations realized as operators in varying Pontryagin spaces were studied.

In Part I, the present paper, we focus on the description of the models for $\mathcal{N}_\kappa^\infty$-functions; it can be considered as a continuation of our previous paper [6]. The minimal model (1.5)–(1.6) is a model which applies to every generalized Nevanlinna function. But when dealing with functions N from the class $\mathcal{N}_\kappa^\infty$ the models can be made more explicit by using analytic representations of N such as (1.2). In this paper we consider two such models. The first one, based on the irreducible representation (1.2), is described in detail in Section 3. This model is minimal and hence is unique up to isomorphisms. For studying convergence properties, the model based on another representation seems more useful. It is described in Section 4. The underlying representation is obtained from N by repeatedly using the difference-quotient operator:

$$(R_\lambda f)(z) = \frac{f(z) - f(\lambda)}{z - \lambda},$$

where λ is a point in the domain of holomorphy of the function f, and is given by

$$N(z) = (z - z_0^*)^\kappa n_r(z)(z - z_0)^\kappa + r_0(z), \tag{1.7}$$

where r_0 is a real polynomial and n_r is the Nevanlinna function $n_r = R_{z_0}^\kappa R_{z_0^*}^\kappa N$. Although the model derived from (1.7) need not be minimal, the representation has the advantage that if a sequence of functions with this representation

$$N^{\{\ell\}}(z) = (z - z_0^*)^\kappa n_r^{\{\ell\}}(z)(z - z_0)^\kappa + r_0^{\{\ell\}}(z)$$

converges then the entries $n_r^{\{\ell\}}$ and $r_0^{\{\ell\}}$ also converge. We think of the representation (1.7) of $N \in \mathcal{N}_\kappa^\infty$ as a regularized version of the irreducible representation (1.2) of N for the following reason. If $m < \kappa$ and n is given by (1.4) then

$$n_r(z) = s(z) + \int_{\mathbb{R}} \frac{1}{t - z} \frac{d\sigma(t)}{|t - z_0|^{2(\kappa - m)}},$$

where $s(z)$ is real polynomial of degree at most 1. Thus the measure in the integral representation of n_r is finite or "more regular" than the measure of n in (1.4). If in (1.2) $m = \kappa$ then the Nevanlinna functions n and n_r have the same measure $d\sigma$ and differ by a real polynomial of degree at most equal to 1. For the details we refer to Section 4.

The state spaces in the models in this paper are reproducing kernel Pontryagin spaces, in the model (1.5)–(1.6) with the Nevanlinna kernel (1.1) and in the models based on the representations (1.2) and (1.7) with matrix-valued kernels constructed from the Nevanlinna function, the factors $(z - z_0)^m$, $(z - z_0^*)^m$, and the real polynomial which appear in these representations of the $\mathcal{N}_\kappa^\infty$-function.

We briefly describe the contents of the paper. In Section 2 we collect the necessary definitions and facts about matrix-valued and scalar generalized Nevanlinna functions and their models (self-adjoint operator representations) to make the paper reasonably self-contained. We prove two lemmas concerning properties of a Nevanlinna function n and the corresponding reproducing kernel space $\mathcal{L}(n)$ with kernel K_n related to the inclusion $n \in \mathcal{L}(n)$. In Sections 3 and 4 we study three different models for $N \in \mathcal{N}_\kappa^\infty$. In the following diagram we give an overview of these models and the mappings $\mathbf{V}$ and $\mathbf{V}_r$ which connect the two models at the bottom with the model at the top. For simplicity, we only name the state spaces (first entry), which are all reproducing kernel Pontryagin spaces, and the representing self-adjoint relations in these spaces (second entry), and omit mentioning the defect functions.

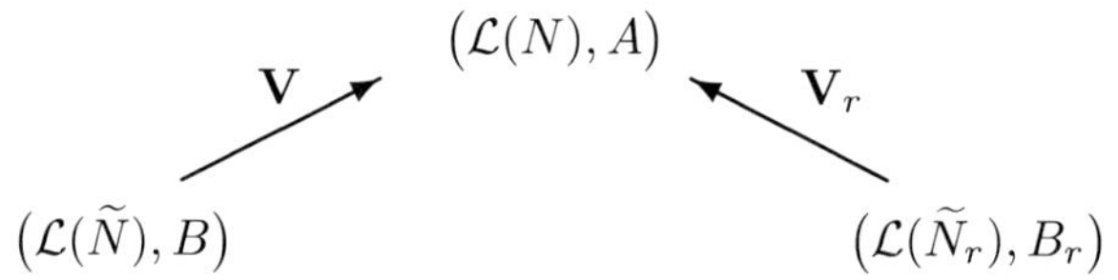

In Section 3, see Theorem 3.1, we first recall the canonical model for a matrix-valued generalized Nevanlinna function N, in the above diagram denoted by $(\mathcal{L}(N), A)$. Then we discuss the minimal model for $N \in \mathcal{N}_\kappa^\infty$ corresponding to the irreducible representation (2.15) of N. In the diagram this is the model denoted by $(\mathcal{L}(\widetilde{N}), B)$, where $\widetilde{N}$ is a matrix-valued generalized Nevanlinna function constructed from the data of the representation. This model was also presented in [6], but in the present paper we consider a slightly more general situation, see Remark 2.7. We complete the proofs given in [6] by adding to Section 3 a generalization of [6, Lemma 4.4] to generalized Nevanlinna functions having the irreducible representation (2.15), where $m = 0$ and the function n satisfies the second condition in (2.16), see Lemma 3.5. Since, for $N \in \mathcal{N}_\kappa^\infty$, both models $(\mathcal{L}(N), A)$ and $(\mathcal{L}(\widetilde{N}), B)$ are minimal, they are isomorphic. The isomorphism, $\mathbf{V}$ in the diagram, is defined by (3.3) where the function v is given in Theorem 3.2. In Section 4 we study the model $(\mathcal{L}(\widetilde{N}_r), B_r)$ mentioned in the diagram corresponding to the regularized representation (4.4) of $N \in \mathcal{N}_\kappa^\infty$, see Theorem 4.3. Here $\widetilde{N}_r$ is a matrix-valued generalized Nevanlinna function with entries obtained from the functions on the right-hand side of (4.4). Lemma 4.1 provides a general form of the regularized representation of a function from $\mathcal{N}_\kappa^\infty$; the representation (4.4) is but a particular case. The mapping $\mathbf{V}_r : \mathcal{L}(\widetilde{N}_r) \to \mathcal{L}(N)$ in the diagram and defined by (4.9) with the function v_r given by (4.10) plays the same role as the mapping $\mathbf{V}$. In Lemma 4.2 we describe its properties: $\mathbf{V}_r$ is a surjective partial isometry, it is unitary if and only if the regularized model is minimal, and if the model is not minimal, then $\mathbf{V}_r$ has a one-dimensional positive kernel.

Finally, in the two models for $N \in \mathcal{N}_\kappa^\infty$ related to the irreducible representation of N and the regularized version of it, there is a natural one-dimensional restriction of the representing self-adjoint relation. This restriction, which is symmetric and has equal defect indices, is described in Theorems 3.2 and 4.3, respectively. The family of all its canonical self-adjoint extensions (with exclusion of the self-adjoint relation given in the models) can be parametrized over the real numbers. They are described in Theorems 3.3 and 4.4 and denoted by $B^{\langle\alpha\rangle}$ and $B_r^{\langle\alpha\rangle}$ respectively with $\alpha \in \mathbb{R}$. It turns out that, as $\alpha \to \infty$, these self-adjoint relations converge in the resolvent sense to B and B_r respectively, whence the notation $B^{\langle\infty\rangle} = B$ and $B_r^{\langle\infty\rangle} = B_r$.

In the follow-up paper, Part II, which will be published elsewhere, we discuss the convergence of the models. This involves the application of the theory of approximating an operator by operators acting in different Pontryagin spaces. Such approximations in Banach and Hilbert space settings have been considered in [12, pp. 512, 513] and [21]; for approximations of this kind in an indefinite setting, see the papers [18] and [17], and also, for special cases, the papers [20] and [19] mentioned above. As a typical example we consider in Part II an approximation problem associated with the singular Bessel differential expression

$$\mathfrak{b}_B\, y(x) := -y''(x) + \frac{2}{x^2}\, y(x), \quad x \in (0,\infty).$$

The minimal and maximal realization of $\mathfrak{b}_B$ in the space $L^2(0,\infty)$ coincide and are equal to a self-adjoint operator which we denote by L. With $\mathfrak{b}_B$ there is also associated a "generalized" Titchmarsh-Weyl coefficient N which belongs to the class $\mathcal{N}_1^\infty$ and is singular, see [9, 15]. The model for N is obtained by lifting L to a self-adjoint relation in a Pontryagin space which contains $L^2(0,\infty)$ as a subspace with finite co-dimension, see [8]. It consists of the self-adjoint lifting and the Pontryagin space in which it acts. Our aim is to construct regular $\mathcal{N}_1^\infty$-functions which approximate N and show that models of these functions converge to the model of N. To this end we consider for $\varepsilon > 0$ the differential expression

$$\mathfrak{b}_B^{\{\varepsilon\}}\, y(x) := -y''(x) + \frac{2}{(x+\varepsilon)^2}\, y(x), \quad x \in (0,\infty),$$

which formally converges to $\mathfrak{b}_B$ as $\varepsilon \downarrow 0$. Since $\mathfrak{b}_B^{\{\varepsilon\}}$ is regular at the left endpoint of the interval $(0,\infty)$, the minimal and maximal realizations in $L^2(0,\infty)$ do not coincide, the first one is a symmetric operator $S^{\{\varepsilon\}}$ with defect numbers both equal to 1 and the second one is its adjoint $S^{\{\varepsilon\}*}$. The self-adjoint extensions $L_\beta^{\{\varepsilon\}}$ of $S^{\{\varepsilon\}}$ in $L^2(0,\infty)$ are the restrictions of $S^{\{\varepsilon\}*}$ determined by the boundary conditions $y'(0) = \beta y(0)$, $\beta \in \mathbb{R} \cup \{\infty\}$; $\beta = \infty$ corresponds to the condition $y(0) = 0$. The Titchmarsh-Weyl coefficients, in this case the Q-functions, associated with $L_\beta^{\{\varepsilon\}}$ and their models do not have the property we aim for: The coefficients belong to the Nevanlinna class $\mathcal{N}_0$ and hence do not converge to N as $\varepsilon \downarrow 0$. As to the models, for any choice of numbers $\beta^{\{\varepsilon\}}$ in $\mathbb{R}\cup\{\infty\}$, the operators $L_{\beta^{\{\varepsilon\}}}^{\{\varepsilon\}}$ converge in the resolvent sense to the operator L in $L^2(0,\infty)$, and hence not to the self-adjoint lifting of L.

We make another substitution, namely in the boundary eigenvalue problems associated with the operators $L_\beta^{\{\varepsilon\}}$, $\varepsilon > 0$, which are given by

$$b_B^{\{\varepsilon\}}\, y = zy, \quad y \in \operatorname{dom} S^{\{\varepsilon\}*}, \quad y'(0) = \beta y(0).$$

We replace the numbers β by linear polynomials $\beta^{\{\varepsilon\}}(z)$ and consider the boundary eigenvalue problems with boundary conditions depending on the eigenvalue parameter z:

$$b_B^{\{\varepsilon\}}\, y = zy, \quad y \in \operatorname{dom} S^{\{\varepsilon\}*}, \quad y'(0) = \beta^{\{\varepsilon\}}(z) y(0).$$

The polynomials are chosen such that the linearizations of these problems are self-adjoint relations in Pontryagin spaces depending on ε and containing $L^2(0,\infty)$ as a subspace with finite co-dimension. In the example we show that the Q-functions $N^{\{\varepsilon\}}$ associated with the self-adjoint linearizations are regular $\mathcal{N}_1^\infty$-functions which converge to N as $\varepsilon \downarrow 0$ and that the self-adjoint linearizations converge to the self-adjoint lifting of L in the sense of compressed resolvents.

2. Preliminaries

We first recall briefly some well-known facts on generalized Nevanlinna functions. A $\nu \times \nu$ matrix function N is called a *generalized Nevanlinna function with* $\kappa = \kappa(N)$ *negative squares* if

(i) it is meromorphic on $\mathbb{C} \setminus \mathbb{R}$,
(ii) it satisfies $N(z) = N(z^*)^*$ for $z \in \operatorname{hol}(N)$, the domain of holomorphy of N, and
(iii) the Nevanlinna kernel K_N defined by

$$K_N(\zeta, z) = \frac{N(\zeta) - N(z)^*}{\zeta - z^*}, \qquad \zeta, z \in \operatorname{hol}(N),$$

has κ negative squares.

Here the expression on the right-hand side of the definition of the kernel K_N is to be understood as $N'(\zeta)$ if $\zeta = z^*$. By the *domain of holomorphy* $\operatorname{hol}(N)$ mentioned in item (ii) we mean the set of all points from $\mathbb{C} \setminus \mathbb{R}$ at which N is holomorphic and of all those real points into which N can be extended by holomorphy. The class of $\nu \times \nu$ matrix functions with κ negative squares is denoted by $\mathcal{N}_\kappa^{\nu\times\nu}$ and by $\mathcal{N}_\kappa$ when the functions are scalar. As an example we mention that a real polynomial q is a generalized Nevanlinna function whose number of negative squares is given by

$$\kappa(q) = \begin{cases} \dfrac{k+1}{2}, & \text{if } k \text{ is odd and the leading coefficient of } q \text{ is } < 0, \\ \left[\dfrac{k}{2}\right], & \text{otherwise,} \end{cases} \tag{2.1}$$

where $k = \deg q$, the degree of q, and $[x]$ stands for the integral (or integer) part of the real number x.

Definition 2.1. A *self-adjoint operator realization*, or *model* for short, of a function $N \in \mathcal{N}_\kappa^{\nu\times\nu}$ is by definition a pair (A, Γ_z) consisting of a self-adjoint relation A in some Pontryagin space $\mathcal{P}$ with nonempty resolvent set $\rho(A)$ and a corresponding Γ-field Γ_z, that is, a family of mappings $\Gamma_z : \mathbb{C}^\nu \to \mathcal{P}$, $z \in \rho(A)$, which satisfy

$$\Gamma_z = \left(I_\mathcal{P} + (z-\zeta)(A-z)^{-1}\right)\Gamma_\zeta, \quad \zeta, z \in \rho(A), \tag{2.2}$$

and

$$\frac{N(\zeta) - N(z)^*}{\zeta - z^*} = \Gamma_z^*\Gamma_\zeta, \quad \zeta,\ z \in \rho(A),\ z \neq \zeta^*. \tag{2.3}$$

If a point $z_0 \in \rho(A)$ is fixed, then (2.2) and (2.3) imply the following *operator representation* of N:

$$N(z) = N(z_0)^* + (z - z_0^*)\Gamma_{z_0}^*\left(I_\mathcal{P} + (z-z_0)(A-z)^{-1}\right)\Gamma_{z_0}, \quad z \in \rho(A). \tag{2.4}$$

The term operator in operator representation refers to A, but it should be noted that A may also be a linear relation with nontrivial multi-valued part. The function N is determined by the self-adjoint relation A and the Γ-field Γ_z up to an additive constant hermitian $\nu \times \nu$ matrix. The space $\mathcal{P}$ is often called the *state space* of the model (A, Γ_z); formula (2.3) implies that the negative index of $\mathcal{P}$ is $\geq \kappa(N)$. A model (A, Γ_z) always exists. Moreover, it can be chosen *minimal* which means that

$$\mathcal{P} = \overline{\operatorname{span}}\,\{\Gamma_z\mathbf{c} \mid z \in \rho(A), \mathbf{c} \in \mathbb{C}^\nu\}.$$

In that case the negative index of the state space $\mathcal{P}$ equals $\kappa(N)$ and $\operatorname{hol}(N) = \rho(A)$; see for example [7, Theorem 1.1]. Two minimal models of N are unitarily equivalent. With a minimal model (A, Γ_z) often a symmetric restriction S of the relation A is associated, defined by

$$S = \{\{f, g\} \in A \,|\, \Gamma_{z_0}^*(g - z_0^* f) = 0\}, \tag{2.5}$$

where $z_0 \in \operatorname{hol}(N)$. The definition of S is independent of the point z_0, S is an operator, and Γ_z maps $\mathbb{C}^\nu$ onto the defect subspace $\operatorname{ran}(S - z^*)^\perp$ of S at z. Sometimes the triples (A, Γ_z, S) or $(\mathcal{P}, A, \Gamma_z)$ are also called models for the function N.

If $\nu = 1$, then the defect subspace $\operatorname{ran}(S - z^*)^\perp$ of S has dimension 1 and the function

$$\varphi(z) := \Gamma_z 1 = \left(I_\mathcal{P} + (z-z_0)(A-z)^{-1}\right)\varphi(z_0),$$

called a *defect function* for S and A, spans the defect subspace of S at z and the representation of N takes the form

$$N(z) = N(z_0)^* + (z - z_0^*)\langle\varphi(z), \varphi(z_0)\rangle_\mathcal{P}.$$

The well-known *canonical model* for $N \in \mathcal{N}_\kappa^{\nu\times\nu}$ is recalled in the next section. In this model the state space is the reproducing kernel Pontryagin space $\mathcal{L}(N)$ associated with the kernel K_N. The elements of $\mathcal{L}(N)$ are ν-vector functions defined and holomorphic on $\operatorname{hol}(N)$, the kernel has the reproducing property:

$$\langle f, K_N(\,\cdot\,, z)\mathbf{c}\rangle_{\mathcal{L}(N)} = \mathbf{c}^* f(z), \quad f \in \mathcal{L}(N),\ z \in \operatorname{hol}(N),\ \mathbf{c} \in \mathbb{C}^\nu,$$

and hence

$$\mathcal{L}(N) = \overline{\operatorname{span}}\,\{K_N(\,\cdot\,, z)\mathbf{c} \mid z \in \operatorname{hol}(N), \mathbf{c} \in \mathbb{C}^\nu\}.$$

These properties determine the space $\mathcal{L}(N)$ uniquely.

Recall that the class $\mathcal{N}_0$ consists of Nevanlinna functions, these are the functions n which are holomorphic in $\mathbb{C} \setminus \mathbb{R}$ (hence $\mathbb{C} \setminus \mathbb{R} \subset \operatorname{hol}(n)$) and such that $n(z)^* = n(z^*)$ and

$$\frac{\operatorname{Im} n(z)}{\operatorname{Im} z} \geq 0, \quad z \in \mathbb{C} \setminus \mathbb{R}.$$

A function n belongs to the class $\mathcal{N}_0$ if and only if it admits the *integral representation*

$$n(z) = \alpha + \beta z + \int_{\mathbb{R}} \left(\frac{1}{t-z} - \frac{t}{t^2+1} \right) d\sigma(t), \tag{2.6}$$

where $\alpha, \beta \in \mathbb{R}$, $\beta \geq 0$ and σ is a nondecreasing function on $\mathbb{R}$ such that

$$\int_{\mathbb{R}} \frac{1}{t^2+1}\, d\sigma(t) < \infty.$$

We shall call σ the *spectral function* and $d\sigma$ the *spectral measure* of n. From (2.6) it follows that

$$\alpha = \operatorname{Re} n(i), \quad \beta = \lim_{y \to \infty} \frac{1}{y} \operatorname{Im} n(iy) = \lim_{y \to \infty} \frac{1}{iy} n(iy). \tag{2.7}$$

In [2] it is shown that if $n \in \mathcal{N}_0$ has the integral representation (2.6), then the reproducing kernel space $\mathcal{L}(n)$ can be described as

$$\mathcal{L}(n) = \left\{ F(\zeta) := \beta c + \int_{\mathbb{R}} \frac{f(t)\, d\sigma(t)}{t - \zeta} \,\middle|\, c \in \mathbb{C},\ f \in L^2_\sigma \right\} \tag{2.8}$$

with norm defined by

$$\|F\|^2 = \beta |c|^2 + \|f\|^2_{L^2_\sigma}, \tag{2.9}$$

where

$$F(\zeta) = \beta c + \int_{\mathbb{R}} \frac{f(t)\, d\sigma(t)}{t - \zeta}, \quad c \in \mathbb{C},\ f \in L^2_\sigma. \tag{2.10}$$

Indeed, the space on the right-hand side of (2.8) with norm (2.9) is a Hilbert space and it contains the functions

$$K_n(\zeta, w) := \beta + \int_{\mathbb{R}} \frac{d\sigma(t)}{(t - w^*)(t - \zeta)}, \qquad w \in \mathbb{C} \setminus \mathbb{R},$$

and they have the reproducing kernel property:

$$\langle F(\,\cdot\,), K_n(\,\cdot\,, w) \rangle = F(w).$$

If $F \in \mathcal{L}(n)$ is given by (2.10), then by the Cauchy-Schwarz inequality

$$|F(iy)| \leq \beta |c| + \|f\|_{L^2_\sigma} \Big(\int_R \frac{1}{t^2 + y^2} d\sigma(t) \Big)^{1/2},$$

hence

$$\lim\nolimits_{y \to \infty} \frac{1}{iy} F(iy) = 0, \quad F \in \mathcal{L}(n), \tag{2.11}$$

and

$$\beta = 0 \Rightarrow \lim_{y\to\infty} F(iy) = 0, \quad F \in \mathcal{L}(n). \tag{2.12}$$

The following two simple lemmas will be referred to later on.

Lemma 2.2. *For $n \in \mathcal{N}_0$ with integral representation* (2.6) *the following statements are equivalent*:

(i) $n \in \mathcal{L}(n)$.
(ii) $\sup_{y>0} y|n(iy)| < \infty$.
(iii) $\beta = 0$, $\int_{\mathbb{R}} d\sigma(t) < \infty$, *and* $\alpha = \int_{\mathbb{R}} \frac{t}{t^2+1} d\sigma(t)$.

Statement (iii) in Lemma 2.2 is equivalent to

$$n(z) = \int_{\mathbb{R}} \frac{d\sigma(t)}{t-z} \quad \text{with} \quad \int_{\mathbb{R}} d\sigma(t) < \infty, \tag{2.13}$$

and hence the equivalence (ii) $\Leftrightarrow$ (iii) is due to I.S. Kac and M.G. Krein (see [11, Theorem S1.4.1]).

Proof of Lemma 2.2. The implication (iii) $\Rightarrow$ (i) follows directly from (2.13) and the description (2.8) of the space $\mathcal{L}(n)$. Now assume (i) holds. Then there is a $c \in \mathbb{C}$ and a function $f \in L^2_\sigma$ such that

$$\alpha + \beta\zeta + \int_{\mathbb{R}} \left(\frac{1}{t-\zeta} - \frac{t}{t^2+1} \right) d\sigma(t) = n(\zeta) = \beta c + \int_{\mathbb{R}} \frac{f(t)\, d\sigma(t)}{t-\zeta}. \tag{2.14}$$

From the formula for β in (2.7) and the last equality in (2.14) it follows that $\beta = 0$. Calculating the difference quotient

$$\frac{n(z) - n(w)}{z-w}$$

in two ways using both sides of (2.14) we obtain for $z, w \in \mathbb{C} \setminus \mathbb{R}$

$$\left\langle \frac{1}{t-z}, \frac{1}{t-w^*} \right\rangle_{L^2_\sigma} = \left\langle \frac{f(t)}{t-z}, \frac{1}{t-w^*} \right\rangle_{L^2_\sigma}.$$

By [22], the functions $1/(t-w)$ with $w \in \mathbb{C}\setminus\mathbb{R}$ are total in L^2_σ and hence $f = 1 \in L^2_\sigma$, that is, $d\sigma$ is a finite measure. Finally it follows from (2.14) that

$$\alpha = \int_{\mathbb{R}} \frac{t}{t^2+1} d\sigma(t).$$

This proves (iii). □

Lemma 2.3. *Assume $n \in \mathcal{N}_0$. Then:*

(i) *If a nonzero polynomial belongs to $\mathcal{L}(n)$ then it is identically equal to a constant.*
(ii) *The space $\mathcal{L}(n)$ contains the constant functions if and only if $\beta \neq 0$ in the integral representation* (2.6) *of n.*
(iii) *If $n \in \mathcal{L}(n)$ then the only constant function in $\mathcal{L}(n)$ is the zero function.*

Proof. (i) follows from (2.11). A function F of the form (2.10) can only be a nonzero constant if $f = 0$ and $\beta \neq 0$. This readily implies (ii). Item (iii) follows from (ii) and Lemma 2.2. □

A function $N \in \mathcal{N}_\kappa$ has κ (generalized) poles of non-positive type in the closed upper half-plane, counted according to their degrees of non-positivity, see [14] and [16].

Remark 2.4. Condition (iii) in Lemma 2.2 can also be interpreted as that n has a generalized zero at ∞, see for example [10, Lemma 3.2] or also [13]. Then the statements in the above lemmas follow alternatively also from the discussions on the space $\mathcal{L}(n)$ in [3].

In the present paper we mainly deal with functions from the subclass $\mathcal{N}_\kappa^\infty$ of $\mathcal{N}_\kappa$ introduced in [4]. In this paper we slightly modify the definition (see Remark 2.7 below):

Definition 2.5. A function N belongs to the class $\mathcal{N}_\kappa^\infty$ if and only if it belongs to $\mathcal{N}_\kappa$ and satisfies the following equivalent conditions:

(i) The point $z = \infty$ is the only (generalized) pole of non-positive type of N, and hence its degree of non-positivity equals κ.

(ii) The root space at ∞ of the self-adjoint relation A in the minimal model (2.4) of N contains a κ-dimensional non-positive subspace.

(iii) N admits the *irreducible representation*

$$N(z) = (z - z_0^*)^m n(z)(z - z_0)^m + p(z), \tag{2.15}$$

where $z_0 \in \operatorname{hol}(N)$, m is an integer ≥ 0, p is a real polynomial, and n belongs to $\mathcal{N}_0$ such that $\operatorname{hol}(n) = \operatorname{hol}(N)$,

$$m \text{ is minimal}, \qquad \lim_{y\to\infty} y^{-1} n(iy) = 0, \tag{2.16}$$

and the normalization condition $\operatorname{Re} n(i) = 0$ holds.

The normalization and the second condition in (2.16) imply that the linear term $\alpha + \beta z$ in the integral representation (2.6) of n in (2.15) is zero, hence

$$n(z) = \int_{\mathbb{R}} \left(\frac{1}{t - z} - \frac{t}{t^2 + 1} \right) d\sigma(t), \tag{2.17}$$

see (2.7). Moreover, if $z_0 \in \operatorname{hol}(N) = \operatorname{hol}(n)$ is real, then σ is constant in a real neighborhood of z_0.

Remark 2.6. (a) If in condition (iii) in Definition 2.5 the Nevanlinna function n satisfies all requirements except the requirement that $\operatorname{hol}(n) = \operatorname{hol}(N)$, then it can be replaced by a Nevanlinna function which does satisfy all requirements. Indeed, if $\operatorname{hol}(n) \neq \operatorname{hol}(N)$, then $m \geq 1$, $z_0 \in \mathbb{R}$, and n has in z_0 an isolated singularity, that is, since n is a Nevanlinna function, a pole of order 1. Hence

$$n(z) = n_0(z) + \frac{\sigma_0}{z_0 - z},$$

where σ_0 is a positive real number and n_0 is a Nevanlinna function with a spectral function which is constant in a real neighborhood of z_0. If follows that

$$N(z) = (z - z_0)^{2m} n_1(z) + p_1(z),$$

where

$$p_1(z) = p(z) - \sigma_0(z - z_0)^{2m-1} + (\operatorname{Re} n_0(i))\,(z - z_0)^{2m}$$

and

$$n_1(z) = n_0(z) - \operatorname{Re} n_0(i), \quad \operatorname{Re} n_0(i) = \frac{z_0\sigma_0}{z_0^2 + 1},$$

satisfies all requirements in condition (iii).
(b) The irreducible representation (2.15) is a special case of the one in [14]. The point z_0 is called a reference point. If N has an irreducible representation of the form (2.15) for some $z_0 \in \operatorname{hol}(N)$, then N admits such a representation for each $z_0 \in \operatorname{hol}(N)$. If z_0 is fixed then this representation is unique. The numbers m and $\deg p$ are independent of the choice of z_0; for the dependence of the functions n and p on z_0, see Remark 1.3 in [4].

If in (2.16) $m > 0$, then the minimality of m is equivalent to the condition

$$\lim_{y \to \infty} y \operatorname{Im} n(iy) = \infty. \tag{2.18}$$

Remark 2.7. The class $\mathcal{N}_\kappa^\infty$ in Definition 2.5 is larger than the one in [4] and [6]. In these papers we require that condition (2.18) is satisfied also in case $m = 0$, and this requirement we no longer impose.

Some properties of $N \in \mathcal{N}_\kappa^\infty$ can be read off directly from its representation (2.15). The number m is the dimension of the isotropic part of the root space at ∞ of the self-adjoint relation A in the minimal model of N, see [4]. If we rewrite (2.15) in the form

$$N(z) = (z - z_0^*)^m (n(z) + q(z))(z - z_0)^m + r(z), \tag{2.19}$$

where q and r are real polynomials such that

$$(z - z_0)^m q(z)(z - z_0^*)^m + r(z) = p(z), \quad \deg r < 2m,$$

then

$$\vartheta(N) = m + \deg q \qquad \text{and} \qquad \kappa(N) = m + \kappa(q). \tag{2.20}$$

Here $\vartheta(N)$ is the dimension of the root space of A at ∞, $\kappa(N)$ is the number of negative squares of N and equal to the dimension of a maximal non-positive subspace of this root space, and $\deg q$ and $\kappa(q)$ are the degree and number of negative squares of q. It follows from the last equality in (2.20) and (2.1) that $m \leq \kappa$ and that

if $m = \kappa$, then $\deg p \leq 2\kappa$ or $\deg p = 2\kappa + 1$ and the leading coefficient of p is > 0, and

if $m < \kappa$, then $\deg p = 2\kappa$ or $\deg p = 2\kappa \pm 1$ and the leading coefficient of p is $\gtrless 0$.

The representation (2.19) plays an important role in the sequel.

3. Canonical and irreducible models

In this section we recall two concrete models. To start with we cite the minimal model for a generalized Nevanlinna function N in which the state space is the reproducing kernel Pontryagin space $\mathcal{L}(N)$ with reproducing Nevanlinna kernel

$$K_N(\zeta, z) = \frac{N(\zeta) - N(z)^*}{\zeta - z^*}$$

and negative index $\kappa(N)$. Whenever defined we denote by R_z the difference-quotient operator and, for later use, by E_ζ the operator of evaluation at the point ζ, that is,

$$(R_z f)(\zeta) := \frac{f(\zeta) - f(z)}{\zeta - z}, \quad E_\zeta\, f := f(\zeta),$$

where f is a scalar or vector function.

Theorem 3.1. *Let $N \in \mathcal{N}_\kappa^{\nu\times\nu}$ be given. Then:*

(i) *The pair (A, Γ_z) is a minimal model of N in $\mathcal{L}(N)$, where*

$$A := \big\{\{f, g\} \in \mathcal{L}(N)^2 \,|\, \exists\, \mathbf{c} \in \mathbb{C}^\nu : \; g(\zeta) - \zeta f(\zeta) \equiv \mathbf{c}\big\}$$

is a self-adjoint relation in $\mathcal{L}(N)$ with $\rho(A) \neq \emptyset$, and

$$\big(\Gamma_z \mathbf{c}\big)(\zeta) := K_N(\zeta, z^*)\mathbf{c}, \quad \mathbf{c} \in \mathbb{C}^\nu,$$

is a corresponding Γ-field.

(ii) *The resolvent of A is the difference-quotient operator in $\mathcal{L}(N)$:*

$$(A - z)^{-1} = R_z, \quad z \in \rho(A).$$

(iii) *The restriction S of A:*

$$S := \big\{\{f, g\} \in A \,|\, \Gamma_{z_0^*}(g - z_0 f) = 0\big\} = \big\{\{f, g\} \in \mathcal{L}(N)^2 \,|\, g(\zeta) - \zeta f(\zeta) \equiv 0\big\}$$

is a symmetric operator in the space $\mathcal{L}(N)$ with equal defect indices $\nu - d$, where $d := \dim\ker\Gamma_z$. Moreover, the point spectrum $\sigma_p(S) = \emptyset$ and the adjoint of S is given by

$$S^* = \overline{\operatorname{span}}\,\big\{\{\Gamma_z \mathbf{h}, z\Gamma_z \mathbf{h}\} \;|\; \mathbf{h} \in \mathbb{C}^\nu, z \in \operatorname{hol}(N)\big\}$$

$$= \big\{\{f, g\} \in \mathcal{L}(N)^2 \,|\, \exists\, \mathbf{c}, \mathbf{d} \in \mathbb{C}^\nu : \; g(\zeta) - \zeta f(\zeta) \equiv \mathbf{c} - N(\zeta)\mathbf{d}\big\}.$$

For the proof of this theorem and remarks concerning its origin we refer to [3, Theorem 2.1]. The minimal model of N described in Theorem 3.1 is called the *canonical model* of N. For these canonical models only, to denote the dependence on N we often write A_N, Γ_{Nz}, S_N etc. instead of A, Γ_z, S etc.

From now on in this section we assume that N belongs to the class $\mathcal{N}_\kappa^\infty$ and admits the irreducible representation (2.19):

$$N(z) = (z - z_0^*)^m (n(z) + q(z))(z - z_0)^m + r(z),$$

where $z_0 \in \operatorname{hol}(N)$ belongs to the possibly smaller set $\operatorname{hol}(n)$. In [6] we have constructed a minimal model of N connected with this irreducible representation. We repeat it here for the convenience of the reader, see Theorem 3.2 below. Of course it is unitarily equivalent to the canonical model of N described in Theorem 3.1.

With the decomposition (2.19) there are associated the matrix-valued generalized Nevanlinna functions $\widetilde{N}$ and M defined by

$$\widetilde{N}(z)=\begin{pmatrix} n(z) & 0 & 0 \\ 0 & q(z) & 0 \\ 0 & 0 & M(z) \end{pmatrix}, \quad M(z)=\begin{pmatrix} \frac{r(z)}{(z-z_0)^m} & (z-z_0^*)^m \\ & 0 \end{pmatrix}.$$

Evidently, the reproducing kernel space $\mathcal{L}(\widetilde{N})$ with kernel $K_{\widetilde{N}}$ can be decomposed as the orthogonal sum $\mathcal{L}(\widetilde{N}) = \mathcal{L}(n) \oplus \mathcal{L}(q) \oplus \mathcal{L}(M)$. Some of the summands can be trivial, more precisely:

(i) if $\deg q > 0$ and $m > 0$, then the elements of $\mathcal{L}(q)$ are the polynomials of degree $< \deg q$ and the elements of $\mathcal{L}(M)$ are 2-vector functions with polynomial entries,

(ii) if $\deg q = 0$ and $m > 0$, then $\mathcal{L}(\widetilde{N}) = \mathcal{L}(n) \oplus \mathcal{L}(M)$, and

(iii) if $\deg q > 0$ and $m = 0$, then $\mathcal{L}(\widetilde{N}) = \mathcal{L}(n) \oplus \mathcal{L}(q)$.

The cases $\deg q = m = 0$ or $n \equiv 0$ will not be considered here since then $N \in \mathcal{N}_0$ or N is a polynomial, respectively.

The minimal model for N in the space $\mathcal{L}(\widetilde{N})$ is given in the following theorem (see [6, Theorem 4.1, Theorem 4.6, Theorem 4.7]).

Theorem 3.2. *For $N \in \mathcal{N}_\kappa^\infty$ a minimal model related to its irreducible representation* (2.19) *is given in the space $\mathcal{L}(\widetilde{N})$ by the triple*

$$(B, K_{\widetilde{N}}(\,\cdot\,, z^*)v(z), \widetilde{S})$$

consisting of the self-adjoint relation

$$B := \{\{\widetilde{f}, \widetilde{g}\} \in \mathcal{L}(\widetilde{N}) \mid \exists \mathbf{h} \in \mathbb{C}^j : \widetilde{g}(\zeta) - \zeta\widetilde{f}(\zeta) = (I_{\mathbb{C}^j} + \widetilde{N}(\zeta)\mathcal{B})\mathbf{h}\}, \tag{3.1}$$

the defect function

$$K_{\widetilde{N}}(\,\cdot\,, z^*)v(z),$$

and the symmetric operator

$$\widetilde{S} := \{\{\widetilde{f}, \widetilde{g}\} \in \mathcal{L}(\widetilde{N}) \mid \exists \mathbf{h} \in \mathbb{C}^j \text{ with } h_{j-1} = 0 : \widetilde{g}(\zeta) - \zeta\widetilde{f}(\zeta) = (\mathcal{A} + \widetilde{N}(\zeta)\mathcal{B})\mathbf{h}\},$$

where

(i) *if* $\deg q > 0$ *and* $m > 0$, *then* $j = 4$ *and*

$$v(z)=\begin{pmatrix} (z-z_0)^m \\ (z-z_0)^m \\ 1 \\ (z-z_0)^m(n(z)+q(z)) \end{pmatrix}, \quad \mathcal{A}=\begin{pmatrix} 1&0&0&0\\0&1&0&0\\0&0&0&0\\0&0&0&1 \end{pmatrix}, \quad \mathcal{B}=-\begin{pmatrix} 0&0&0&1\\0&0&0&1\\0&0&0&0\\1&1&0&0 \end{pmatrix},$$

(ii) *if* $\deg q = 0$ *and* $m > 0$, *then* $j = 3$ *and*

$$v(z)=\begin{pmatrix} (z-z_0)^\kappa \\ 1 \\ (z-z_0)^\kappa(n(z)+q_0) \end{pmatrix}, \quad \mathcal{A}=\begin{pmatrix} 1&0&0\\0&0&0\\0&0&1 \end{pmatrix}, \quad \mathcal{B}=-\begin{pmatrix} 0&0&1\\0&0&0\\1&0&0 \end{pmatrix},$$

(iii) *if* $\deg q > 0$ *and* $m = 0$, *then* $j = 2$ *and*

$$v(z) = \begin{pmatrix} 1 \\ 1 \end{pmatrix}, \quad \mathcal{A} = \begin{pmatrix} 0 & 1 \\ 0 & -1 \end{pmatrix}, \quad \mathcal{B} = \begin{pmatrix} 0 & 0 \\ 0 & 0 \end{pmatrix}.$$

Theorem 3.3. *The family of all self-adjoint extensions of the symmetric operator* $\widetilde{S}$ *in Theorem* 3.2 *consists of* B *and*

$$B^{\langle\alpha\rangle} := \{\{\widetilde{f}, \widetilde{g}\} \in \mathcal{L}(\widetilde{N}) \mid \exists \mathbf{h} \in \mathbb{C}^j : \widetilde{g}(\zeta) - \zeta \widetilde{f}(\zeta) = (\mathcal{A}_\alpha + \widetilde{N}(\zeta)\mathcal{C})\mathbf{h}\}, \ \alpha \in \mathbb{R}, \quad (3.2)$$

where

(i) *if* $\deg q > 0$ *and* $m > 0$, *then* $j = 4$ *and*

$$\mathcal{A}_\alpha = \begin{pmatrix} 1 & 0 & 0 & 0 \\ 0 & 1 & 0 & 0 \\ 0 & 0 & \alpha & 0 \\ 0 & 0 & 0 & 1 \end{pmatrix}, \quad \mathcal{C} = -\begin{pmatrix} 0 & 0 & 0 & 1 \\ 0 & 0 & 0 & 1 \\ 0 & 0 & 1 & 0 \\ 1 & 1 & 0 & 0 \end{pmatrix},$$

(ii) *if* $\deg q = 0$ *and* $m > 0$, *then* $j = 3$ *and*

$$\mathcal{A}_\alpha = \begin{pmatrix} 1 & 0 & 0 \\ 0 & \alpha & 0 \\ 0 & 0 & 1 \end{pmatrix}, \quad \mathcal{C} = -\begin{pmatrix} 0 & 0 & 1 \\ 0 & 1 & 0 \\ 1 & 0 & 0 \end{pmatrix},$$

(iii) *if* $\deg q > 0$ *and* $m = 0$, *then* $j = 2$ *and*

$$\mathcal{A}_\alpha = \begin{pmatrix} 1 & 0 \\ -1 & \alpha \end{pmatrix}, \quad \mathcal{C} = -\begin{pmatrix} 0 & 1 \\ 0 & 1 \end{pmatrix}.$$

Remark 3.4. The operators $B^{\langle\alpha\rangle}$, $\alpha \in \mathbb{R}$, are rank-one perturbations of $B^{\langle 0\rangle}$:

$$B^{\langle\alpha\rangle} = B^{\langle 0\rangle} + \alpha\langle \cdot, \widetilde{w}\rangle_{\mathcal{L}(\widetilde{N})}\widetilde{w},$$

where $\widetilde{w} = \begin{pmatrix} 0 & 0 & 1 & 0 \end{pmatrix}^\top$, $\widetilde{w} = \begin{pmatrix} 0 & 1 & 0 \end{pmatrix}^\top$, and $\widetilde{w} = \begin{pmatrix} 0 & 1 \end{pmatrix}^\top$ according to the cases (i), (ii), and (iii). Moreover, the relation B is the limit of $B^{\langle\alpha\rangle}$ in the resolvent sense for $\alpha \to \infty$. For that reason we define $B^{\langle\infty\rangle} = B$.

By Remark 2.7, in the theorems in case (iii) the assumptions on N are less restrictive than in the corresponding theorems in [6]. The proofs of the theorems in [6] are based on the unitarity of the mapping $\mathbf{V} : \mathcal{L}(\widetilde{N}) \to \mathcal{L}(N)$ defined by

$$(\mathbf{V}\widetilde{f})(\zeta) = v^{\#}(\zeta)\,\widetilde{f}(\zeta), \qquad (3.3)$$

where $v^{\#}(\zeta) := v(\zeta^*)^*$ and v is the vector-valued function defined in the formulation of Proposition 3.2. Thus to complete the proofs of Theorem 3.2 (iii) and Theorem 3.3 (iii) it suffices to show that $\mathbf{V}$ is also unitary in the case $m = 0$ (and $N \in \mathcal{N}_\kappa^\infty$ does not necessarily satisfy the additional condition (2.18)).

Lemma 3.5. *Assume* $N \in \mathcal{N}_\kappa^\infty$ *and in* (2.19) $m = 0$, *that is,*

$$N(z) = n(z) + q(z),$$

where n is a Nevanlinna function satisfying the second condition in (2.16). *Then* $\mathcal{L}(\widetilde{N}) = \mathcal{L}(n) \oplus \mathcal{L}(q)$ *and the mapping* $\mathbf{V} : \mathcal{L}(\widetilde{N}) \to \mathcal{L}(N)$ *defined by*

$$(\mathbf{V}\widetilde{f})(\zeta) = \begin{pmatrix} 1 & 1 \end{pmatrix} \widetilde{f}(\zeta)$$

is unitary.

Proof. In a similar way as in the proof of Lemma 4.2 in Section 4 below, or as in the proof of Lemma 4.4 in [6], it can be shown that $\mathbf{V}$ is a surjective partial isometry. We show that $\ker \mathbf{V} = \{0\}$. An element $\widetilde{f} := \begin{pmatrix} f & a \end{pmatrix}^\top$ belongs to $\ker \mathbf{V}$ if and only if

$$f(\zeta) + a(\zeta) \equiv 0.$$

The last equality implies that $a = -f \in \mathcal{L}(n)$. But a is an element of $\mathcal{L}(q)$ and, therefore, a is a polynomial of degree $< \deg q$. Then according to Lemma 2.3 the assumption $a \not\equiv 0$ would imply that $\beta \neq 0$ in the integral representation (2.6) of n, which contradicts (2.16). Hence $\ker \mathbf{V}$ is trivial, and $\mathbf{V}$ is unitary. □

4. Regularization

In this section we introduce a representation for a function $N \in \mathcal{N}_\kappa^\infty$ which is different from (2.15) and (2.19) and which need not be irreducible. We also present the model related to this representation. Although this model need not be minimal, it is useful in the approximation of models with varying state spaces.

Lemma 4.1. *Let $N \in \mathcal{N}_\kappa^\infty$, let $\Lambda = (\lambda_1, \lambda_2, \dots, \lambda_\kappa)$ be a sequence in* hol(N), *and set*

$$c_\Lambda(z) = (z - \lambda_1)(z - \lambda_2) \dots (z - \lambda_\kappa).$$

Then N can be written as

$$N(z) = c_\Lambda^\#(z) n_\Lambda(z) c_\Lambda(z) + r_\Lambda(z), \tag{4.1}$$

where

(i) *the function $n_\Lambda := R_{\lambda_1} R_{\lambda_2} \dots R_{\lambda_\kappa} R_{\lambda_1^*} R_{\lambda_2^*} \dots R_{\lambda_\kappa^*} N$ is a Nevanlinna function with* hol$(n_\Lambda) =$ hol(N) *and*

(ii) *r_Λ is a real polynomial of degree $\leq 2\kappa - 1$.*

If N has representation (2.15), *then*

$$n_\Lambda(z) = \alpha_\Lambda + p_{2\kappa} + p_{2\kappa+1} \sum_{j=1}^{\kappa} (\lambda_j + \lambda_j^*) + p_{2\kappa+1} z + \int_{\mathbb{R}} \left(\frac{1}{t - z} - \frac{t}{t^2 + 1} \right) \frac{|t - z_0|^{2m}}{|c_\Lambda(t)|^2} \, d\sigma(t), \tag{4.2}$$

where σ is the spectral function of n, p_j is the coefficient of z^j in $p(z)$, and $\alpha_\Lambda = 0$ if $m = \kappa$ and if $m < \kappa$ then

$$\alpha_\Lambda = \int_{\mathbb{R}} \frac{t}{t^2 + 1} \frac{|t - z_0|^{2m}}{|c_\Lambda(t)|^2} d\sigma(t).$$

Proof. The Nevanlinna function n in (2.15) has the integral representation (2.17):

$$n(z) = \int_{\mathbb{R}} \left(\frac{1}{t-z} - \frac{t}{1+t^2} \right) d\sigma(t),$$

where the spectral function σ is constant on a real neighborhood of the real points in Λ. It follows that the integrals in the lemma are all finite. Hence

$$\begin{aligned} N(z) &= (z-z_0)^m (z-z_0^*)^m \int_{\mathbb{R}} \left(\frac{1}{t-z} - \frac{t}{1+t^2} \right) d\sigma(t) + p(z) \\ &= c_\Lambda(z) c_\Lambda^{\#}(z) \int_{\mathbb{R}} \left(\frac{1}{t-z} - \frac{t}{1+t^2} \right) \frac{|t-z_0|^{2m}\, d\sigma(t)}{|c_\Lambda(t)|^2} + p(z) \qquad (4.3) \\ &\quad + \int_{\mathbb{R}} \frac{tz+1}{t^2+1} \frac{(z-z_0)^m (z-z_0^*)^m |c_\Lambda(t)|^2 - |t-z_0|^{2m} c_\Lambda(z) c_\Lambda^{\#}(z)}{t-z} \frac{d\sigma(t)}{|c_\Lambda(t)|^2}. \end{aligned}$$

The last integral is a real polynomial in z of degree $< 2\kappa$ if $m = \kappa$ and of degree $= 2\kappa$ if $m < \kappa$, and the coefficient of $z^{2\kappa}$ is given by α_Λ from the lemma. If we apply $R_{\lambda_1} R_{\lambda_2} \ldots R_{\lambda_\kappa} R_{\lambda_1^*} R_{\lambda_2^*} \ldots R_{\lambda_\kappa^*}$ to both sides of this equality and use

$$R_{\lambda_1} R_{\lambda_2} \ldots R_{\lambda_\kappa} R_{\lambda_1^*} R_{\lambda_2^*} \ldots R_{\lambda_\kappa^*} p(z) = p_{2\kappa+1} \Big(z + \sum_{j=1}^{\kappa} (\lambda_j + \lambda_j^*) \Big) + p_{2\kappa},$$

we obtain formula (4.2) for n_Λ defined in item (i) of the lemma. The formula implies that $n_\Lambda \in \mathcal{N}_0$. Replacing the middle integral in (4.3) by

$$n_\Lambda(z) - \alpha_\Lambda - p_{2\kappa} - p_{2\kappa+1} \sum_{j=1}^{\kappa} (\lambda_j + \lambda_j^*) - p_{2\kappa+1} z$$

and using that

$$c_\Lambda(z) c_\Lambda^{\#}(z) = z^{2\kappa} - \sum_{j=1}^{\kappa} (\lambda_j + \lambda_j^*)\, z^{2\kappa-1} + \text{terms with lower powers of } z,$$

we obtain formula (4.1) for N with r_Λ satisfying (ii). □

In the sequel we take $\Lambda = (z_0, \ldots, z_0)$, where z_0 is the reference point of N in (2.15), and we write n_r for n_Λ and r_0 for r_Λ, hence in this case (4.1) and (4.2) become

$$N(z) = (z - z_0^*)^\kappa n_r(z) (z-z_0)^\kappa + r_0(z) \qquad (4.4)$$

and

$$n_r(z) = \begin{cases} p_{2\kappa} + \kappa(z_0 + z_0^*) p_{2\kappa+1} + p_{2\kappa+1} z + \int_{\mathbb{R}} \dfrac{1}{t-z} \dfrac{d\sigma(t)}{|t-z_0|^{2(\kappa-m)}}, & m < \kappa, \\ p_{2\kappa} + \kappa(z_0 + z_0^*) p_{2\kappa+1} + p_{2\kappa+1} z + \int_{\mathbb{R}} \left(\dfrac{1}{t-z} - \dfrac{t}{t^2+1} \right) d\sigma(t), & m = \kappa. \end{cases} \qquad (4.5)$$

Note that if $m < \kappa$, then the measure in the integral representation of n_r is finite. Therefore, in this case, n_r is said to be "more regular" than n. Evidently, (4.4) and (4.5) can also be obtained directly by applying the operator $R_{z_0}^\kappa R_{z_0^*}^\kappa$ to both sides of the representation (2.15) of N.

With representation (4.4) we associate the matrix-valued generalized Nevanlinna functions $\widetilde{N}_r$ and M_r defined by

$$\widetilde{N}_r(z) = \begin{pmatrix} n_r(z) & 0 \\ 0 & M_r(z) \end{pmatrix}, \quad M_r(z) = \begin{pmatrix} r_0(z) & (z - z_0^*)^\kappa \\ (z - z_0)^\kappa & 0 \end{pmatrix}. \tag{4.6}$$

Again the reproducing kernel space $\mathcal{L}(\widetilde{N}_r)$ with the kernel $K_{\widetilde{N}_r}$ can be decomposed as the orthogonal sum $\mathcal{L}(\widetilde{N}_r) = \mathcal{L}(n_r) \oplus \mathcal{L}(M_r)$. The space $\mathcal{L}(M_r)$ is a Pontryagin space with basis $u_1, \dots, u_\kappa, v_1, \dots, v_\kappa$ consisting of the 2-vector polynomials

$$u_j(\zeta) = R_{z_0}^j \begin{pmatrix} r_0(\zeta) \\ (\zeta - z_0)^\kappa \end{pmatrix}, \; v_j(\zeta) = R_{z_0^*}^{\kappa-j+1} \begin{pmatrix} (\zeta - z_0^*)^\kappa \\ 0 \end{pmatrix}, \quad j = 1, \dots, \kappa, \tag{4.7}$$

and equipped with an inner product such that the Gram matrix $\widetilde{G}$ associated with this basis has the form

$$\widetilde{G} = \begin{pmatrix} G & I_{\mathbb{C}^\kappa} \\ I_{\mathbb{C}^\kappa} & 0 \end{pmatrix},$$

where

$$\begin{aligned} G_{ij} &:= \langle u_j, u_i \rangle_{\mathcal{L}(\widetilde{N})} \\ &= \frac{1}{(j-1)!} \frac{1}{(i-1)!} \left(\frac{d}{dz}\right)^{j-1} \left(\frac{d}{dw^*}\right)^{i-1} \frac{r_0(z) - r_0(w^*)}{z - w^*} \Big|_{z=w=z_0}, \end{aligned} \tag{4.8}$$

see [6, Lemma 5.1].

We now define a (not necessarily minimal) model for N in the space $\mathcal{L}(\widetilde{N}_r)$. The idea is very similar to the model corresponding to the irreducible representation presented in the preceding section, however, the proofs here become slightly more technical. To relate the canonical model in $\mathcal{L}(N)$ and the model in $\mathcal{L}(\widetilde{N}_r)$ we use an analog of the mapping $\mathbf{V}$ which now is a surjective partial isometry with a possibly nontrivial kernel, namely the mapping $\mathbf{V}_r : \mathcal{L}(\widetilde{N}_r) \to \mathcal{L}(N)$ defined by

$$(\mathbf{V}_r \widetilde{f})(\zeta) := v_r^\#(\zeta)\, \widetilde{f}(\zeta), \quad f \in \mathcal{L}(\widetilde{N}_r), \tag{4.9}$$

where v_r is the vector function

$$v_r(\zeta) := \begin{pmatrix} (\zeta - z_0)^\kappa & 1 & (\zeta - z_0)^\kappa n_r(\zeta) \end{pmatrix}^\top. \tag{4.10}$$

That $\mathbf{V}_r$ is well defined follows from the next lemma.

Lemma 4.2. *Let $N \in \mathcal{N}_\kappa^\infty$ be given and let n_r and $\widetilde{N}_r$ be defined by (4.4)–(4.6). Then:*

(a) *The mapping $\mathbf{V}_r$ is a partial isometry from $\mathcal{L}(\widetilde{N}_r)$ onto $\mathcal{L}(N)$.*

(b) *The following statements are equivalent:*

  (i) *In the representation (2.15) of N: $m < \kappa$ and $\deg p = 2\kappa - 1$.*

  (ii) *$n_r \in \mathcal{L}(n_r)$.*

  (iii) *$\ker \mathbf{V}_r \neq \{0\}$.*

If these conditions hold, then $\ker \mathbf{V}_r$ *is a one-dimensional positive subspace of* $\mathcal{L}(\widetilde{N}_r)$ *spanned by the element*

$$\eta_r := \begin{pmatrix} -n_r & 0 & 1 \end{pmatrix}^\top .$$

(c) *The following statements are equivalent:*
   (1) *In the representation* (2.15) *of* N : $m = \kappa$ *or* $m < \kappa$ *and* $\deg p \in \{2\kappa, 2\kappa+1\}$.
   (2) $n_r \notin \mathcal{L}(n_r)$.
   (3) V_r *is a unitary mapping.*

Proof. A straightforward calculation shows

$$v_r^{\#}(\zeta) K_{\widetilde{N}_r}(\zeta, z) v_r(z^*) = K_N(\zeta, z).$$

Since the number of negative squares of the kernel $K_{\widetilde{N}_r}$ is equal to that of the kernel K_{M_r} which is κ, the numbers of negative squares of $K_{\widetilde{N}_r}$ and K_N coincide. Hence, by [1, Theorem 1.5.7], $\mathbf{V}_r$ is a well-defined partial isometry from $\mathcal{L}(\widetilde{N}_r)$ onto $\mathcal{L}(N)$. This proves (a).

We prove the equivalence between (i) and (ii): If (i) holds, then from (4.5) it follows that

$$n_r(z) = \int_{\mathbb{R}} \frac{1}{t-z} d\tau(t),$$

where

$$d\tau(t) := \frac{d\sigma(t)}{|t - z_0|^{2(\kappa - m)}}$$

is a finite measure. Hence, on account of (2.13) and Lemma 2.2, (ii) holds. Now assume (ii). Then Lemma 2.2 applied to n_r in (4.5) yields that

$$p_{2\kappa+1} = p_{2\kappa} = 0$$

and also that if $m = \kappa$, then $\int_{\mathbb{R}} d\sigma(t) < \infty$, which is in contradiction with (2.18). Thus $m < \kappa$ and deg $p = 2\kappa - 1$ according to the list of possible values of m and $\deg p$ at the end of Section 2. So (i) holds.

Since (i) and (1) are each others complement in the list of possible values of m and $\deg p$ at the end of Section 2 and statements (i) and (ii) are equivalent, also statements (1) and (2) are equivalent, and it remains to prove the implication (2) $\Rightarrow$ (3), the implication (ii) $\Rightarrow$ (iii), and the statement about the $\ker \mathbf{V}_r$ in (b).

We begin by calculating $\ker \mathbf{V}_r$. Assume $\widetilde{f} := \begin{pmatrix} f & a & b \end{pmatrix}^\top$ belongs to $\ker \mathbf{V}_r$. Then

$$(\zeta - z_0^*)^\kappa f(\zeta) + a(\zeta) + (\zeta - z_0^*)^\kappa n_r(\zeta) b(\zeta) \equiv 0. \tag{4.11}$$

Since n_r is holomorphic at z_0, also $f \in \mathcal{L}(n_r)$ is holomorphic at z_0, and the equality (4.11) implies that the polynomial a has a zero at z_0^* of order at least κ, that is,

$$a(\zeta) = (\zeta - z_0^*)^\kappa a_1(\zeta) \tag{4.12}$$

with some polynomial a_1. Then equality (4.11) implies

$$a_1 + n_r b = -f \in \mathcal{L}(n_r). \tag{4.13}$$

The element $\begin{pmatrix} a & b \end{pmatrix}^\top \in \mathcal{L}(M_r)$ can be written as

$$\begin{aligned}\begin{pmatrix} a(\zeta) \\ b(\zeta) \end{pmatrix} &= \sum_{i=1}^{\kappa} a_i R_{z_0^*}^{\kappa-i+1} \begin{pmatrix} (\zeta - z_0^*)^\kappa \\ 0 \end{pmatrix} + \sum_{j=1}^{\kappa} b_j R_{z_0}^j \begin{pmatrix} r_0(\zeta) \\ (\zeta - z_0)^\kappa \end{pmatrix} \\ &= \begin{pmatrix} \sum_{i=1}^\kappa a_i(\zeta - z_0^*)^{i-1} + \sum_{j=1}^\kappa b_j R_{z_0}^j r_0(\zeta) \\ \sum_{j=1}^\kappa b_j(\zeta - z_0)^{\kappa-j}. \end{pmatrix}. \end{aligned} \tag{4.14}$$

Applying the operator $R_{z_0^*}^\kappa$ to the first component and using (4.12) we find

$$a_1(\zeta) = \sum_{j=1}^{\kappa} b_j R_{z_0^*}^\kappa R_{z_0}^j r_0(\zeta). \tag{4.15}$$

Inserting this in (4.13) gives

$$\begin{aligned} a_1(\zeta) + n_r(\zeta) b(\zeta) &= \sum_{j=1}^{\kappa} b_j \left(R_{z_0^*}^\kappa R_{z_0}^j r_0(\zeta) + (\zeta - z_0)^{\kappa-j} n_r(\zeta) \right) \\ &= \sum_{j=1}^{\kappa} b_j R_{z_0^*}^\kappa R_{z_0}^j \left(r_0(\zeta) + (\zeta - z_0)^\kappa (\zeta - z_0^*)^\kappa n_r(\zeta) \right) = \sum_{j=1}^{\kappa} b_j R_{z_0^*}^\kappa R_{z_0}^j N(\zeta). \end{aligned}$$

Hence (4.13) can be written as

$$\sum_{j=1}^{\kappa} b_j R_{z_0^*}^\kappa R_{z_0}^j N = -f \in \mathcal{L}(n_r). \tag{4.16}$$

Note that for $l > \kappa$ it holds $R_{z_0^*}^\kappa R_{z_0}^l N = R_{z_0}^{l-\kappa} n_r \in \mathcal{L}(n_r)$. Hence applying $R_{z_0}^{\kappa-1}$ to (4.16) gives

$$b_1 n_r \in \mathcal{L}(n_r).$$

There are two cases. Let us first assume $n_r \notin \mathcal{L}(n_r)$, hence $b_1 = 0$. Then applying $R_{z_0}^{\kappa-2}$ to (4.16) and so on gives $b_j = 0$ for $j = 2, \ldots, \kappa$. Together with (4.15) this implies $\widetilde{f} = 0$ and hence in this case the mapping $\mathbf{V}_r$ is unitary. This proves that (2) $\Rightarrow$ (3).

Now assume (ii), that is, $n_r \in \mathcal{L}(n_r)$. Then (i) holds: $m < \kappa$, $\deg p = 2\kappa - 1$, and $p_{2\kappa-1} < 0$. Applying the operator $R_{z_0}^{\kappa-2}$ to (4.16) yields

$$b_1 R_{z_0^*}^\kappa R_{z_0}^{\kappa-1} N(\zeta) \in \mathcal{L}(n_r),$$

or, inserting here the representation (2.15) for N,

$$b_1 \left(\int_{\mathbb{R}} \frac{d\sigma(t)}{(t-z)(t-z_0)^{\kappa-m-1}(t-z_0^*)^{\kappa-m}} + p_{2\kappa-1} \right) \in \mathcal{L}(n_r),$$

where $d\sigma$ is the measure in the integral representation of n in (2.15). In the integral representation (4.5) of n_r we have $p_{2k+1} = 0$. Hence, by (2.12) applied to n_r, for every $F \in \mathcal{L}(n_r)$ it holds $\lim_{y\to\infty} F(iy) = 0$, but the corresponding limit of the function within the brackets is $p_{2\kappa-1} \neq 0$, and hence $b_1 = 0$. Multiplying now (4.16) by $R_{z_0}^{\kappa-3}$ and repeating the above argument yield $b_2 = 0$. In the same way one obtains $b_3 = \cdots = b_{\kappa-1} = 0$ and finally $b_\kappa n_r \in \mathcal{L}(n_r)$. It follows from (4.14), (4.15), and (4.16) that

$$\widetilde{f} = \begin{pmatrix} f & a & b \end{pmatrix}^\top = b_\kappa \begin{pmatrix} -n_r & 0 & 1 \end{pmatrix}^\top = b_\kappa \eta_r,$$

and hence in this case $\ker \mathbf{V}_r = \operatorname{span}\{\eta_r\}$. This proves (ii) $\Rightarrow$ (iii).
To calculate the sign of the subspace $\ker \mathbf{V}_r$ recall that η_r belongs to the orthogonal sum $\mathcal{L}(\widetilde{N}_r) = \mathcal{L}(n_r) \oplus \mathcal{L}(M_r)$ and hence

$$\langle \eta_r, \eta_r \rangle_{\mathcal{L}(\widetilde{N}_r)} = \langle n_r, n_r \rangle_{\mathcal{L}(n_r)} + \langle \begin{pmatrix} 0 \\ 1 \end{pmatrix}, \begin{pmatrix} 0 \\ 1 \end{pmatrix} \rangle_{\mathcal{L}(M_r)}.$$

According to (2.9),

$$\langle n_r, n_r \rangle_{\mathcal{L}(n_r)} = \int \frac{d\sigma(t)}{|t - z_0|^{2(\kappa - m)}}$$

and, by [6] or using directly the basis in (4.7),

$$\langle \begin{pmatrix} 0 \\ 1 \end{pmatrix}, \begin{pmatrix} 0 \\ 1 \end{pmatrix} \rangle_{\mathcal{L}(M_r)} = -r_{0,2\kappa-1},$$

where $r_{0,2\kappa-1}$ is the coefficient of $z^{2\kappa-1}$ in the polynomial $r_0(z)$, and hence

$$\langle \eta_r, \eta_r \rangle_{\mathcal{L}(\widetilde{N}_r)} = \int \frac{d\sigma(t)}{|t - z_0|^{2(\kappa - m)}} - r_{0,2\kappa-1}.$$

Calculating the limit

$$\lim_{y \to \infty} \frac{1}{(iy)^{2\kappa-1}} N(iy)$$

by using on the one hand (4.4) and on the other hand (2.15) gives the equality

$$-\int \frac{d\sigma(t)}{|t - z_0|^{2(\kappa - m)}} + r_{0,2\kappa-1} = p_{2\kappa-1}$$

and hence $\langle \eta_r, \eta_r \rangle_{\mathcal{L}(\widetilde{N}_r)} = -p_{2\kappa-1} > 0$. □

In the case $m < \kappa$ and $\deg p = 2\kappa - 1$ we introduce the space $\mathcal{P}'$ by

$$\mathcal{P}' = \mathcal{L}(\widetilde{N}_r) \ominus \{\eta_r\}$$

and we denote by P' the orthogonal projection in $\mathcal{L}(\widetilde{N}_r)$ onto $\mathcal{P}'$. Then by Lemma 4.2 the mapping $\mathbf{V}'_r := \mathbf{V}_r \,|_{\mathcal{P}'} \colon \mathcal{P}' \to \mathcal{L}(N)$ is unitary.

Theorem 4.3. *For $N \in \mathcal{N}_\kappa^\infty$ a model related to its representation (4.4) is given in the space $\mathcal{L}(\widetilde{N}_r)$ by the triple*

$$(B_r, K_{\widetilde{N}_r}(\,\cdot\,, z^*) v_r(z), \widetilde{S}_r)$$

consisting of the self-adjoint relation

$$B_r := \{\{\widetilde{f}, \widetilde{g}\} \in \mathcal{L}(\widetilde{N}_r) \mid \exists \mathbf{h} \in \mathbb{C}^3 : \widetilde{g}(\zeta) - \zeta \widetilde{f}(\zeta) = (I_{\mathbb{C}^3} + \widetilde{N}_r(\zeta)\mathcal{B})\mathbf{h}\}, \tag{4.17}$$

the defect function

$$K_{\widetilde{N}_r}(\,\cdot\,, z^*) v_r(z),$$

and the symmetric relation

$$\widetilde{S}_r := \{\{\widetilde{f},\widetilde{g}\} \in \mathcal{L}(\widetilde{N}_r) \mid \exists \mathbf{h} \in \mathbb{C}^3 \text{ with } h_2 = 0 : \widetilde{g}(\zeta) - \zeta\widetilde{f}(\zeta) = (\mathcal{A} + \widetilde{N}_r(\zeta)\mathcal{B})\mathbf{h}\},$$

where the 3×3 *matrices* $\mathcal{A}$ *and* $\mathcal{B}$ *are as in Theorem* 3.2(ii).

The following statements hold.

(i) *The model is minimal if and only if either* $m = \kappa$ *or* $m < \kappa$ *and* $\deg p \geq 2\kappa$.

(ii) *If* $m < \kappa$ *and* $\deg p = 2\kappa - 1$, *then*

$$\begin{aligned} \mathcal{L}(\widetilde{N}_r) &= \mathcal{P}' \oplus \operatorname{span}\{\eta_r\}, & \mathcal{P}' &= \overline{\operatorname{span}}\,\{K_{\widetilde{N}_r}(\,\cdot\,, z^*)v_r(z) \mid z \in \operatorname{hol}(N)\}, \\ \widetilde{S}_r &= \widetilde{S}'_r \oplus \operatorname{span}\{\{0, \eta_r\}\}, & \widetilde{S}'_r &:= P'\widetilde{S}_r\,|_{\mathcal{P}'} = \mathbf{V}'^{-1}_r S_N \mathbf{V}'_r, \\ B_r &= B'_r \oplus \operatorname{span}\{\{0, \eta_r\}\}, & B'_r &:= P'B_r\,|_{\mathcal{P}'} = \mathbf{V}'^{-1}_r A_N \mathbf{V}'_r \end{aligned}$$

and the triple

$$(B'_r, K_{\widetilde{N}_r}(\,\cdot\,, z^*)v_r(z), \widetilde{S}'_r)$$

is a minimal model of N *in* $\mathcal{P}'$.

Proof. Assume $m = \kappa$ or $m < \kappa$ and $\deg p \geq 2\kappa$. Then, according to Lemma 4.2, the mapping $\mathbf{V}_r$ is unitary. The same reasoning as in the proof of [6, Theorem 4.1 (i)] yields that the triple $(B_r, K_{\widetilde{N}_r}(\,\cdot\,, z^*)v_r(z), \widetilde{S}_r)$ defined in the first part of the theorem is unitarily equivalent to the canonical model of N under $\mathbf{V}_r$, that is,

$$(B_r, K_{\widetilde{N}_r}(\,\cdot\,, z^*)v_r(z), \widetilde{S}_r) = (\mathbf{V}_r^{-1}A_N\mathbf{V}_r, \mathbf{V}_r^{-1}K_N(\,\cdot\,, z^*), \mathbf{V}_r^{-1}S_N\mathbf{V}_r).$$

Hence this triple is a minimal model for N. This proves the "if" part of item (i). According to the list (at the end of Section 2) of possible values for m and $\deg p$, the case that remains is when $m < \kappa$ and $\deg p = 2\kappa - 1$. This case we consider now. Then, again by Lemma 4.2, $\mathbf{V}_r$ is a surjective partial isometry with a one-dimensional kernel. We claim that

$$\mathcal{P}' = \operatorname{ran} \mathbf{V}_r^* = \overline{\operatorname{span}}\,\{K_{\widetilde{N}_r}(\,\cdot\,, z^*)v_r(z) \mid z \in \operatorname{hol}(N)\}. \tag{4.18}$$

Since

$$\mathcal{L}(\widetilde{N}_r) = \operatorname{ran} \mathbf{V}_r^* \oplus \ker \mathbf{V}_r, \tag{4.19}$$

the second equality in (4.18) implies that in this case the triple is not minimal, which shows the "only if" part of item (i).

The first equality in (4.18) follows from the definition of the space $\mathcal{P}'$, the orthogonal sum decomposition (4.19) and the equality $\ker \mathbf{V}_r = \operatorname{span}\{\eta_r\}$, see Lemma 4.2(b). To verify the second equality it suffices to show that

$$\mathbf{V}_r^* K_N(\,\cdot\,, z^*) = K_{\widetilde{N}_r}(\,\cdot\,, z^*)v_r(z). \tag{4.20}$$

But this follows from the reproducing kernel property of the kernels K_N and $K_{\widetilde{N}_r}$ and the following chain of equalities with $w \in \operatorname{hol}(\widetilde{N}_r)$ and $\mathbf{c} \in \mathbb{C}^3$:

$$\begin{aligned} \langle \mathbf{V}_r^* K_N(\,\cdot\,, z^*), K_{\widetilde{N}_r}(\,\cdot\,, w)\mathbf{c}\rangle_{\mathcal{L}(\widetilde{N}_r)} &= \langle K_N(\,\cdot\,, z^*), \mathbf{V}_r K_{\widetilde{N}_r}(\,\cdot\,, w)\mathbf{c}\rangle_{\mathcal{L}(N)} \\ &= \langle K_N(\,\cdot\,, z^*), v_r^{\#}(\,\cdot\,)K_{\widetilde{N}_r}(\,\cdot\,, w)\mathbf{c}\rangle_{\mathcal{L}(N)} = \left(v_r(z)^* K_{\widetilde{N}_r}(z^*, w)\mathbf{c}\right)^* \\ &= \langle K_N(\,\cdot\,, z^*)v_r(z), K_{\widetilde{N}_r}(\,\cdot\,, w)\mathbf{c}\rangle_{\mathcal{L}(\widetilde{N}_r)}. \end{aligned}$$

From

$$\operatorname{rank}\begin{pmatrix} I_{\mathbb{C}^3} \\ \mathcal{B} \end{pmatrix} = 3, \qquad \mathcal{B} - \mathcal{B}^* = 0,$$

and [3, Theorem 2.4] it follows that the right-hand side in (4.17) describes a self-adjoint relation. Again as in the proof of [6, Theorem 4.1 (i)] one can show the equality

$$\mathbf{V}_r' B_r \mid_{\mathcal{P}'} = A_N \mathbf{V}_r'. \tag{4.21}$$

The multi-valued part of B_r is given by

$$B_r(0) = \{\widetilde{g} \in \mathcal{L}(\widetilde{N}_r) \mid \exists \mathbf{h} \in \mathbb{C}^3 : \widetilde{g}(\zeta) = (I_{\mathbb{C}^3} + \widetilde{N}_r(\zeta)\mathcal{B})\mathbf{h}\}.$$

Since

$$(I_{\mathbb{C}^3} + \widetilde{N}_r(\zeta)\mathcal{B})\mathbf{h} = \begin{pmatrix} h_1 - h_3 n_r(\zeta) \\ h_2 - h_1(\zeta - z_0^*)^\kappa \\ h_3 \end{pmatrix},$$

this element belongs to $\mathcal{L}(\widetilde{N}_r)$ if and only if the first component $h_1 - h_3 n_r$ belongs to $\mathcal{L}(n_r)$ and the second component $\begin{pmatrix} h_2 - h_1(\zeta - z_0^*)^\kappa & h_3 \end{pmatrix}^\top$ to $\mathcal{L}(M_r)$. These inclusions hold if and only if $h_1 = 0$. On the one hand this follows from the fact that, by Lemma 4.2, $n_r \in \mathcal{L}(n_r)$ and hence, on account of Lemma 2.3(iii), $h_1 = 0$. On the other hand $h_1 = 0$ also because $\mathcal{L}(M_r)$ is spanned by functions of the form (4.7), recall (4.6). We conclude that

$$B_r(0) = \operatorname{span}\{\eta_r, \begin{pmatrix} 0 & 1 & 0 \end{pmatrix}^\top\}.$$

As B_r is a self-adjoint relation the manifolds $\operatorname{dom} B_r$ and $B_r(0)$ are orthogonal. Together with the equality (4.21) this implies the decomposition

$$B_r = B_r' \oplus \operatorname{span}\{\{0, \eta_r\}\}$$

with $B_r' = \mathbf{V}_r'^{-1} A_N \mathbf{V}_r'$. The formula $\widetilde{S}_r = \widetilde{S}_r' \oplus \operatorname{span}\{\{0, \eta_r\}\}$ is obtained similarly. It follows that the triple $(B_r', K_{\widetilde{N}_r}(\,\cdot\,, z^*)v_r(z), \widetilde{S}_r')$ is a minimal model for N in $\mathcal{P}'$. This proves (ii).

It remains to show that the first triple in the theorem is also a model for N if $m < \kappa$ and $\deg p = 2\kappa - 1$. But this follows from formulas (2.2), (2.3), and (2.5). These formulas hold with $A = B_r$ and $\Gamma_z = K_{\widetilde{N}_r}(\,\cdot\,, z^*)v_r(z)$, because they hold with $A = B_r'$ and $\Gamma_z = K_{\widetilde{N}_r}(\,\cdot\,, z^*)v_r(z)$, and because

$$\operatorname{dom} B_r \subset \mathcal{P}', (B_r - z)^{-1} = (B_r' - z)^{-1} P', \ K_{\widetilde{N}_r}(\,\cdot\,, z^*)v_r(z) = P' K_{\widetilde{N}_r}(\,\cdot\,, z^*)v_r(z).$$

□

In the following theorem we describe the family all of self-adjoint extensions of the symmetry $\widetilde{S}_r$. For this we use the mapping

$$\mathbf{W} := \mathbf{V}^{-1}\mathbf{V}_r : \mathcal{L}(\widetilde{N}_r) \to \mathcal{L}(\widetilde{N}) \quad (\mathbf{V} : \mathcal{L}(\widetilde{N}) \to \mathcal{L}(N),\ \mathbf{V}_r : \mathcal{L}(\widetilde{N}_r) \to \mathcal{L}(N)),$$

which is unitary if $\mathbf{V}_r$ is unitary and a surjective partial isometry with

$$\ker \mathbf{W} = \ker \mathbf{V}_r = \operatorname{span}\{\eta_r\}$$

otherwise. In the last case we use the restricted mapping

$$\mathbf{W}' := \mathbf{V}^*\mathbf{V}_r' : \mathcal{P}' \to \mathcal{L}(\widetilde{N}) \quad (\mathbf{V}_r' : \mathcal{P}' \to \mathcal{L}(N))$$

which is unitary.

Theorem 4.4. *The family of all self-adjoint extensions of the symmetric relation $\widetilde{S}_r$ in Theorem 4.3 consists of B_r and $B_r^{\langle\alpha\rangle}$, $\alpha \in \mathbb{R}$, where*

$$B_r^{\langle\alpha\rangle} := \{\{\widetilde{f},\widetilde{g}\} \in \mathcal{L}(\widetilde{N}_r) \mid \exists \mathbf{h} \in \mathbb{C}^3 : \widetilde{g}(\zeta) - \zeta\widetilde{f}(\zeta) = (\mathcal{A}_\alpha + \widetilde{N}_r(\zeta)\mathcal{C})\mathbf{h}\}, \tag{4.22}$$

and the 3×3 matrices $\mathcal{A}_\alpha$ and $\mathcal{C}$ are as in Theorem 3.3 (ii).
The following statements hold:

(i) *If either $m = \kappa$ or $m < \kappa$ and $\deg p \geq 2\kappa$ then $B_r^{\langle\alpha\rangle}$, $\alpha \in \mathbb{R}$, is unitarily equivalent to the operator $B^{\langle\alpha\rangle}$ from Theorem 3.3 under the mapping $\mathbf{W}$:*

$$B_r^{\langle\alpha\rangle} = \mathbf{W}^{-1}B^{\langle\alpha\rangle}\mathbf{W}.$$

(ii) *If $m < \kappa$ and $\deg p = 2\kappa - 1$ then $B_r^{\langle\alpha\rangle}$, $\alpha \in \mathbb{R}$, is a linear relation whose multi-valued part is spanned by the vector η_r and*

$$B_r^{\langle\alpha\rangle} = \mathbf{W}'^{-1}B^{\langle\alpha\rangle}\mathbf{W}' \oplus \operatorname{span}\{\{0,\eta_r\}\}.$$

Proof. The matrices $\mathcal{A}_\alpha$ and $\mathcal{C}$ satisfy

$$\operatorname{rank}\begin{pmatrix}\mathcal{A}_\alpha \\ \mathcal{C}\end{pmatrix} = 3, \qquad \mathcal{A}_\alpha^*\mathcal{C} - \mathcal{C}^*\mathcal{A}_\alpha = 0.$$

Hence by [3, Theorem 2.4] the right-hand side in (4.22) describes a self-adjoint relation. That B_r and $B_r^{\langle\alpha\rangle}$ with $\alpha \in \mathbb{R}$ are all self-adjoint extensions of S_r follows from Theorem 3.3 and the unitariy equivalences described in items (i) and (ii).
To prove (i) we only need to show that the mapping $\mathbf{W} : \mathcal{L}(\widetilde{N}_r) \to \mathcal{L}(\widetilde{N})$ intertwines $\widetilde{S}$ and $\widetilde{S}_r$ as well as $B_r^{\langle\alpha\rangle}$ and $B^{\langle\alpha\rangle}$:

$$\widetilde{S} = \{\{\mathbf{W}\widetilde{f},\mathbf{W}\widetilde{g}\} \mid \{\widetilde{f},\widetilde{g}\} \in \widetilde{S}_r\}, \quad B^{\langle\alpha\rangle} = \{\{\mathbf{W}\widetilde{f},\mathbf{W}\widetilde{g}\} \mid \{\widetilde{f},\widetilde{g}\} \in B_r^{\langle\alpha\rangle}\}.$$

We only prove the last equality. Using the definitions of $\mathbf{V}$ and $B^{\langle\alpha\rangle}$ we obtain that the relation $\{\{\mathbf{V}\widetilde{f},\mathbf{V}\widetilde{g}\} \mid \{\widetilde{f},\widetilde{g}\} \in B^{\langle\alpha\rangle}\}$ is equal to

$$\{\{f,g\} \in \mathcal{L}(N) \mid \exists h \in \mathbb{C} : g(\zeta) - \zeta f(\zeta) = (\alpha - N(\zeta))h\}.$$

Similarly the relation $\{\{\mathbf{V}_r\widetilde{f},\mathbf{V}_r\widetilde{g}\} \mid \{\widetilde{f},\widetilde{g}\} \in B_r^{\langle\alpha\rangle}\}$ can be written as

$$\{\{f,g\} \in \mathcal{L}(N) \mid \exists h \in \mathbb{C} : g(\zeta) - \zeta f(\zeta) = (\alpha - N(\zeta))h\}.$$

Hence

$$\{\{\mathbf{V}\widetilde{f},\mathbf{V}\widetilde{g}\} \mid \{\widetilde{f},\widetilde{g}\} \in B^{\langle\alpha\rangle}\} = \{\{\mathbf{V}_r\widetilde{f},\mathbf{V}_r\widetilde{g}\} \mid \{\widetilde{f},\widetilde{g}\} \in B_r^{\langle\alpha\rangle}\},$$

which completes the proof of (i).
As to (ii), assume $m < \kappa$ and $\deg p = 2\kappa - 1$. We show that the multi-valued part of $B_r^{\langle\alpha\rangle}$ is given by

$$B_r^{\langle\alpha\rangle}(0) = \operatorname{span}\{\eta_r\}. \tag{4.23}$$

From (4.22) we obtain

$$B_r^{\langle\alpha\rangle}(0) = \{\widetilde{g} \in \mathcal{L}(\widetilde{N}_r) \mid \exists \mathbf{h} \in \mathbb{C}^3 : \widetilde{g}(\zeta) = (\mathcal{A}_\alpha + \widetilde{N}_r(\zeta)\mathcal{C})\mathbf{h}\}.$$

Here

$$(\mathcal{A}_\alpha + \widetilde{N}_r(\zeta)\mathcal{C})\mathbf{h} = \begin{pmatrix} h_1 - h_3 n_r(\zeta) \\ h_2(\alpha - r_0(z)) - h_1(\zeta - z_0^*)^\kappa \\ h_3 - h_2(\zeta - z_0)^\kappa \end{pmatrix}.$$

The right-hand side of this equality belongs to $\mathcal{L}(\widetilde{N}_r)$ if and only if $h_1 = h_2 = 0$. This can be shown as in the proof of Theorem 4.3 by means of Lemma 4.2 and Lemma 2.3(iii) which imply that $h_1 = 0$ and from the fact that $\mathcal{L}(M_r)$ is spanned by the functions in (4.7) which implies that deg $(h_3 - h_2(\zeta - z_0)^\kappa) < \kappa$ and hence that $h_2 = 0$. This proves (4.23). The formulas in (ii) now easily follow from the fact that, since B_r is self-adjoint, dom $B_r^{\langle\alpha\rangle}$ is orthogonal to η_r. □

Remark 4.5. (a) The relation $B_r^{\langle\alpha\rangle}$, $\alpha \in \mathbb{R}$, can also (see Remark 3.4) be seen as a rank-one perturbation of $B_r^{\langle 0\rangle}$:

$$B_r^{\langle\alpha\rangle} = B_r^{\langle 0\rangle} + \alpha\langle\,\cdot\,, \widetilde{w}_r\rangle_{\mathcal{L}(\widetilde{N}_r)}\widetilde{w}_r,$$

where $\widetilde{w}_r := \begin{pmatrix}0 & 1 & 0\end{pmatrix}^\top \in \mathcal{L}(\widetilde{N}_r)$. Moreover, B_r is the limit of $B_r^{\langle\alpha\rangle}$ in the resolvent sense as $\alpha \to \infty$. For this reason we set $B_r^{\langle\infty\rangle} = B_r$.
(b) In the case $m < \kappa$ and $\deg p = 2\kappa - 1$ it holds that

$$\langle\widetilde{w}_r, \eta_r\rangle = \begin{cases} 0 & \text{if } \kappa > 1, \\ 1 & \text{if } \kappa = 1. \end{cases}$$

To prove this, observe that

$$\langle\widetilde{w}_r, \eta_r\rangle_{\mathcal{L}(\widetilde{N}_r)} = \langle P_{\mathcal{L}(M_r)}\widetilde{w}_r, P_{\mathcal{L}(M_r)}\eta_r\rangle_{\mathcal{L}(M_r)}, \quad P_{\mathcal{L}(M_r)}\eta_r = \begin{pmatrix}0 & 1\end{pmatrix}^\top,$$

and

$$P_{\mathcal{L}(M_r)}\widetilde{w}_r = \begin{pmatrix}1 & 0\end{pmatrix}^\top = \frac{1}{(\kappa-1)!}\frac{d^{\kappa-1}}{dw^{*\kappa-1}}K_{M_r}(\cdot, w)|_{w=z_0}\begin{pmatrix}0 & 1\end{pmatrix}^\top.$$

Hence

$$\begin{aligned}\langle P_{\mathcal{L}(M_r)}\widetilde{w}_r, P_{\mathcal{L}(M_r)}\eta_r\rangle_{\mathcal{L}(\widetilde{M}_r)} &= \frac{1}{(\kappa-1)!}\frac{d^{\kappa-1}}{dw^{*\kappa-1}}\langle K_{M_r}(\cdot, w)\begin{pmatrix}0\\1\end{pmatrix}, \begin{pmatrix}0\\1\end{pmatrix}\rangle_{\mathcal{L}(M_r)}|_{w=z_0} \\ &= \frac{1}{(\kappa-1)!}\frac{d^{\kappa-1}}{dw^{*\kappa-1}}\begin{pmatrix}0 & 1\end{pmatrix}\begin{pmatrix}0\\1\end{pmatrix} = \begin{cases} 0 & \text{if } \kappa > 1, \\ 1 & \text{if } \kappa = 1. \end{cases}\end{aligned}$$

(c) A straightforward calculation of the resolvents of $B_r^{\langle\alpha\rangle}$ and B_r leads to formulas which are similar to those in [6, Corollary 4.5]:

$$(B_r^{\langle\alpha\rangle} - z)^{-1} = (A_{\widetilde{N}_r} - z)^{-1} - \frac{1}{N(z) - \alpha}\Gamma_{\widetilde{N}_r z}K_{r\alpha}(z)E_z, \tag{4.24}$$

$$(B_r - z)^{-1} = (A_{\widetilde{N}_r} - z)^{-1} + \Gamma_{\widetilde{N}_r z}K_r(z)E_z, \tag{4.25}$$

where $(A_{\widetilde{N}_r} - z)^{-1}$ is the difference-quotient operator in the space $\mathcal{L}(\widetilde{N}_r)$,

$$\Gamma_{\widetilde{N}z} := \operatorname{diag}\{K_{n_r}(\,\cdot\,, z^*), K_{M_r}(\,\cdot\,, z^*)\},$$

$E_z := (\Gamma_{\widetilde{N}z^*})^*$ is the evaluation operator at the point z, and

$$\begin{aligned} K_{r\alpha}(z) := \;& \mathcal{C}(\mathcal{A}_\alpha + \widetilde{N}_r(z)\mathcal{C})^{-1} \\ =\;& \begin{pmatrix} (z-z_0)^\kappa(z-z_0^*)^\kappa & (z-z_0)^\kappa & \alpha-r_0(z) \\ (z-z_0^*)^\kappa & 1 & n_r(z)(z-z_0^*)^\kappa \\ \alpha-r_0(z) & n_r(z)(z-z_0)^\kappa & n_r(z)(\alpha-r_0(z)) \end{pmatrix}, \\ K_r(z) := \;& \lim_{\alpha\to\infty}\frac{1}{\alpha}K_{r\alpha}(z) = \begin{pmatrix} 0 & 0 & 1 \\ 0 & 0 & 0 \\ 1 & 0 & n_r(z) \end{pmatrix}. \end{aligned}$$

References

[1] D. Alpay, A. Dijksma, J. Rovnyak, and H. de Snoo, *Schur functions, operator colligations, and reproducing kernel Pontryagin spaces*, Operator Theory: Adv. Appl. **96**, Birkhäuser Verlag, Basel, 1997.

[2] L. de Branges, *Hilbert spaces of entire functions*, Prentice-Hall, Inc., Englewood Cliffs, N.J., 1968.

[3] A. Dijksma, H. Langer, A. Luger, and Yu. Shondin, *Minimal realizations of scalar generalized Nevanlinna functions related to their basic factorization*, Operator Theory: Adv. Appl. **154**, Birkhäuser Verlag, Basel, 2004, 69–90.

[4] A. Dijksma, H. Langer, and Yu. Shondin, *Rank one perturbations at infinite coupling in Pontryagin spaces*, J. of Functional Analysis **209** (2004), 206–246.

[5] A. Dijksma, H. Langer, Yu. Shondin, and C. Zeinstra, *Self-adjoint operators with inner singularities and Pontryagin spaces*, Operator Theory: Adv., Appl., **118**, Birkhäuser Verlag, Basel, 2000, 105–175.

[6] A. Dijksma, A. Luger, and Yu. Shondin, *Minimal models for $\mathcal{N}_\kappa^\infty$-functions*, Operator Theory: Adv., Appl. **163**, Birkhäuser Verlag, Basel, 2005, 97–134.

[7] A. Dijksma, H. Langer, and H.S.V. de Snoo, *Eigenvalues and pole functions of Hamiltonian systems with eigenvalue depending boundary conditions*, Math. Nachr. **161** (1993), 107–154.

[8] A. Dijksma and Yu. Shondin, *Singular point-like perturbations of the Bessel operator in a Pontryagin space*, J. Differential Equations **164** (2000), 49–91.

[9] C. Fulton, *Titchmarsh-Weyl m-functions for Second-order Sturm-Liouville Problems with two singular endpoints*, to appear in Math. Nachr.

[10] S. Hassi and A. Luger, *Generalized zeros and poles of $\mathcal{N}_\kappa$-functions: On the underlying spectral structure*, Methods Funct. Anal. Topology **12** (2006), no. 2, 131–150.

[11] I.S. Kac and M.G. Krein, *R-functions – analytic functions mapping the upper halfplane into itself*, Amer. Math. Soc. Transl. (2) **103** (1974), 1–18.

[12] T. Kato, *Perturbation theory for linear operators*, Die Grundlehren der mathematischen Wissenschaften, Band 132, Springer-Verlag, Heidelberg, 1966.

[13] M.G. Krein and H. Langer, *Über einige Fortsetzungsprobleme, die eng mit der Theorie hermitescher Operatoren im Raume Π_κ zusammenhängen. I. Einige Funktionenklassen und ihre Darstellungen*, Math. Nachr. **77** (1977), 187–236.

[14] M.G. Krein and H. Langer, *Some propositions on analytic matrix functions related to the theory of operators on the space Π_κ*, Acta Sci. Math. (Szeged) **43** (1981), 181–205.

[15] P. Kurasov and A. Luger, *An operator theoretic interpretation of the generalized Titchmarsh-Weyl coefficient for a singular Sturm-Liouville problem*, submitted. (published as Preprint 2007:8, Lund University, Center for Mathematical Sciences, *Singular differential operators: Titchmarsh–Weyl coefficients and operator models*).

[16] H. Langer, *A characterization of generalized zeros of negative type of functions of the class $\mathcal{N}_\kappa$*, Operator Theory: Adv. Appl. **17**, Birkhäuser Verlag, Basel, 1986, 201–212.

[17] H. Langer and B. Najman, *Perturbation theory for definizable operators in Krein spaces* J. Operator Theory **9** (1983) 247–317.

[18] B. Najman, *Perturbation theory for selfadjoint operators in Pontrjagin spaces*, Glasnik Mat. **15** (1980) 351–370.

[19] Yu. Shondin, *On approximation of high order singular perturbations*, J. Phys. A: Math. Gen. **38** (2005), 5023–5039.

[20] O.Yu. Shvedov, *Approximations for strongly singular evolution equations*, J. Funct. Analysis. **210**(2) (2004), 259–294.

[21] F. Stummel, *Diskrete Konvergenz linearer Operatoren* I, Math. Annal. **190** (1970), 45-92; II, Math. Z. **141** (1975), 231–264.

[22] J. L. Walsh, *Interpolation and approximation by rational functions in the complex domain*, Amer. Math. Soc. Coll. Publ. XX, 3-rd ed., Amer. Math. Soc., Providence, R.I., 1960.

Aad Dijksma
Department of Mathematics
University of Groningen
P.O. Box 407
9700 AK Groningen, The Netherlands
e-mail: `dijksma@math.rug.nl`

Annemarie Luger
Department of Mathematics
Lund Institute of Technology
Box 118
SE-221 00 Lund, Sweden
e-mail: `luger@maths.lth.se`

Yuri Shondin
Department of theoretical Physics
State Pedagogical University
Str. Ulyanova 1
Nizhny Novgorod GSP 37, 603950 Russia
e-mail: `shondin@sinn.ru`

Operator Theory:
Advances and Applications, Vol. 188, 113–133

On the Block Numerical Range of Nonnegative Matrices

K.-H. Förster and N. Hartanto

Abstract. We present a Perron-Frobenius Theory for the block numerical range of entrywise nonnegative square matrices similar to that known for the special cases of the spectrum and of the standard numerical range. For irreducible matrices we prove a corresponding version of Wielandt's Lemma. With help of the Frobenius Form we study the block numerical range of a nonnegative matrix and its peripheral part. Finally we give an application to the numerical range of Perron Polynomials.

1. Introduction

The spectral theory of nonnegative matrices is well developed, especially the Perron-Frobenius Theory for irreducible nonnegative matrices, and it is standard nowadays, see the books [1], [7], [13]. This theory has been extended to the numerical range of a nonnegative matrix by N. Issos in his unpublished Ph.D. thesis [9] and got recently much attention in the papers [11] and [12].

The classical notion of the numerical range has been generalized in various ways in the literature (see [5], [8]). One of these generalizations is the block numerical range which was introduced in its full generality in [18] and [19] for bounded linear operators in Hilbert spaces. In this article we present the Perron-Frobenius Theory for the block numerical range of a nonnegative matrix; we obtain corresponding generalizations of results known for the spectrum and/or the standard numerical range of such matrices.

The paper is organized as follows. In Section 2 we use the concept of isometric projections to define the block numerical range of a matrix, modify the definitions of some concepts and simplify proofs of some results concerning the block numerical range of an arbitrary matrix. In Section 3 we prove basic results on the radius of a block numerical range of a nonnegative matrix; they are generalizations of the corresponding facts for the spectral radius. Section 4 is considered as the core of this paper, it contains the Perron-Frobenius Theory for the block numerical ranges of nonnegative irreducible matrices; as for the spectrum, we will prove a version of Wielandt's Lemma and draw corresponding conclusions, for example that the number of peripheral points in each block numerical range is equal to the index of imprimitivty of the matrix. We use these results and the Frobenius Form of a nonnegative matrix to extend the results of Section 3. In Section 5 we consider the numerical range of Perron Polynomials (for which the sum of its coefficients is irreducible) using the fact that the nonzero part of it is equal to the nonzero part of a specific block numerical range of its companion matrix. In the last Section 6 we consider block numerical ranges of the nilpotent shift, the cyclic shift and some other nonnegative matrices to illustrate the results of Section 4.

Notation

$\mathbb{R}_+$	the interval $[0, \infty)$
$\Re(\lambda)$	the real part of $\lambda \in \mathbb{C}$
$\mathbb{C}^m$	the set of all complex column vectors with m components
$\mathbf{1}_m$	the vector with m components which are all 1
$\|x\|_2$	the euclidian norm of the vector x
$\mathbb{C}^{m\times n}$	the set of all $m \times n$ complex matrices
I_m	the $m \times m$ identity matrix
$\|A\|$	the entrywise absolute value of the matrix (or vector)A
A^t	the transpose of the matrix (or vector) A
A^*	the conjugate transpose of the matrix (or vector) A
$\Sigma(A)$	the spectrum of the matrix A
$r(A)$	the spectral radius of the matrix A
$\Theta(A)$	the numerical range of the matrix A
$w(A)$	the numerical radius of the matrix A
$A \leq B$	entrywise (weak) inequality for matrices (or vectors) of same size
$A < B$	$A \leq B$ and $A \neq B$
$A \ll B$	entrywise strict inequality for matrices (or vectors) of same size
$\langle n\rangle$	the set $\{1, 2, \ldots, n\}$
$\#M$	the number of elements in the set M

2. The block numerical range of a matrix, revisited

The block numerical range was introduced in its full generality in [18] and [19] for bounded linear operators in Hilbert spaces. In this paper we consider matrices and then it is possible to define the block numerical range with help of certain classes of isometric projections. The purpose of this section is to present this definition and to work with it. For the use of isometrc projections in generalizations of the numerical range of matrices see [5] and [8, Chapter 1].

Let $p, n \in \mathbb{N}, p \leq n$. We denote by $\mathcal{Z}^{n\times p}$ the set of all $n \times p$ matrices with entries from $\{0, 1\}$ such that they have exactly one 1 in each row and at least one

1 in each column; for example $I_n \in \mathcal{Z}^{n\times n}$ and $\mathbf{1}_m \in \mathcal{Z}^{n\times 1}$. For $Z \in \{0,1\}^{n\times p}$ we have $Z \in \mathcal{Z}^{n\times p}$ if and only if $Z^t Z$ is a nonsingular $p\times p$ diagonal matrix.

Let $Z = (\zeta_{rs}) \in \mathcal{Z}^{n\times p}$. We define

$$\mathcal{U}(Z) = \{X = (\xi_{rs}) \in \mathbb{C}^{n\times p} : X^*X = I_p, \quad \zeta_{rs} = 0 \quad \text{implies} \quad \xi_{rs} = 0 \\ for \quad 1 \le r \le n, 1 \le s \le p\}, \tag{2.1}$$

and for $A \in \mathbb{C}^{n\times n}$ we define the **block numerical range of A with respect to** Z by

$$\Theta(A, Z) = \bigcup_{X\in\mathcal{U}(Z)} \Sigma(X^*AX). \tag{2.2}$$

This set is a compact subset of $\mathbb{C}$, since we are working in finite dimensions. The **block numerical radius of A with respect to** Z is defined as follows

$$w(A, Z) = \max\{|\lambda| : \lambda \in \Theta(A, Z)\}; \tag{2.3}$$

Example 2.1.
1. $\Theta(A, I_n) = \Sigma(A)$; note that $\mathcal{U}(I_n)$ is the set of the unitary diagonal matrices. More general we have $\Theta(A, Z) = \Sigma(A)$ for all $Z \in \mathcal{Z}^{n\times n}$.
2. $\Theta(A, \mathbf{1}_n) = \Theta(A)$; note that $\mathcal{U}(\mathbf{1}_n) = \{X \in \mathbb{C}^{n\times 1} = \mathbb{C}^n : \|X\|_2 = 1\}$.

Let $Z = (\zeta_{rs}) \in \mathcal{Z}^{n\times p}$. For $s = 1, 2, \dots, p$ define $\langle Z, s\rangle = \{r \in \langle n\rangle : \zeta_{rs} = 1\}$. The family $\{\langle Z, s\rangle : s = 1, 2, \dots, p\}$ is a decomposition of $\langle n\rangle = \{1, 2, \dots, n\}$; i.e., the sets of the family are nonempty and pairwise disjoint and their union is equal to $\langle n\rangle$.

Conversely, let $\{\mathbf{M}_j : j = 1, 2, \dots, p\}$ be a decomposition of $\langle n\rangle$. Define $Z = (\zeta_{rs}) \in \mathbb{C}^{n\times p}$ by

$$\zeta_{rs} = 1 \quad \text{if} \quad r \in \mathbf{M}_s \quad \text{and} \quad \zeta_{rs} = 0 \quad \text{if} \quad r \notin \mathbf{M}_s,$$

then $Z \in \mathcal{Z}^{n\times p}$.

For a vector $x = (\xi_j) \in \mathbb{C}^n$ we denote by $x_{\langle Z,s\rangle}$ its subvector with entries in $\langle Z, s\rangle$. We define

$$\mathcal{S}(Z) = \{x \in \mathbb{C}^n : \|x_{\langle Z,s\rangle}\|_2 = 1 \quad \text{for} \quad s = 1, 2, \dots, p\}. \tag{2.4}$$

Remark 2.2. Next we will show that the block numerical range in the sense of [18] coincides with that defined above.

There exists an one-to-one correspondence between the vectors $x \in \mathcal{S}(Z)$ and the matrices $X \in \mathcal{U}(Z)$ given by

$$\xi_{rs} = \xi_r \quad \text{if} \quad \zeta_{rs} = 1 \quad \text{and} \quad \xi_{rs} = 0 \quad \text{otherwise},$$

for $1 \le r \le n, \quad 1 \le s \le p$. For $X \in \mathcal{U}(Z)$ the corresponding $x \in \mathcal{S}(Z)$ is the sum of the column of X.

Then we have with the notation in [18]

$$A_x = X^*AX = \left(x_{\langle Z,r\rangle}{}^* A_{\langle r,Z,s\rangle} x_{\langle Z,s\rangle}\right) \in \mathbb{C}^p, \tag{2.5}$$

where $A_{\langle r,Z,s\rangle}$ denotes the submatrix of A with entries of A at the row-indices in $\langle Z, r\rangle$ and the column-indices in $\langle Z, s\rangle$.

This implies the equivalence of both definitions; in the following we will use both of these characterisations of the block numerical range.

Let $1 \leq p \leq \widetilde{p} \leq n$. We call $\widetilde{Z} \in \mathcal{Z}^{n\times\widetilde{p}}$ a *refinement* of $Z \in \mathcal{Z}^{n\times p}$ if each column of Z is a sum of some columns of $\widetilde{Z}$. The following assertions are equivalent

1. $\widetilde{Z} \in \mathcal{Z}^{n\times\widetilde{p}}$ is a refinement of $Z \in \mathcal{Z}^{n\times p}$;
2. for all $\widetilde{r} \in \langle\widetilde{p}\rangle$ there exists a $r \in \langle p\rangle$ with $\langle\widetilde{Z},\widetilde{r}\rangle \subset \langle Z,r\rangle$;
3. for all $r \in \langle p\rangle$ we have $\bigcup_{\langle\widetilde{Z},\widetilde{r}\rangle\subset\langle Z,r\rangle}\langle\widetilde{Z},\widetilde{r}\rangle = \langle Z,r\rangle$.

If $\widetilde{Z} \in \mathcal{Z}^{n\times\widetilde{p}}$ is a refinement of $Z \in \mathcal{Z}^{n\times p}$ then by [18], Theorem 3.5

$$\Sigma(A) \subset \Theta(A,\widetilde{Z}) \subset \Theta(A,Z) \subset \Theta(A) \tag{2.6}$$

$$r(A) \leq w(A,\tilde{Z}) \leq w(A,Z) \leq w(A) \tag{2.7}$$

For sake of completeness we include a proof of these inclusions which in the finite-dimensional case is simpler than the original one in [18] but uses the same main ideas.

Proof. The third inclusion follows from the second since each $Z \in \mathcal{Z}^{n\times p}$ is a refinement of $\mathbf{1}_n$ and $\Theta(A,\mathbf{1}_n) = \Theta(A)$.

We prove the first. Let λ be an eigenvalue of A and $u \in \mathbb{C}^n$ be a corresponding eigenvector. For $j = 1,2,\ldots,\tilde{p}$ choose $\tilde{x}_j \in \mathbb{C}^{|\langle Z,j\rangle|}$ such that

$$u_{\langle\tilde{Z},j\rangle} = \|u_{\langle\tilde{Z},j\rangle}\|_2\,\tilde{x}_j, \quad \|\tilde{x}_j\|_2 = 1.$$

Define $\tilde{x} \in \mathbb{C}^n$ by $\tilde{x}_{\langle\tilde{Z},j\rangle} = \tilde{x}_j$ and $v = (\|u_{\langle\tilde{Z},j\rangle}\|_2) \in \mathbb{C}^{\tilde{p}}$. Then $\tilde{x} \in \mathcal{S}(\tilde{Z})$. For the corresponding $\tilde{X} \in \mathcal{U}(\tilde{Z})$ we obtain $\tilde{X}v = u$ and $\tilde{X}^*A\tilde{X}v = \lambda v$. Therefore, $\lambda \in \Sigma(\tilde{X}^*A\tilde{X}) \subset \Theta(A,\widetilde{Z})$.

We prove the second inclusion. We have the decomposition $\{\{k : \langle\tilde{Z},k\rangle \subset \langle Z,j\rangle\} : j = 1,2,\ldots,p\}$ of $\langle\tilde{p}\rangle$. Let $\hat{Z} = (\hat{\zeta}_{rs})$ be the corresponding matrix in $\mathcal{Z}^{\tilde{p}\times p}$, then $Z = \tilde{Z}\hat{Z}$. Let $Z = (\zeta_{rs}) \in \mathcal{Z}^{n\times p}$and $\tilde{Z} = (\tilde{\zeta}_{rs}) \in \tilde{\mathcal{Z}}^{n\times\tilde{p}}$ then $Z = \tilde{Z}\hat{Z}$ implies for $r = 1,2,\ldots,n$ and $s = 1,2,\ldots,p$

$$\zeta_{rs} = 0 \Leftrightarrow \quad \tilde{\zeta}_{rj} = 0 \quad \text{or} \quad \hat{\zeta}_{js} = 0, \quad j = 1,2,\ldots,\tilde{p}.$$

For $\tilde{X} \in \mathcal{U}(\tilde{Z})$ and $\hat{X} \in \mathcal{U}(\hat{Z})$ now $\tilde{X}\hat{X} \in \mathcal{U}(Z)$ follows easily.

With the first inclusion we get for all $\tilde{X} \in \mathcal{U}(\tilde{Z})$

$$\Sigma(\tilde{X}^*A\tilde{X}) \subset \Theta((\tilde{X}^*A\tilde{X},\hat{Z}) = \bigcup_{\hat{X}\in\mathcal{U}(\hat{Z})} \Sigma(\hat{X}^*(\tilde{X}^*A\tilde{X})\hat{X}) \subset \Theta(A,Z).$$

The second inclusion now follows immediately.

The proof of the inequalities in (2.7) follows from the inclusions in (2.6) and the definition of the relevant quantities. □

Finally we state some simple facts which we will refer to later.

Let $Z = (\zeta_{rs}) \in \mathcal{Z}^{n \times p}$ and P and Q permutation matrices in $\mathbb{C}^{n\times n}$ and $\mathbb{C}^{p\times p}$, respectively, then we have

$$X \in \mathcal{U}(Z) \quad \Leftrightarrow \quad PXQ \in \mathcal{U}(PZQ). \tag{2.8}$$

Indeed, $X^*X = I_p$ if and only if $(PXQ)^*(PXQ) = I_p$ and if $X \in \mathcal{U}(Z)$ then

$$(PZQ)_{rs} = 0 \Leftrightarrow \zeta_{\pi_P(r)\pi_Q(s)} = 0 \Rightarrow \xi_{\pi_P(r)\pi_Q(s)} = 0 \Leftrightarrow (PXQ)_{rs} = 0,$$

finally if $PXQ \in \mathcal{U}(PZQ)$ then

$$\zeta_{rs} = 0 \Leftrightarrow (PZQ)_{\pi_P(r)\pi_Q(s)} = 0 \Rightarrow (PXQ)_{\pi_P(r)\pi_Q(s)} = 0 \Leftrightarrow X_{rs} = 0$$

for all $1 \le r \le n,\ 1 \le s \le p$.

Furthermore note that for all $X \in \mathcal{U}(Z)$ and all $A \in \mathbb{C}^{n\times n}$ clearly

$$\Sigma((PX)^*APX) = \Sigma((PXQ)^*A(PXQ)) \tag{2.9}$$

and also

$$\Theta(P^tAP; Z) = \Theta(A; PZQ) \tag{2.10}$$

are true, where (2.10) follows from the definition of the block numerical range and from (2.8) and (2.9).

The block numerical range of A can be shifted by $\mu \in \mathbb{C}$ by adding μI_n:

$$\Theta(A + \mu I_n; Z) = \Theta(A; Z) + \mu. \tag{2.11}$$

$$\Theta(\alpha A,Z) = \alpha\Theta(A,Z) \quad \text{for all } \alpha \in \mathbb{C}. \tag{2.12}$$

We call a matrix $D \in \mathbb{C}^{n\times n}$ Z-block diagonal if the block matrices $D_{\langle r,Z,s\rangle}$ are zero for $r \neq s$. These are the off diagonal blocks. A Z-block diagonal matrix D is unitary if and only if the diagonal blocks $D_{\langle r,Z,r\rangle}$ are unitary.

Finally we will prove that for all Z-block diagonal unitary matrices $D \in \mathbb{C}^{n\times n}$ the equality

$$\Theta(A,Z) = \Theta(DAD^{-1},Z) \tag{2.13}$$

holds.

For the proof let $D = (d_{rs})$ be Z-block diagonal and unitary. We first show that $DX \in \mathcal{U}(Z)$ if and only if $X \in \mathcal{U}(Z)$. Take $X = (\xi_{rs}) \in \mathcal{U}(Z)$. Suppose $\zeta_{kl} = 0$ and consider in $(DX)_{kl} = \sum_{j=1}^{n} d_{kj}\xi_{jl}$ those ξ_{jl} which are not zero. For those consequently $\zeta_{jl} \neq 0$ and therefore, $j \in \langle Z, l\rangle$ by definition. This implies that $d_{kj} = 0$. To see this assume the opposite. Then d_{kj} has to be an element of any Z-diagonal block $D_{\langle s_0,Z,s_0\rangle}$ of D, $s_0 \in \{1,\dots,p\}$, which means $j,k \in \langle Z, s_0\rangle$ But since $j \in \langle Z,l\rangle$, also $k \in \langle Z,l\rangle$. Again by definition of $\langle Z,l\rangle$ this means that $\zeta_{kl} \neq 0$, which is a contradiction to our assumption.

Therefore, $(DX)_{kl} = 0$ and since $(DX)^*(DX) = I_n$ we obtain $DX \in \mathcal{U}(Z)$.

If we assume the other way around $DX \in \mathcal{U}(Z)$, where D is Z-block diagonal unitary and suppose $\zeta_{kl} = 0$, we set $DX = (\eta_{rs})$ and write $\xi_{kl} = \sum_{j=1}^{n} \bar{d}_{jk}\eta_{jl}$. With the same arguments as above we obtain $\xi_{kl} = 0$. And with $X^*X = X^*D^{-1}DX = I_n$ follows $X \in \mathcal{U}(Z)$. Therefore,

$$\begin{aligned}\Theta(DAD^{-1},Z) &= \bigcup_{X \in \mathcal{U}(Z)} \Sigma(X^*DAD^{-1}X) \\ &= \bigcup_{D^{-1}X \in \mathcal{U}(Z)} \Sigma\big((D^{-1}X)A(D^{-1}X)\big) \\ &= \bigcup_{Y \in \mathcal{U}(Z)} \Sigma(Y^*AY) \\ &= \Theta(A,Z).\end{aligned}$$

3. The block numerical range of a nonnegative matrix

In this section, we prove some basic facts on the block numerical range of a nonnegative matrix which are known for the spectrum and the standard numerical range of such matrices. We continue these considerations at the end of the next section using the Frobenius Form and fundamental results on the block numerical range of irreducible nonnegative matrices which will be proved in the next section.

For this section, let A, B be entrywise nonnegative $n \times n$ square matrices and $Z = (\zeta_{rs}) \in \mathcal{Z}^{n \times p}$, $p, n \in \mathbb{N}$, $p \leq n$.

Proposition 3.1. *Let* $A \geq 0$.

1. $w(A,Z) = \max\{r(X^*AX) : X \in \mathcal{U}(Z),\ X \geq 0\}$
   $= \max\{r(A_x) : x \in \mathcal{S}(Z),\ x \geq 0\}$
2. $w(A,Z) \in \Theta(A,Z)$

Proof. 1. Since $X \in \mathcal{U}(Z) \Leftrightarrow |X| \in \mathcal{U}(Z)$ the assertion follows from the inequality

$$r(X^*AX) \leq r(|X^*AX|) \leq r(|X^*|A|X|),$$

note that A is nonnegative.

2. Due to 1. there exists a nonnegative $X \in \mathcal{U}(Z)$ such that $w(A,Z) = r(X^*AX)$. Since $A \geq 0$, also $X^*AX \geq 0$ and therefore, by the Perron Frobenius Theorem $w(A,Z) = r(X^*AX) \in \Sigma(X^*AX) \subset \Theta(A,Z)$. □

Proposition 3.2.

1. *Let* $C \in \mathbb{C}^{n \times n}$. *Then*
$$w(C,Z) \leq w(|C|,Z).$$
2. *Suppose* $0 \leq A \leq B$. *Then*
$$w(A,Z) \leq w(B,Z).$$

Proof. 1. By definition of $w(C,Z)$ this follows from $r(X^*CX) \leq r(|X^*||C||X|)$.

2. Since $O \leq A \leq B$ for all $X \geq 0$ follows $O \leq X^*AX \leq X^*BX$. With Proposition 3.1.1 and the monotonicity of the spectral radius the assertion follows. □

We now generalize Proposition 6 of [17], which states that the spectrum of a nonnegative matrix is located in a disc with radius strictly less than the spectral radius of the matrix, provided all diagonal elements of the matrix are positive.

Let $A = (\alpha_{rs}) \in \mathbb{C}^{n\times n}$ be a nonnegative matrix, $Z \in \mathcal{Z}^{n\times p}$, $1 \le p \le n$ and set

$$\varepsilon(A) = \min_{j\in\langle n\rangle} \alpha_{jj}. \tag{3.1}$$

Then

$$\beta \ge -\varepsilon(A) \text{ implies } \quad w(A+\beta I_n, Z) = w(A,Z) + \beta \tag{3.2}$$

holds. Indeed if $\beta \ge -\varepsilon(A)$ then $A+\beta I_n$ is a nonnegative matrix and by Proposition 3.1 and (2.11) we obtain

$$0 \le w(A+\beta I_n, Z) \in \Theta(A+\beta I_n, Z) = \Theta(A,Z) + \beta,$$

and thus, $w(A+\beta I_n, Z) \le w(A,Z) + \beta$. On the other hand

$$w(A,Z) + \beta \in \Theta(A,Z) + \beta = \Theta(A+\beta I_n, Z),$$

and thus, $w(A+\beta I_n, Z) \ge w(A,Z)+\beta$.

Note that

$$w(A,Z) = \max\{\Re(\lambda) \; : \; \lambda \in \Theta(A,Z)\}.$$

We define

$$\mu(A,Z) = \min\{\Re(\lambda) \; : \; \lambda \in \Theta(A,Z)\}.$$

Since there exists a $\lambda \in \Theta(A,Z)$ with $|\mu(A,Z)| = |\Re(\lambda)| \le |\lambda| \le w(A,Z)$, the inequalities

$$-w(A,Z) \le \mu(A,Z) \le w(A,Z) \tag{3.3}$$

hold. Furthermore we have

$$\frac{\mu(A,Z) + w(A,Z)}{2} \ge \varepsilon(A). \tag{3.4}$$

To see that (3.4) holds assume the opposite and choose $\eta > 0$ such that

$$0 \le \frac{\mu(A,Z)+w(A,Z)}{2} = \frac{\mu + w}{2} < \eta < \varepsilon(A).$$

By (3.2), it follows $w(A-\eta I_n, Z) = w - \eta$. Consider a $\lambda \in \Theta(A,Z)$ with $\Re(\lambda) = \mu$. Then

$$|\lambda - \eta| \ge |\mu - \eta| > \left|\mu - \frac{\mu+w}{2}\right| = \frac{w-\mu}{2} = w - \frac{w+\mu}{2} \ge w - \eta$$
$$= w(A - \eta I_n, Z).$$

That contradicts $\lambda - \eta \in \Theta(A - \eta I_n, Z)$.

Note that from (3.3) and (3.4) follows that $0 \le \varepsilon(A) \le w(A,Z)$.

We can now formulate the generalization of Proposition 6 in [17].

Proposition 3.3. *Let $A \geq 0$, $Z \in \mathcal{Z}^{n\times p}$, $\varepsilon(A)$ as in (3.1). Then*

$$\Theta(A,Z) \subset \{\, \lambda \in \mathbb{C} \;:\; |\lambda - \varepsilon(A)| \leq w(A,Z) - \varepsilon(A)\,\}.$$

Especially, $\varepsilon(A) > 0$, implies that the the peripheral part of block numerical range of A consists only of the point $w(A,Z)$, i.e.,

$$\Theta_{\rm per}(A,Z) = \{\lambda \in \Theta(A,Z) : |\lambda| = w(A,Z)\} = \{w(A,Z)\}.$$

Proof. Consider $\lambda \in \Theta(A,Z)$. Then $\lambda - \varepsilon(A) \in \Theta(A - \varepsilon I_n, Z)$ and therefore,

$$|\lambda - \varepsilon(A)| \leq w(A - \varepsilon(A)I_n, Z) = w(A,Z) - \varepsilon(A). \qquad \square$$

4. The block numerical range of an irreducible matrix

In this section we present the important Perron-Frobenius Theory for the block numerical range of an irreducible nonnegative matrix as known for the spectrum (see [1], [7], [13]) and the standard numerical range (see [9], [11], [12]). We prove a corresponding version of Wielandt's Lemma. Later we use these results and the Frobenius Form of a nonnegative matrix to extent the results of Section 3.

Proposition 4.1. *Let $A \geq 0$ be irreducible, $Z \in \mathcal{Z}^{n\times p}$ and $x \in \mathcal{S}(Z)$.*

1. *If $x \gg 0$ then the matrix A_x is irreducible.*
2. *If $w(A,Z) = r(A_x)$, $x \in \mathcal{S}(Z)$ then $x \gg 0$.*

Proof. 1. Let us assume A_x is reducible. Then there is a $p\times p$ permutation matrix P and a corresponding permutation π of $\langle n\rangle$ such that

$$PA_xP^T = \left[\begin{array}{c|c} B_1 & C \\ \hline 0 & B_2 \end{array}\right] = \left(x_{\langle\pi(r),Z\rangle}{}^* A_{\langle\pi(r),Z,\pi(s)\rangle} x_{\langle\pi(s),Z\rangle}\right),$$

with the square matrices $B_1 \in \mathbb{C}^{k\times k}$, $B_2 \in \mathbb{C}^{l\times l}$ where $k,l > 1$ and $k + l = p$.

Thus $x_{\langle\pi(r),Z\rangle}{}^* A_{\langle\pi(r),Z,\pi(s)\rangle} x_{\langle\pi(s),Z\rangle} = 0$ for all r and s such that $\pi(r) > k$ $\pi(s) \leq k$, respectively. As $x \gg 0$ and $A \geq 0$, we conclude

$$\left(A_{\pi(r),Z,\pi(s)} x_{\langle\pi(s),Z\rangle}\right)_j = \sum_{i\in\langle Z,\pi(s)\rangle} (A_{\langle\pi(r),Z\pi(s)\rangle})_{ji} (x_{\langle Z,\pi(s)\rangle})_i = 0$$

for all $j \in \langle Z,\pi(r)\rangle$. Again because of $x \gg 0$ we get that $A_{\langle\pi(r),Z,\pi(s)\rangle} = 0$ for all r and s such that $\pi(r) > k$ and $\pi(s) \leq k$, respectively. This is a contradiction to the irreducibility of A.

2. We will prove the equivalent assertion: If $x \in \mathcal{S}(Z)$ is nonnegative but not strictly positive, then $r(A_x) < w(A,Z)$. We proceed in the following way: We construct a refinement $\tilde{Z}$ of Z and a $\tilde{x} \in \mathcal{S}(\tilde{Z})$ such that $r(A_x) < r(A_{\tilde{x}})$. Then by (2.7) we obtain the assertion, namely $r(A_x) < r(A_{\tilde{x}}) \leq w(A,\tilde{Z}) \leq w(A,Z)$.

Let $Z = (\zeta_{kl}) = (z_1,\ldots,z_p)$, where $z_j \in \mathbb{C}^n$, $j = 1,\ldots,p$ are the columns of Z and $X = (\xi_{kl}) = (x_1,\ldots,x_p) \in \mathcal{U}(Z)$ the matrix corresponding to x according to Remark 2.2, with the columns $x_j \in \mathbb{C}^n, j = 1,2,\ldots,p$. Then $x = \sum_{j=1}^p x_j$ and $X > 0$. Since x is not strictly positive and $\|x_{\langle Z,s\rangle}\|_2 = 1$, there exist $s \in \langle p\rangle$ such that $\{r \in \langle Z, s\rangle : \xi_{rs} = 0\}$ is a nonempty and proper subset of $\langle Z, s\rangle$. For such s we define the vectors $\tilde{z}_s^0$ and $\tilde{z}_s^1 \in \{0,1\}^n \subset \mathbb{C}^n$ by

$$\tilde{z}_s^0 = (\tilde{\zeta}_{rs}^0) \quad \text{with} \quad \tilde{\zeta}_{rs}^0 = \begin{cases} 1 & \text{iff} \quad r \in \langle Z, s\rangle \text{ and } \xi_{rs} = 0 \\ 0 & \text{otherwise} \end{cases}$$

$$\tilde{z}_s^1 = (\tilde{\zeta}_{rs}^1) \quad \text{with} \quad \tilde{\zeta}_{rs}^1 = \begin{cases} 1 & \text{iff} \quad r \in \langle Z, s\rangle \text{ and } \xi_{rs} > 0 \\ 0 & \text{otherwise} \end{cases} .$$

Then $z_s = \tilde{z}_s^0 + \tilde{z}_s^1$.

If ξ_{rs} is positive for all $r \in \langle Z, s\rangle$,then $\tilde{z}_s^0$ is the empty vector, i.e., it does not appear. We define the matrix

$$\tilde{Z} = (\tilde{z}_1^0, \tilde{z}_1^1, \tilde{z}_2^0, \tilde{z}_2^1, \ldots, \tilde{z}_p^0, \tilde{z}_p^1) = (\tilde{z}_1, \tilde{z}_2, \ldots, \tilde{z}_{\tilde{p}}) \in \{0,1\}^{n \times \tilde{p}}.$$

Accordingly, we define for $s = 1, 2, \ldots, \tilde{p}$ the nonnegative vectors

$$\tilde{x}_s^0 = (\tilde{\xi}_{rs}^0) \quad \text{with} \quad \tilde{\xi}_{rs}^0 \begin{cases} > 0 & \text{iff } r \in \langle Z, s\rangle \text{ and } \xi_{rs} = 0 \\ = 0 & \text{otherwise} \end{cases} ,$$

such that $\|\tilde{x}_s^0\|_2 = 1$ and

$$\tilde{x}_s^1 = (\tilde{\xi}_{rs}^1) = x_s.$$

If $\tilde{z}_s^0$ is the empty vector, then also $\tilde{x}_s^0$ is the empty vector. We define the matrix

$$\tilde{X} = (\tilde{x}_1^0, \tilde{x}_1^1, \tilde{x}_2^0, \tilde{x}_2^1, \ldots, \tilde{x}_p^0, \tilde{x}_p^1) = (\tilde{x}_1, \tilde{x}_2, \ldots, \tilde{x}_{\tilde{p}}) \in \mathbb{R}_+^{n \times \tilde{p}}.$$

Clearly $\tilde{Z} \in \tilde{\mathcal{Z}}^{n \times \tilde{p}}$, $\tilde{Z}$ is a refinement of Z, $\tilde{X} \in \mathcal{U}(\tilde{Z})$ and $\tilde{x} = \sum_{1 \le j \le \tilde{p}} \tilde{x}_j \in \mathcal{S}(\tilde{Z})$ is strictly positive. $A_x = X^*AX$ is a proper principle submatrix of A_x, since X^*AX corresponds to the submatrix of $\tilde{X}^*A\tilde{X}$ which is obtained by deleting in $\tilde{X}$ the columns $\tilde{x}_s^0$. Now $A_{\tilde{x}} = \tilde{X}^*A\tilde{X}$ is irreducible by 1. and we obtain $r(A_x) < r(A_{\tilde{x}})$ from Theorem 5.1 of [13]. □

Theorem 4.2. *Let $A, C \in \mathbb{C}^{n \times n}$, $A \ge 0$, $Z \in \mathcal{Z}^{n \times p}$ and assume $|C| \le A$. Then*

1. *$w(C,Z) \le w(A,Z)$*
2. *Suppose in addition that A is irreducible.*

   Then $w(C,Z) = w(A,Z)$ if and only if C is of the form

$$C = \zeta DAD^{-1}, \tag{4.1}$$

where $\zeta \in \mathbb{C}$ is such that $|\zeta| = 1$ and $\zeta w(A,Z) \in \Theta(C,Z)$ and D is a unitary diagonal matrix, i.e., $|D| = I_n$.

Proof. 1. For any $x \in \mathbb{C}_1 \times \cdots \times \mathbb{C}_n$ we have by triangle inequality

$$|C_x| \le |C|_{|x|} \le A_{|x|}.$$

Therefore, with $r(C_x) \le r(|C_x|)$ we obtain

$$\begin{aligned} w(C,Z) &= \max\{r(C_x) \mid x \in \mathcal{S}(Z)\} \\ &\le \max\{r(|C_x|) \mid x \in \mathcal{S}(Z)\} \\ &\le \max\{r(|C|_{|x|}) \mid x \in \mathcal{S}(Z)\} \\ &\le \max\{r(A_{|x|}) \mid x \in \mathcal{S}(Z)\} \\ &= w(A,Z) \end{aligned}$$

2. Let (4.1) hold. Then $\Theta(C,Z) = \zeta\Theta(DAD^{-1},Z) = \zeta\Theta(A,Z)$; this of course implies $w(C,Z) = w(A,Z)$.

For the necessary part, note that $w(C,Z) = w(A,Z)$ implies that there exists a $\zeta \in \mathbb{C}$, with $|\zeta| = 1$ and $\zeta w(A,Z) \in \Theta(C,Z)$. This means that there is a $y \in \mathcal{S}(Z)$ such that $\zeta w(A,Z) \in \Sigma(C_y)$. By the same arguments as in 1. we get that

$$|C_y| \le |C|_{|y|} \le A_{|y|}$$

and

$$r(A_{|y|}) \le w(A,Z) \le r(C_y) = r(|C_y|) \le r(A_{|y|}),$$

which implies $r(A_{|y|}) = w(A,Z)$. Since A is irreducible Proposition 4.1.2 implies that $|y| \gg 0$, thus from the first part of the same proposition follows that $A_{|y|}$ is irreducible.

Let $v \in \mathbb{C}^p$ be an eigenvector of C_y such that $\zeta\, r(A_{|y|})v = \zeta w(A,Z)v = C_y v$ then we have

$$A_{|y|}|v| \ge |C_y||v| \ge |C_y v| = |\zeta r(A_{|y|})v| = r(A_{|y|})|v|.$$

Since $A_{|y|}$ is irreducible we obtain $r(A_{|y|}) = A_{|v|}v$ (see [7]) and $|v| \gg 0$, since v is nonzero.

So we have a nonnegative irreducible matrix $A_{|y|}$ which dominates a matrix C_y and there exists a unit complex number ζ such that $\zeta\, r(A_{|y|}) \in \Sigma(C_y)$ and the eigenvector v for that eigenvalue of $C_{|y|}$ has no zero entry.

By Wielandt's Lemma (see [13]) we get

$$C_y = \zeta D_v A_{|y|} D_v^{-1}, \tag{4.2}$$

where $D_v = \mathrm{Diag}(\frac{v_1}{|v_1|}, \dots, \frac{v_p}{|v_p|}) \in \mathbb{C}^{p\times p}$. Therefore, $|C_y| = A_{|y|}$.

We now show that this implies $|C| = A$. Let $A = (\alpha_{kl})$ and $C = (\gamma_{kl})$. By the assumption of the theorem we have $|C| \le A$. Assume that $|C| < A$. Then there exist $k, l \in \langle n\rangle$ with $|\gamma_{kl}| < \alpha_{kl}$. Furthermore, there exist $r, s \in \langle p\rangle$ with $k \in \langle Z, r\rangle$ and $l \in \langle s, Z\rangle$, respectively. Since $|y|$ is strictly positive and $|C| \le A$ it follows that

$$|y^*_{\langle Z,r\rangle} C_{\langle r,Z,s\rangle} y_{\langle s,Z\rangle}| \quad < \quad |y|^*_{\langle Z,r\rangle} A_{\langle r,Z,s\rangle} |y|_{\langle s,Z\rangle},$$

which implies the contradiction $|C_y| < A_{|y|}$.

For $y = (\eta_j) \in \mathcal{S}(Z) \subset \mathbb{C}^n$ define the unitary diagonal matrix

$$D_y = \mathrm{Diag}\Big(\frac{\eta_1}{|\eta_1|}, \dots, \frac{\eta_d}{|\eta_d|}\Big) \in \mathbb{C}^{n \times n}$$

and the corresponding matrix $Y \in \mathcal{U}(Z) \subset \mathbb{C}^{n \times p}$. Then $D_y|y| = y$ and $D_y|Y| = Y$.

Define the unitary diagonal matrix $\widehat{D}_v = \mathrm{Diag}(\delta_1, \dots, \delta_n)$ by

$$\delta_j = \frac{v_s}{|v_s|} \quad \text{if} \quad j \in \langle Z, s\rangle.$$

We now show that $C = \zeta(D_y\widehat{D}_v)^* A(D_y\widehat{D}_v)$, which proves formula (4.1) with $D = D_y\widehat{D}_v$.

First note that

$$YD_v = \widehat{D}_v Y \quad \text{and} \quad |Y|D_v = \widehat{D}_v|Y|.$$

Then

$$\begin{aligned}
(|Y||v|)^* \, \widehat{D}_v^* D_y^* \, \bar{\zeta} C \, D_y \widehat{D}_v \, |Y||v| &= |v|^* D_v^* \, |Y^*| D_y^* \, \bar{\zeta} C \, D_y |Y| \, D_v |v| \\
&= D_v^* Y^* \, \bar{\zeta} C \, Y D_v \\
&= |v|^* D_v^* \bar{\zeta} C_y D_v |v| \\
&\overset{(4.2)}{=} |v|^* A_{|y|} |v| \\
&= (|Y||v|)^* A \, (|Y||v|).
\end{aligned}$$

Now $|\widehat{D}_v^* D_y^* \, \bar{\zeta} C \, D_y \widehat{D}_v| = |C| = A$ due to the fact that D_y and $\widehat{D}_v$ are unitary diagonal matrices, therefore,

$$(|Y||v|)^* (D_y\widehat{D}_v)^* \bar{\zeta} C (D_y\widehat{D}_v) |Y||v| = (|Y||v|)^* A \, (|Y||v|).$$

Since $|Y||v|$ is strictly positive we obtain from the two last equations that

$$\widehat{D}_v^* D_y^* \, \bar{\zeta} C \, D_y \widehat{D}_v = |\widehat{D}_v^* D_y^* \, \bar{\zeta} C \, D_y \widehat{D}_v| = A. \tag{4.3}$$

Remember that if we have $\alpha_1, \dots, \alpha_p \in \mathbb{C}$ such that $\sum_{j=1}^p \alpha_j = \sum_{j=1}^p |\alpha_j|$ then for all $j = 1, \dots, n$ we already have that $\alpha_j = |\alpha_j|$. □

Example 4.3. Recall Example 2.1 for irreducible A.

1. In the case $Z = I_p$, i.e., $\Theta(A,Z) = \Sigma(A)$, the matrix Y in the proof of Theorem 4.2,2 is unitary since $p = n$. Thus $\Sigma(C_y) = \Sigma(C) = \Sigma(C_{\mathbf{1}_n})$. Therefore, in the proof we can start with $y = \mathbf{1}_n$ which means that $D_y = I_n$.

   So, in the case $\Theta(A,Z) = \Sigma(A)$ Theorem 4.2 is the well-known result of Wielandt for the spectral radii of A and C with $D = \widehat{D}_v$. See, e.g., [13].

2. Suppose the case $Z = \mathbf{1}_n$, i.e., $\Theta(A,Z) = \Theta(A)$. Then $p = 1$ and the matrix D_v in the proof of Theorem 4.2, 2 is a unit complex number τ, thus $\widehat{D}_v = \tau I_n$.

   Therefore, equation (4.3) reduces to $D_y^* \bar{\zeta} C D_y = A$, which is the Perron-Frobenius type result for the numerical range of A in [11] with $D = D_y$.

Since equation (4.1) implies $|C| = A$ we immediately get

Corollary 4.4. *Assume $0 \leq C \leq A$, where A is irreducible. Then the following assertions are equivalent*

1. *$C = A$.*
2. *There exist $p \in \langle n \rangle$ and $Z \in \mathcal{Z}^{n\times p}$ such that $w(C,Z) = w(A,Z)$.*
3. *For all $p \in \langle n \rangle$ and all $Z \in \mathcal{Z}^{n\times p}$ we have $w(C,Z) = w(A,Z)$.*

The following theorem is well known for the spectrum and the standard numerical range of a nonnegative irreducible matrix A; see for example [7], [13] and [11], [12].

Theorem 4.5. *Let $A \geq 0$ be irreducible, let $p_0 \in \langle n \rangle$ and $Z_0 \in \mathcal{Z}^{n\times p_0}$, let $\zeta \in \mathbb{C}$, $|\zeta| = 1$ and $\zeta \in \Theta(A, Z_0)$. Then for all $p \in \langle n \rangle$ and for all $Z \in \mathcal{Z}^{n\times p}$ the rotation condition*

$$\zeta\Theta(A,Z) = \Theta(A,Z)$$

holds.

Proof. In Theorem 4.2 set $C = A$. Then A has the form $A = \zeta DAD^{-1}$ where ζ is a unit complex number such that $\zeta w(A;Z_0) \in \Theta(A;Z_0)$ and D is a unitary diagonal matrix. By (2.13) we have

$$\zeta\Theta(A,Z) = \zeta\Theta(DAD^{-1},Z) = \Theta(\zeta DAD^{-1},Z) = \Theta(A,Z). \qquad \square$$

It is well known from Perron-Frobenius Theory that the spectrum of a nonnegative irreducible matrix with index of imprimitivity m (it is defined to be the number of peripheral eigenvalues of this matrix, see [13]) is rotation invariant with respect to the angles $\frac{2k}{m}\pi$, $k = 0, \ldots, m-1$ and only to these.

As we mentioned in Example 2.1, for $p = n$ and $Z_0 = I_n$ the block numerical range of A coincides with the spectrum of A. Therefore, Theorem 4.5 implies that for a nonnegative irreducible matrix A with index of imprimitivity m the complex unit numbers ζ from that theorem are the mth roots of unity

$$\zeta = e^{\frac{2k}{m}\pi i}, \ k = 0, \ldots, m-1.$$

We have proved the following.

Theorem 4.6. *Let $A \geq 0$ be irreducible with index of imprimitivity m. Then for all block numercial ranges $\Theta(A, Z)$ of A*

$$\Theta_{\text{per}}(A,Z) = \big\{\lambda \in \Theta(A,Z) : |\lambda| = w(A,Z)\big\} = \big\{w(A,Z)e^{\frac{2k}{m}\pi i}, k = 0, \ldots, m-1\big\},$$

and they are invariant under rotation of the angle $(2\pi)/m$, but not invariant under rotation with a smaller angle.

It is well known that a nonnegative matrix is permutation similar to a block triangular matrix with irreducible blocks on the diagonal, see [16]; this (not unique) block triangular form is usually called the **Frobenius form** of the matrix. In the following we will use the Frobenius form of a nonnegative matrix to discuss properties of its block numerical ranges.

Let A be a $n \times n$ nonnegative matrix. Two numbers $k, l \in \langle n \rangle$ belong to the same **class of** A if there exist nonnegative integers q_1, q_2 such that in A^{q_1} the entry in position (k,l) is positive and in A^{q_2} the entry in position (l,k) is positive, see [1]; there are other possibilities to define the classes of A, for example with help of the directed graph associated with A.

The family of all classes of A is a decomposition of $\langle n \rangle$. If A has p classes let $Z_F \in \mathcal{Z}^{n\times p}$ the corresponding matrix, see Section 2. It is possible (and we will do that) to enumerate the family of classes $\{\langle Z_F, s\rangle : s = 1, 2., \dots, p\}$ in such a way that $(A_{\langle r, Z_F, s\rangle})$ is a $p \times p$ lower block triangular matrix, i.e., $A_{\langle r, Z_F, s\rangle} = 0$ if $1 \le r < s \le p$. Note that the diagonal blocks $A_{\langle s, Z_F, s\rangle}$ are irreducible matrices, for $s = 1, 2, \dots, p$.

Then by Proposition 1.9 of [19],

$$\Theta(A, Z_F) = \bigcup_{s \in \langle p \rangle} \Theta(A_{\langle s, Z_F, s\rangle}),$$

and

$$\Theta_{\text{per}}(A, Z_F) = \bigcup \{\Theta_{\text{per}}(A_{\langle s, Z_F, s\rangle}) : w(A_{\langle s, Z_F, s\rangle}) = w(A, Z_F)\}.$$

The last equality and Theorem 4.6 imply that the peripheral numerical range $\Theta_{\text{per}}(A, Z_F)$ is of the form $w(A, Z_F)\Gamma$, where Γ is a finite union of finite groups of unity.

These statements have natural generalizations to refinements of Z_F. Let $\tilde{Z}$ be a refinenment of Z_F. Then $\{\langle \tilde{Z}, j\rangle : \langle \tilde{Z}, j\rangle \subset \langle Z_F, s\rangle\}$ is a decomposition of $\langle Z_F, s\rangle$, $s = 1, 2, \dots, p$. With the notations $n_s = \#\langle Z_F, s\rangle$ and $\tilde{p}_s = \#\{j : \langle \tilde{Z}, j\rangle \subset \langle Z_F, s\rangle\}$ and the corresponding matrices $\tilde{Z}_s \in \mathcal{Z}^{n_s \times \tilde{p}_s}$ for $s = 1, 2, \dots, p$ we obtain

$$\Theta(A, \tilde{Z}) = \bigcup_{s \in lsp\rangle} \Theta(A_{\langle s, Z_F, s\rangle}, \tilde{Z}_s)$$

and

$$\Theta_{\text{per}}(A, \tilde{Z}) = \bigcup \{\Theta_{\text{per}}(A_{\langle s, Z_F, s\rangle}, \tilde{Z}_s) : w(A_{\langle s, Z_F, s\rangle}, \tilde{Z}_s) = w(A, \tilde{Z})\}.$$

The last equality and Theorem 4.6 imply that the peripheral numerical range $\Theta_{\text{per}}(A, \tilde{Z})$ is of the form $w(A, \tilde{Z})\Gamma$, where Γ is a finite union of finite groups of unity.

For a proof of these equalities note that for $x \in \mathbb{C}^n$

$$x \in \mathcal{S}(\tilde{Z}) \Leftrightarrow x_{\langle Z_F, s\rangle} \in \mathcal{S}(\tilde{Z}_s) \quad \text{for} \quad s = 1, 2, \dots, p$$

and for $x \in \mathcal{S}(\tilde{Z})$ the matrix $A_x \in \mathbb{C}^{\tilde{p}\times\tilde{p}}$ is permutationaly similar to a $p \times p$ block matrix, which is lower block triangular and $(A_{\langle s, Z_F, s\rangle})_{x_{\langle Z_F, s\rangle}}$ are its diagonal blocks for $s = 1, 2, \dots, p$.

For $\tilde{Z} = I_n$ these are well-known results for the spectrum of A.

The following result was proved in Theorem 1.2 of [14] for the peripheral spectra of the matrices A and B.

Proposition 4.7. *Let A and B be nonnegative square matrices of the same size, Let B be irreducible and $A \leq \gamma B$ for some nonnegative constant γ.*

Then:

1. *The index of imprimitivity of B divides the index of imprimitivity of each $A_{\langle s,Z_F,s\rangle}$, $s = 1,2,\dots,p$.*

   If $\tilde{Z}$ is a refinement of the Frobenius decomposition $Z_{A,F}$ of A, then

2. $$\Theta_{\text{per}}(A,\tilde{Z}) \subset \Theta_{\text{per}}(A_{\langle s,Z_{A,F},s\rangle},\tilde{Z}_s)$$
   for all $s = 1,2,\dots,p$ with $w(B,\tilde{Z}) = w(A_{\langle s,Z_{A,F},s\rangle},\tilde{Z}_s)$; here we use the notation from above.

3. $$w(A,\tilde{Z}) = w(B,\tilde{Z}) \Longleftrightarrow \Theta_{\text{per}}(A,\tilde{Z}) \subset \Theta_{\text{per}}(B,\tilde{Z}).$$

Proof. 1. From Theorem 2.30 of [1] we know that the index of imprimitivity of a nonnegative irreducible $n \times n$ matrix C is the gcd (= greatest common divisor) of all nonnegative integers k, such that $(C^k)_{jj}$ is positive for an arbitrary $j \in \langle n\rangle$.

For all $s \in \langle p\rangle$, $j \in \langle Z_{A,F},s\rangle$ and all $k \in \mathbb{N}$ the lower block triangular structure of $(A_{\langle r,Z_{A,F},s\rangle})$ implies

$$((A_{\langle s,Z_{A,F},s\rangle})^k)_{jj} \leq (A^k)_{jj} \leq (\gamma)^k(B^k)_{jj},$$

so 1. follows immediately.

2. The assertion foffows from 1, and Theorem 4.6.

3. Note that the peripheral parts of the block numerical ranges of matrices are nonempty. The assertion follows from 2. ☐

In Section 6, we will give an example which shows that assertions 2. and 3. in the last proposition do not hold if $\tilde{Z}$ is not a refinement of the Frobenius decomposition $Z_{A,F}$ of A.

5. The numerical range of a Perron polynomial

A monic $n \times n$ matrix polynomial of degree l

$$L(\lambda) = \lambda^{l+1}I_n - \lambda^l A_l - \dots \lambda A_1 - A_0 \tag{5.1}$$

is called a Perron polynomial if all coefficients A_j are entrywise nonnegative $n \times n$ matrices, $j = 0,1,\dots,l$; see [15].

Its numerical range is defined as (see [10], [15])

$$\Theta(L(\cdot)) = \{\lambda \in \mathbb{C} : \quad x^*L(\lambda)x = 0 \quad \text{for some} \quad x \in \mathbb{C}^n,\, x \neq 0\}. \tag{5.2}$$

Assume that the coefficients of $L(\cdot)$ are nonnegative and their sum is irreducible then in ([2], p.15) it is proved that there exist a $h \in \mathbf{N}$ such that

$$\Sigma_{\text{per}}(L(\cdot)) = \{r(L(\cdot))e^{(i2\pi j)/h} : j = 0,1,\dots,h\}, \tag{5.3}$$

and $\Sigma(L(\cdot))$is invariant under rotation of the angle $(2\pi)/h$, but not invariant under rotation with a smaller angle. In this section we will prove that the numerical range of $L(\cdot)$ has the corresponding properties with the same h. For a similar result see ([15], Theorem 5.5, Corollary 5.6)

In the proof we will use ([18], Theorem 5.2)

$$\Theta(L(\cdot)) \cup \{0\} = \Theta(C_L, \hat{Z}) \cup \{0\}, \tag{5.4}$$

where C_L denotes the companion matrix of $L(\cdot)$ and $\hat{Z}$ is the $n(l+1) \times n(l+1)$ matrix

$$\begin{pmatrix} I_{nl} & 0_{nl} \\ 0_{n\times nl} & \mathbf{1}_n \end{pmatrix}.$$

The companion matrix of the polynomial in (5.1) is not irreducible, if the coefficients are nonnegative and their sum is irreducible; for example, if A_0 has a zero-column then C_L has a zero-column and cannot be irreducible.

Let $A_j \geq O$ for $j = 1, 2, \ldots, l$ and $S = A_l + A_{l-1} + \cdots + A_1 + A_0$ be irreducible. From Theorem 3.2 of [2] it follows that after an appropriate permutation of the first nl row and columns of C_L we obtain the cogredient matrix

$$P^t C_L P = \hat{C}_L = \begin{pmatrix} N & B \\ 0 & C \end{pmatrix}, \tag{5.5}$$

where N is an upper diagonal square matrix with zero diagonal, 0 is the zero matrix of appropriate size and C is an irreducible nonnegative square matrix. It follows

$$\Sigma(L(\cdot)) \cup \{0\} = \Sigma(C_L) \cup \{0\} = \Sigma(\hat{C}_L) \cup \{0\} = = \Sigma(C) \cup \{0\}, \tag{5.6}$$

$$\Theta(L(\cdot)) \cup \{0\} = \Theta(C_L, \hat{Z}) \cup \{0\} = \Theta(\hat{C}_L, \hat{Z}) \cup \{0\} = \Theta(C) \cup \{0\}, \tag{5.7}$$

for the second equality in (5.7) note that the permutation matrix P used to go from $\mathbb{C}_L$ to $\hat{\mathbb{C}}_L$ does not affect $\hat{Z}$, i.e., $\hat{Z} = P\hat{Z}$.

From (5.7) and Theorem 4.6 follows the

Theorem 5.1. *Let h be the index of imprimitivity of the matrix C in* (5.5). *Then*

1. $\Sigma_{per}(L(\cdot)) = \{r(L(\cdot))e^{(i2\pi j)/h} : j = 0, 1, \ldots, h\}$,
2. $\Theta_{\mathrm{per}}(L(\cdot)) = \Theta_{\mathrm{per}}(C) = \{w(L(\cdot))e^{(i2\pi j)/h} : j = 0, 1, \ldots, h\}$;

further $\Sigma(L(\cdot))$ and $\Theta(L(\cdot))$ are invariant under rotation of the angle $(2\pi)/h$, but not invariant under rotation with a smaller angle.

We mention that a relation between this h and the index of imprimitivity of $S = A_0 + A_{l-1} + \cdots + A_1 + A_0$ does not exists, see [3], Section 4.

6. Examples

In this last section we give descriptions and pictures of block numerical ranges of some nonnegative matrices.

Example 6.1 (Reducible matrix in Frobenius Form). We consider the reducible nonnegative matrix

$$A_1 = \begin{bmatrix} 0 & 1 & 0 & 0 \\ 1 & 0 & 0 & 0 \\ 0 & 0 & 0 & 0 \\ 1 & 0 & 2 & 0 \end{bmatrix}$$

with numerical range $\Theta(A_1)$ as in Figure 1 and its spectrum $\Sigma(A_1) = \{-1, 0, 1\}$.

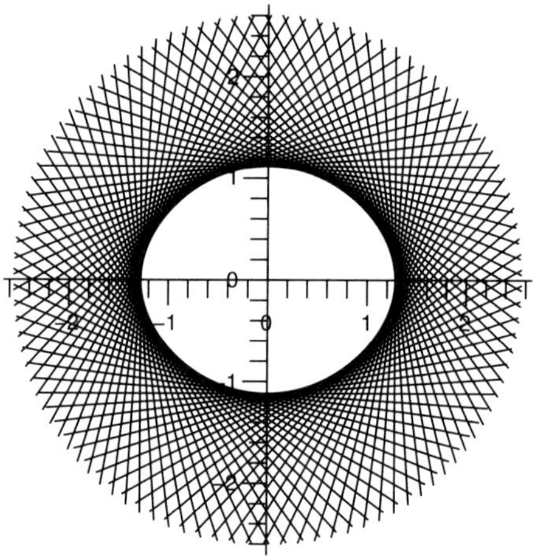

FIGURE 1. $\langle Z, 1\rangle = \{1, 2, 3, 4\}$

In the following, we will see the block numerical range for different partitions of A_1.

- Let $Z_F \in \mathcal{Z}^{4\times 3}$ correspond to the Frobenius decomposition of A_1, i.e., $\langle Z_F, 1\rangle = \{1, 2\}$, $\langle Z_F, 2\rangle = \{3\}$ and $\langle Z_F, 3\rangle = \{4\}$. Let $X \in \mathcal{U}(Z_F)$. Then

$$X = \begin{bmatrix} r_1 e^{i\varphi_1} & 0 & 0 \\ r_2 e^{i\varphi_2} & 0 & 0 \\ 0 & e^{i\varphi_3} & 0 \\ 0 & 0 & e^{i\varphi_4} \end{bmatrix},$$

  where $r_2 = \sqrt{1 - r_1^2}$, $r_1 \in [0, 1]$ and $\varphi_j \in [0, 2\pi]$, $j = 1, \dots, 4]$. The spectrum of $(A_1)_x = X^*AX$ is $\Sigma(X^*AX) = \{0, \frac{r_1 r_2}{2}\cos(\varphi_1 - \varphi_2)\}$, and thus,

$$\Theta(A_1, Z_F) = [-1, 1]$$

  is the block numrical range according to the Frobenius decomposition.

- Let us assume that $Z_a \in \mathcal{Z}^{4\times 2}$ is such that

$$\langle Z_a, 1\rangle = \{1,2\} \text{ and } \langle Z_a, 2\rangle = \{3,4\}.$$

Then every $X \in \mathcal{U}(Z_a)$ is of the form

$$X = \begin{bmatrix} r_1 e^{i\varphi_1} & 0 \\ r_2 e^{i\varphi_2} & 0 \\ 0 & r_3 e^{i\varphi_3} \\ 0 & r_4 e^{i\varphi_4} \end{bmatrix},$$

where $r_j \in [0,1]$, $\varphi_j \in [0,2\pi]$ for $j = 1,\dots,4$ and $r_2 = \sqrt{1-r_1^2}$ and $r_4 = \sqrt{1-r_3^2}$. Hence, $(A_1)_x$ is a 2×2 triangular matrix and its spectrum is

$$\Sigma(A_x) = \{2r_1r_2\cos(\varphi_1-\varphi_2),\ 2r_3r_4e^{\varphi_3-\varphi_4}\}.$$

So the block numerical range $\Theta(A_1, Z_a)$ of A_1 is the closed unit disc $\overline{\mathbb{D}}$.

- Let now $Z_b \in \mathcal{Z}^{4\times 2}$ be such that

$$\langle Z_b, 1\rangle = \{1,2,4\} \text{ and } \langle Z_b, 2\rangle = \{3\}.$$

Then $X \in \mathcal{U}(Z_b)$ has the form

$$X = \begin{bmatrix} \sin\vartheta\,\cos\varphi\, e^{i\varphi_1} & 0 \\ \sin\vartheta\,\sin\varphi\, e^{i\varphi_2} & 0 \\ 0 & e^{i\varphi_4} \\ \cos\vartheta\, e^{i\varphi_3} & 0 \end{bmatrix},$$

where $\vartheta, \varphi, \varphi_j \in [0,2\pi]$ for $j = 1,\dots,4$. The matrix $(A_1)_x$ has one nonzero eigenvalue, which determines the block numerical range

$$\Theta(A_1, Z_b) = \{\sin^2\vartheta\,\sin 2\varphi\,\cos(\varphi_1-\varphi_2) + \frac{1}{2}\sin 2\vartheta\,\cos\varphi\, e^{i(\varphi_1-\varphi_3)} : \vartheta, \varphi, \varphi_1, \varphi_2, \varphi_3 \in [0,2\pi]\}.$$

Since we can choose φ_2 and φ_3 independently from φ_1, we can drop φ_1 and $\Theta(A_1, Z_b)$ reduces to

$$\Theta(A_1, Z_b) = \{\sin^2\vartheta\,\sin 2\varphi\,\cos\varphi_2 + \frac{1}{2}\sin 2\vartheta\,\cos\varphi\, e^{i\varphi_3} : \vartheta, \varphi, \varphi_2, \varphi_3 \in [0,2\pi]\},$$

see Figure 2 (The vertical lines mark $x = \pm 1$, the horizontal ones $y = \pm 0.5$).

Note that $\Theta_{\text{per}}(A_1, Z_b) \not\subset \Theta_{\text{per}}(A_1, Z_F)$. This example shows that if $\tilde{Z}$ is no refinement of the Frobenius decomposition, the peripheral block numerical range of A_1 with respect to $\tilde{Z}$ does not need to be contained in the peripheral block numerical range of A_1 with respect to Z_F.

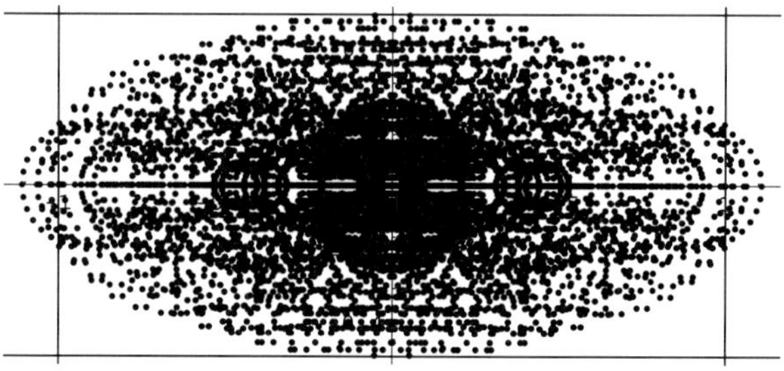

FIGURE 2. $\langle Z_b, 1\rangle = \{1, 2, 4\}$, $\langle Z_b, 2\rangle = \{3\}$

Example 6.2 (Nilpotent matrices). Let

$$S_n = \begin{bmatrix} 0 & 0 & \dots & \dots & 0 & 0 \\ 1 & 0 & \dots & \dots & 0 & 0 \\ 0 & 1 & \dots & \dots & 0 & 0 \\ \vdots & \vdots & \dots & \dots & \vdots & \vdots \\ \vdots & \vdots & \dots & \dots & \vdots & \vdots \\ 0 & 0 & \dots & \dots & 1 & 0 \end{bmatrix}, \tag{6.1}$$

be the matrix of the n-dimensional, nilpotent right shift. The numerical range of S_n is the closed circular disc at the origin with radius $cos(\pi/(n+1)$, see [4], [5], [6]. Let $p \in \langle n\rangle$ and let $n_1, n_2, \dots, n_p$ be natural numbers which sum up to n. For the decomposition

$$\{1, \dots, n_1\}, \dots, \{n_1 + \dots + n_{p-1} + 1, \cdots, n_1 + \dots + n_p\}$$

of $\langle n\rangle$ and the corresponding matrix $Z \in \mathcal{Z}^{n\times p}$ we obtain

$$\Theta(S_n, Z) = \Theta(S_{max\{n_1,\dots,n_p\}})$$

and

$$w(S_n, Z) = cos(\pi/(max\{n_1, \dots, n_p\} + 1)).$$

Note that S_n is a $p\times p$ a lower block triangular matrix with S_{n_k} as diagonal blocks, $k = 1, \dots, p$ and apply [19], Proposition 1.9.

Example 6.3 (Cyclic irreducible matrices). Let

$$C_n = \begin{bmatrix} 0 & 1 & 0 & \dots & \dots & 0 & 0 \\ 0 & 0 & 1 & \dots & \dots & 0 & 0 \\ 0 & 0 & 0 & \dots & \dots & 0 & 0 \\ \vdots & \vdots & \dots & \dots & \vdots & \vdots & \\ \vdots & \vdots & \dots & \dots & \vdots & \vdots & \\ 0 & 0 & 0 & \dots & \dots & 0 & 1 \\ 1 & 0 & 0 & \dots & \dots & 0 & 0 \end{bmatrix}, \tag{6.2}$$

be the matrix of the n-dimensional cyclic shift. This matrix is irreducible with n as index of imprimitivity and the eigenvalues of C_n are exactly the nth roots of unity. This matrix is unitary and therefore, it is normal, thus its numerical range is the convex hull of its spectrum, i.e., $\Theta(C_n)$ is the regular polygon centered at the origin with the nth roots of unity as the vertices. From Theorem 4.6 we obtain that for all $p \in \langle n \rangle$ and all $Z \in \mathcal{Z}^{n \times p}$

$$\Theta_{\text{per}}(C_n, Z) = \{\exp(2\pi ik/n) : k = 0, 1, \ldots, n\}.$$

The block numerical ranges of C_n can change dramatically as the following example shows.

We consider the irreducible matrix

$$A_2 = \begin{bmatrix} 0 & 1 & 0 \\ 0 & 0 & 1 \\ 1 & 0 & 0 \end{bmatrix}.$$

Its spectrum is $\Sigma(A_2) = \{1, e^{i\frac{2\pi}{3}}, e^{i\frac{4\pi}{3}}\}$ and its numerical range $\Theta(A_2)$ as illustrated in Figure 3.

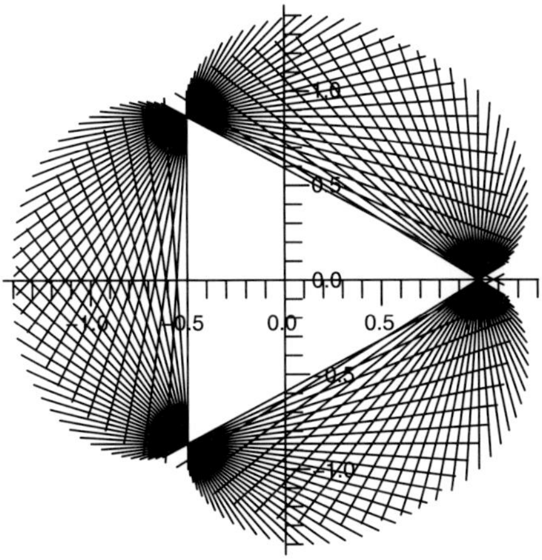

FIGURE 3. $\langle Z, 1 \rangle = \{1, 2, 3, 4\}$

Let $Z_a \in \mathcal{Z}^{3 \times 2}$ be such that $\langle Z_a, 1 \rangle = \{1, 2\}$ and $\langle Z_a, 2 \rangle = \{3\}$ and

$$X = \begin{bmatrix} r_1 e^{i\varphi_1} & 0 \\ r_2 e^{i\varphi_2} & 0 \\ 0 & e^{i\varphi_3} \end{bmatrix}.$$

Calculating the spectrum of $A_x \in \mathbb{C}^{2 \times 2}$ one obtains the block numerical range

$$\Theta(A_2, Z_a) = \{re^{i\varphi} \pm \sqrt{r^2 e^{2i\varphi} + 2re^{-i\varphi}} \; : \; r \in [0, \tfrac{1}{2}], \; \varphi \in [0, 2\pi]\}.$$

Its boundary is shown in Figure 4 (we thank Prof. Dr. Dirk Ferus for this picture).

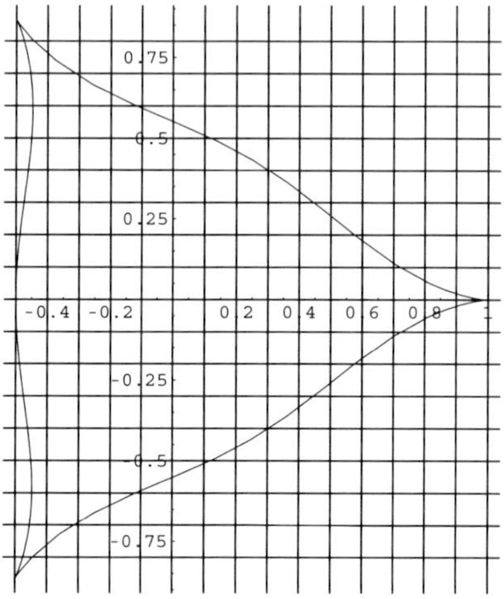

FIGURE 4. $\langle Z,1\rangle = \{1\}$, $\langle Z,2\rangle = \{2,3\}$

References

[1] Berman, A., Plemmons, R.J.,*Nonnegative Matrices in the Mathematical Sciences*, Classiscs in Appied Mathematics. vol. 9, SIAM; Philadelphia, 1994 (Revised reprint of the 1979 original).

[2] Foerster, K.-H., Nagy, B, *Spektraleigenschaften von Matrix- und Operatorpolynomen.* Sitzungsberichte d. Berliner Math. Gesellschaft, 1988–1992, 243–262. Berlin, Vorstand der Gesellschaft, 1994. MR 96b:47009.

[3] Foerster, K.-H., Nagy, B., *On Spectra of Expansion Graphs and Matrix Polynomials.* Linear Algebra and its Applications **363** (2003), 89–101.

[4] Gau, H.L., Wu, P.Y., *Companion matrices: reducibility, numerical ranges and similarity to contractions.* Linear Algebra and its Applications **383** (2004), 127–142.

[5] Gustafson, K.E., Rao, D.K.M., *Numerical Range – The Field of Values of Linear Operators and Matrices*, Springer New York, 1997.

[6] Haagerup, P., de la Harpe, P., *The numerical radius of a nilpotent operator on a Hilbert space.* Proc. Amer. Math. Soc.**115** (1992), 171–179.

[7] Horn, R.A., Johnson, C.R., *Matrix Analysis.* Cambridge University Press: Cambridge, 1985.

[8] Horn, R.A., Johnson, C.R., *Topics in Matrix Analysis.* Cambridge University Press: Cambridge, 1991.

[9] Issos, J.N., *The Field of Values of non-Negative Irreducible Matrices.* Ph.D. Thesis, Auburn University, 1966.

[10] Li, C.-K., Rodman, L., *Numerical Range of Matrix Polynomials.* SIAM J. Matrix Anal. Appl. **15**(4) (1994), 1256–1265.

[11] Li, C.-K., Tam, B.-S., Wu, P.Y., *The Numerical Range of a Nonnegative Matrix.* Linear Algebra and its Applications **350** (2002), 1–23.

[12] Maroulas, J., Psarrakos, P.J., Tsatsomeros, M.J., *Perron-Frobenius Type Results on the Numerical Range.* Linear Algebra and its Applications **348** (2002), 49–62.

[13] Minc, H., *Nonnegative Matrices.* Wiley New York, 1988.

[14] Moustakas, U., *Majorisierung und Spektraleigenschaften positiver Operatoren in Banachverbänden.* Dissertation, Universität Tübingen, 1984

[15] Psarrakos, P.J., Tsatsomeros M.J., *A Primer of Perron-Frobenius Theory for Matrix Polynomials.* Linear Algebra and its Applications **393** (2004), 333–351.

[16] Schneider, H., *The Influence of Marked Reduced Graph of a Nonnegative Matrix on the Jordan Form and on related Properties: A Survey.* Linear Algebra and its Applications **84** (1986), 161–189.

[17] Smyth, M.R.F., West, T.T., *The Minimum Diagonal Element of a Positive Matrix.* Studia Mathematica **131**(1) (1998), 95–99.

[18] Tretter, Ch., Wagenhofer, M., *The Block Numerical Range of an $n \times n$ Block Operator Matrix.* SIAM J. Matrix Anal. Appl. **24**(4) (2003), 1003–1017.

[19] Wagenhofer, M., *Block Numerical Ranges.* Dissertation, Universität Bremen, 2007.

K.-H. Förster and N. Hartanto
Technische Universität Berlin
Institut für Mathematik, MA 6-4
D-10623 Berlin, Germany
e-mail: `foerster@math.tu-berlin.de`

Operator Theory:
Advances and Applications, Vol. 188, 135–148

A Quantum Dot with Impurity in the Lobachevsky Plane

V. Geyler, P. Šťovíček and M. Tušek

Abstract. The curvature effect on a quantum dot with impurity is investigated. The model is considered on the Lobachevsky plane. The confinement and impurity potentials are chosen so that the model is explicitly solvable. The Green function as well as the Krein Q-function are computed.

Keywords. Quantum dot, Lobachevsky plane, point interaction, spectrum.

1. Introduction

Physically, quantum dots are nanostructures with a charge carriers confinement in all space directions. They have an atom-like energy spectrum which can be modified by adjusting geometric parameters of the dots as well as by the presence of an impurity. Thus the study of these dependencies may be of interest from the point of view of the nanoscopic physics.

A detailed analysis of three-dimensional quantum dots with a short-range impurity in the Euclidean space can be found in [1]. Therein, the harmonic oscillator potential was used to introduce the confinement, and the impurity was modeled by a point interaction (δ-potential). The starting point of the analysis was derivation of a formula for the Green function of the unperturbed Hamiltonian (i.e., in the impurity free case), and application of the Krein resolvent formula jointly with the notion of the Krein Q-function.

The current paper is devoted to a similar model in the hyperbolic plane. The nontrivial hyperbolic geometry attracts regularly attention, and its influence on the properties of quantum-mechanical systems has been studied on various models (see, for example, [2, 3, 4]). Here we make use of the same method as in [1] to investigate a quantum dot with impurity in the Lobachevsky plane. We will introduce an appropriate Hamiltonian in a manner quite analogous to that of [1] and derive an explicit formula for the corresponding Green function. In this sense, our model is solvable, and thus its properties may be of interest also from the mathematical point of view.

During the computations to follow, the spheroidal functions appear naturally. Unfortunately, the notation in the literature concerned with this type of special functions is not yet uniform (see, e.g., [5] and [6]). This is why we supply, for the reader's convenience, a short appendix comprising basic definitions and results related to spheroidal functions which are necessary for our approach.

2. A quantum dot with impurity in the Lobachevsky plane

2.1. The model

Denote by (ϱ, ϕ), $0 < \varrho < \infty$, $0 \le \phi < 2\pi$, the geodesic polar coordinates on the Lobachevsky plane. Then the metric tensor is diagonal and reads

$$(g_{ij}) = \operatorname{diag}\left(1, a^2 \sinh^2 \frac{\varrho}{a}\right)$$

where a, $0 < a < \infty$, denotes the so-called curvature radius which is related to the scalar curvature by the formula $R = -2/a^2$. Furthermore, the volume form equals $dV = a \sinh(\varrho/a) \mathrm{d}\varrho \wedge \mathrm{d}\phi$. The Hamiltonian for a free particle of mass $m = 1/2$ takes the form

$$H^0 = -\left(\Delta_{LB} + \frac{1}{4a^2}\right) = -\frac{1}{\sqrt{g}}\frac{\partial}{\partial x^i}\sqrt{g}g^{ij}\frac{\partial}{\partial x^j} - \frac{1}{4a^2}$$

where Δ_{LB} is the Laplace-Beltrami operator and $g = \det g_{ij}$. We have set $\hbar = 1$.

The choice of a potential modeling the confinement is ambiguous. We naturally require that the potential takes the standard form of the quantum dot potential in the flat limit ($a \to \infty$). This is to say that, in the limiting case, it becomes the potential of the isotropic harmonic oscillator $V_\infty = \frac{1}{4}\omega^2\varrho^2$. However, this condition clearly does not specify the potential uniquely. Having the freedom of choice let us discuss the following two possibilities:

$$\text{a)} \quad V_a(\varrho) = \tfrac{1}{4}\, a^2\omega^2 \tanh^2 \tfrac{\varrho}{a}, \tag{2.1}$$

$$\text{b)} \quad U_a(\varrho) = \tfrac{1}{4}\, a^2\omega^2 \sinh^2 \tfrac{\varrho}{a}. \tag{2.2}$$

Potential V_a is the same as that proposed in [7] for the classical harmonic oscillator on the Lobachevsky plane. With this choice, it has been demonstrated in [7] that the model is superintegrable, i.e., there exist three functionally independent constants of motion. Let us remark that this potential is bounded, and so it represents a bounded perturbation to the free Hamiltonian. On the other hand, the potential U_a is unbounded. Moreover, as shown below, the stationary Schrödinger equation for this potential leads, after the partial wave decomposition, to the differential equation of spheroidal functions. The current paper concentrates exclusively on case b).

The impurity is modeled by a δ-potential which is introduced with the aid of self-adjoint extensions and is determined by boundary conditions at the base point. We restrict ourselves to the case when the impurity is located in the center of the dot ($\varrho = 0$). Thus we start from the following symmetric operator:

$$H = -\left(\frac{\partial^2}{\partial\varrho^2} + \frac{1}{a}\coth\left(\frac{\varrho}{a}\right)\frac{\partial}{\partial\varrho} + \frac{1}{a^2}\sinh^{-2}\left(\frac{\varrho}{a}\right)\frac{\partial^2}{\partial\phi^2} + \frac{1}{4a^2}\right) + \frac{1}{4}\,a^2\omega^2\sinh^2\left(\frac{\varrho}{a}\right),$$
$$\mathrm{Dom}(H) = C_0^\infty((0,\infty)\times S^1) \subset L^2\left((0,\infty)\times S^1, a\,\sinh\left(\frac{\varrho}{a}\right)\mathrm{d}\varrho\,\mathrm{d}\phi\right). \tag{2.3}$$

2.2. Partial wave decomposition

Substituting $\xi = \cosh(\varrho/a)$ we obtain

$$H = \frac{1}{a^2}\left[(1-\xi^2)\frac{\partial^2}{\partial\xi^2} - 2\xi\frac{\partial}{\partial\xi} + (1-\xi^2)^{-1}\frac{\partial^2}{\partial\phi^2} + \frac{a^4\omega^2}{4}(\xi^2-1) - \frac{1}{4}\right] =: \frac{1}{a^2}\tilde{H},$$
$$\mathrm{Dom}(H) = C_0^\infty((1,\infty)\times S^1) \subset L^2\left((1,\infty)\times S^1, a^2\mathrm{d}\xi\,\mathrm{d}\phi\right). \tag{2.4}$$

Using the rotational symmetry which amounts to a Fourier transform in the variable ϕ, $\tilde{H}$ may be decomposed into a direct sum as follows

$$\tilde{H} = \bigoplus_{m=-\infty}^{\infty} \tilde{H}_m,$$
$$\tilde{H}_m = -\frac{\partial}{\partial\xi}\left((\xi^2-1)\frac{\partial}{\partial\xi}\right) + \frac{m^2}{\xi^2-1} + \frac{a^4\omega^2}{4}(\xi^2-1) - \frac{1}{4},$$
$$\mathrm{Dom}(\tilde{H}_m) = C_0^\infty(1,\infty) \subset L^2((1,\infty),\mathrm{d}\xi).$$

Note that $\tilde{H}_m$ is a Sturm-Liouville operator.

Proposition 2.1. *$\tilde{H}_m$ is essentially self-adjoint for $m \neq 0$, $\tilde{H}_0$ has deficiency indices* $(1,1)$.

Proof. The operator $\tilde{H}_m$ is symmetric and semibounded, and so the deficiency indices are equal. If we set

$$\mu = |m|,\ 4\theta = -\frac{a^4\omega^2}{4},\ \lambda = -z - \frac{1}{4},$$

then the eigenvalue equation

$$\left(-\frac{\partial}{\partial\xi}\left((\xi^2-1)\frac{\partial}{\partial\xi}\right) + \frac{m^2}{\xi^2-1} + \frac{a^4\omega^2}{4}(\xi^2-1) - \frac{1}{4}\right)\psi = z\psi \tag{2.5}$$

takes the standard form of the differential equation of spheroidal functions (A.1). According to Chapter 3.12, Satz 5 in [6], for $\mu = m \in \mathbb{N}_0$ a fundamental system $\{y_{\mathrm{I}},\ y_{\mathrm{II}}\}$ of solutions to equation (2.5) exists such that

$$y_{\mathrm{I}}(\xi) = (1-\xi)^{m/2}\,\mathfrak{P}_1(1-\xi),\quad \mathfrak{P}_1(0) = 1,$$
$$y_{\mathrm{II}}(\xi) = (1-\xi)^{-m/2}\mathfrak{P}_2(1-\xi) + A_m\,y_{\mathrm{I}}(\xi)\log(1-\xi),$$

where, for $|\xi - 1| < 2$, $\mathfrak{P}_1, \mathfrak{P}_2$ are analytic functions in ξ, λ, θ; and A_m is a polynomial in λ and θ of total order m with respect to λ and $\sqrt{\theta}$; $A_0 = -1/2$.

Suppose that $z \in \mathbb{C}\backslash\mathbb{R}$. For $m = 0$, every solutions to (2.5) is square integrable near 1; while for $m \neq 0$, y_{I} is the only one solution, up to a factor, which is square integrable in a neighborhood of 1. On the other hand, by a classical analysis due to Weyl, there exists exactly one linearly independent solution to (2.5) which is square integrable in a neighborhood of ∞, see Theorem XIII.6.14 in [8]. In the case of $m = 0$ this obviously implies that the deficiency indices are $(1, 1)$. If $m \neq 0$ then, by Theorem XIII.2.30 in [8], the operator $\tilde{H}_m$ is essentially self-adjoint. □

Define the maximal operator associated to the formal differential expression

$$L = -\frac{\partial}{\partial\xi}\left((\xi^2 - 1)\frac{\partial}{\partial\xi}\right) + \frac{a^4\omega^2}{4}(\xi^2 - 1) - \frac{1}{4}$$

as follows

$$\begin{aligned}\mathrm{Dom}(H_{\mathrm{max}}) = \Big\{ & f \in L^2((1,\infty), \mathrm{d}\xi) :\ f, f' \in AC((1,\infty)),\\ & -\frac{\partial}{\partial\xi}\left((\xi^2 - 1)\frac{\partial f}{\partial\xi}\right) + \frac{a^4\omega^2}{4}(\xi^2 - 1)f \in L^2((1,\infty), \mathrm{d}\xi)\Big\},\\ H_{\mathrm{max}} f = Lf. &\end{aligned}$$

According to Theorem 8.22 in [9], $H_{\mathrm{max}} = \tilde{H}_0^\dagger$.

Proposition 2.2. *Let $\kappa \in (-\infty, \infty]$. The operator $\tilde{H}_0(\kappa)$ defined by the formulae*

$$\mathrm{Dom}(\tilde{H}_0(\kappa)) = \{f \in \mathrm{Dom}(H_{\mathrm{max}}) :\ f_1 = \kappa f_0\},\ \tilde{H}_0(\kappa)f = H_{\mathrm{max}} f,$$

where

$$f_0 := -4\pi a^2 \lim_{\xi\to 1+} \frac{f(\xi)}{\log(2a^2(\xi - 1))},\ f_1 := \lim_{\xi\to 1+} f(\xi) + \frac{1}{4\pi a^2} f_0 \log\bigl(2a^2(\xi - 1)\bigr),$$

is a self-adjoint extension of $\tilde{H}_0$. There are no other self-adjoint extensions of $\tilde{H}_0$.

Proof. The methods to treat δ like potentials are now well established [10]. Here we follow an approach described in [11], and we refer to this source also for the terminology and notations. Near the point $\xi = 1$, each $f \in \mathrm{Dom}(H_{\mathrm{max}})$ has the asymptotic behavior

$$f(\xi) = f_0\, F(\xi, 1) + f_1 + o(1) \quad \text{as } \xi \to 1+$$

where $f_0, f_1 \in \mathbb{C}$ and $F(\xi, \xi')$ is the divergent part of the Green function for the Friedrichs extension of $\tilde{H}_0$. By formula (2.11) which is derived below, $F(\xi, 1) = -1/(4\pi a^2)\log\bigl(2a^2(\xi - 1)\bigr)$. Proposition 1.37 in [11] states that $(\mathbb{C}, \Gamma_1, \Gamma_2)$, with $\Gamma_1 f = f_0$ and $\Gamma_2 f = f_1$, is a boundary triple for H_{max}.

According to Theorem 1.12 in [11], there is a one-to-one correspondence between all self-adjoint linear relations κ in $\mathbb{C}$ and all self-adjoint extensions of $\tilde{H}_0$

given by $\kappa \longleftrightarrow \tilde{H}_0(\kappa)$ where $\tilde{H}_0(\kappa)$ is the restriction of $H_{\max}$ to the domain of vectors $f \in \mathrm{Dom}(H_{\max})$ satisfying

$$(\Gamma_1 f, \Gamma_2 f) \in \kappa. \tag{2.6}$$

Every self-adjoint relation in $\mathbb{C}$ is of the form $\kappa = \mathbb{C}v \subset \mathbb{C}^2$ for some $v \in \mathbb{R}^2$, $v \neq 0$. If (with some abuse of notation) $v = (1, \kappa)$, $\kappa \in \mathbb{R}$, then relation (2.6) means that $f_1 = \kappa f_0$. If $v = (0, 1)$ then (2.6) means that $f_0 = 0$ which may be identified with the case $\kappa = \infty$. □

Remark. Let $\mathfrak{q}_0$ be the closure of the quadratic form associated to the semibounded symmetric operator $\tilde{H}_0$. Only the self-adjoint extension $\tilde{H}_0(\infty)$ has the property that all functions from its domain have no singularity at the point $\xi = 1$ and belong to the form domain of $\mathfrak{q}_0$. It follows that $\tilde{H}_0(\infty)$ is the Friedrichs extension of $\tilde{H}_0$ (see, for example, Theorem X.23 in [12] or Theorems 5.34 and 5.38 in [9]).

2.3. The Green function

Let us consider the Friedrichs extension of the operator $\tilde{H}$ in $L^2\left((1,\infty)\times S^1, \mathrm{d}\xi\,\mathrm{d}\phi\right)$ which was introduced in (2.4). The resulting self-adjoint operator is in fact the Hamiltonian for the impurity free case. The corresponding Green function $\mathcal{G}_z$ is the generalized kernel of the Hamiltonian, and it should obey the equation

$$(\tilde{H} - z)\mathcal{G}_z(\xi, \phi; \xi', \phi') = \delta(\xi - \xi')\delta(\phi - \phi') = \frac{1}{2\pi} \sum_{m=-\infty}^{\infty} \delta(\xi - \xi')\mathrm{e}^{im(\phi-\phi')}.$$

If we suppose $\mathcal{G}_z$ to be of the form

$$\mathcal{G}_z(\xi, \phi; \xi', \phi') = \frac{1}{2\pi} \sum_{m=-\infty}^{\infty} \mathcal{G}_z^m(\xi, \xi')\mathrm{e}^{im(\phi-\phi')}, \tag{2.7}$$

then, for all $m \in \mathbb{Z}$,

$$(\tilde{H}_m - z)\mathcal{G}_z^m(\xi, \xi') = \delta(\xi - \xi'). \tag{2.8}$$

Let us consider an arbitrary fixed ξ', and set

$$\mu = m,\ 4\theta = -\frac{a^4\omega^2}{4},\ \lambda = -z - \frac{1}{4}.$$

Then for all $\xi \neq \xi'$ equation (2.8) takes the standard form of the differential equation of spheroidal functions (A.1). As one can see from (A.8), the solution which is square integrable near infinity equals $S_\nu^{|m|(3)}(\xi, -a^4\omega^2/16)$. Furthermore, the solution which is square integrable near $\xi = 1$ equals $Ps_\nu^{|m|}(\xi, -a^4\omega^2/16)$ as one may verify with the aid of the asymptotic formula

$$P_\nu^m(\xi) \sim \frac{\Gamma(\nu + m + 1)}{2^{m/2}\, m!\, \Gamma(\nu - m + 1)}\, (\xi - 1)^{m/2} \quad \text{as } \xi \to 1+,\ \text{for } m \in \mathbb{N}_0.$$

We conclude that the mth partial Green function equals

$$\mathcal{G}_z^m(\xi,\xi') = -\frac{1}{(\xi^2-1)\mathscr{W}(Ps_\nu^{|m|}, S_\nu^{|m|(3)})}\, Ps_\nu^{|m|}\left(\xi_<, -\frac{a^4\omega^2}{16}\right) S_\nu^{|m|(3)}\left(\xi_>, -\frac{a^4\omega^2}{16}\right) \tag{2.9}$$

where the symbol $\mathscr{W}(Ps_\nu^{|m|}, S_\nu^{|m|(3)})$ denotes the Wronskian, and $\xi_<$, $\xi_>$ are respectively the smaller and the greater of ξ and ξ'. By the general Sturm-Liouville theory, the factor $(\xi^2-1)\mathscr{W}(Ps_\nu^{|m|}, S_\nu^{|m|(3)})$ is constant. Since $\mathcal{G}_z^m = \mathcal{G}_z^{-m}$ decomposition (2.7) may be simplified,

$$\mathcal{G}_z(\xi,\phi;\xi',\phi') = \frac{1}{2\pi}\,\mathcal{G}_z^0(\xi,\xi') + \frac{1}{\pi}\sum_{m=1}^{\infty}\mathcal{G}_z^m(\xi,\xi')\cos\left[m(\phi-\phi')\right]. \tag{2.10}$$

2.4. The Krein Q-function

The Krein Q-function plays a crucial role in the spectral analysis of impurities. It is defined at a point of the configuration space as the regularized Green function evaluated at this point. Here we deal with the impurity located in the center of the dot ($\xi = 1$, ϕ arbitrary), and so, by definition,

$$Q(z) := \mathcal{G}_z^{\mathrm{reg}}(1,0;1,0).$$

Due to the rotational symmetry,

$$\mathcal{G}_z(\xi) := \mathcal{G}_z(\xi,\phi;1,0) = \mathcal{G}_z(\xi,\phi;1,\phi) = \mathcal{G}_z(\xi,0;1,0) = \frac{1}{2\pi}\,\mathcal{G}_z^0(\xi,1),$$

and hence

$$(\tilde{H}_0 - z)\mathcal{G}_z(\xi) = 0, \quad \text{for } \xi \in (1,\infty).$$

Let us note that from the explicit formula (2.9), one can deduce that the coefficients $\mathcal{G}_z^m(\xi,1)$ in the series in (2.10) vanish for $m = 1,2,3,\ldots$. The solution to this equation is

$$\mathcal{G}_z(\xi) \propto S_\nu^{0(3)}\left(\xi, -\frac{a^4\omega^2}{16}\right).$$

The constant of proportionality can be determined with the aid the following theorem which we reproduce from [13].

Theorem 2.3. *Let $d(x,y)$ denote the geodesic distance between points x,y of a two-dimensional manifold X of bounded geometry. Let*

$$U \in \mathcal{P}(X) := \left\{U\colon\, U_+ := \max(U,0) \in L^{p_0}_{\mathrm{loc}}(X),\; U_- := \max(-U,0) \in \sum_{i=1}^{n} L^{p_i}(X)\right\}$$

for an arbitrary $n \in \mathbb{N}$ and $2 \le p_i \le \infty$. Then the Green function $\mathcal{G}_U$ of the Schrödinger operator $H_U = -\Delta_{LB} + U$ has the same on-diagonal singularity as that for the Laplace-Beltrami operator itself, i.e.,

$$\mathcal{G}_U(\zeta;x,y) = \frac{1}{2\pi}\log\frac{1}{d(x,y)} + \mathcal{G}_U^{\mathrm{reg}}(\zeta;x,y)$$

where $\mathcal{G}_U^{\mathrm{reg}}$ is continuous on $X \times X$.

Let us denote by $\mathcal{G}_z^H$ and $Q^H(z)$ the Green function and the Krein Q-function for the Friedrichs extension of H, respectively. Since $\tilde{H} = a^2 H$ and $(\tilde{H}-z)\mathcal{G}_z = \delta$, we have

$$\mathcal{G}_z^H(\xi,\phi;\xi',\phi') = a^2\mathcal{G}_{a^2z}(\xi,\phi;\xi',\phi'),\ Q^H(z) = a^2Q(a^2z).$$

One may verify that

$$\log d(\varrho,0;\vec{0}) = \log\varrho = \log(a\operatorname{arg}\cosh\xi) = \frac{1}{2}\log\bigl(2a^2(\xi-1)\bigr) + O(\xi-1)$$

as $\varrho \to 0+$ or, equivalently, $\xi \to 1+$. Finally, for the divergent part

$$F(\xi,\xi') := \mathcal{G}_z(\xi,\phi;\xi',\phi) - \mathcal{G}_z^{\mathrm{reg}}(\xi,\phi;\xi',\phi) = \mathcal{G}_z(\xi,0;\xi',0) - \mathcal{G}_z^{\mathrm{reg}}(\xi,0;\xi',0)$$

of the Green function $\mathcal{G}_z$ we obtain the expression

$$F(\xi,1) = -\frac{1}{4\pi a^2}\log\bigl(2a^2(\xi-1)\bigr). \tag{2.11}$$

From the above discussion, it follows that the Krein Q-function depends on the coefficients α, β in the asymptotic expansion

$$S_\nu^{0(3)}\left(\xi, -\frac{a^4\omega^2}{16}\right) = \alpha\log(\xi-1) + \beta + o(1) \quad \text{as } \xi \to 1+, \tag{2.12}$$

and equals

$$Q(z) = -\frac{\beta}{4\pi a^2\alpha} + \frac{\log(2a^2)}{4\pi a^2}. \tag{2.13}$$

To determine α, β we need relation (A.10) for the radial spheroidal function of the third kind. For ν and $\nu+1/2$ being non-integer, formula (A.12) implies that

$$\begin{aligned} S_\nu^{0(1)}(\xi,\theta) &= \frac{\sin(\nu\pi)}{\pi}\,\mathrm{e}^{-i\pi(\nu+1)}K_\nu^0(\theta)Qs_{-\nu-1}^0(\xi,\theta),\\ S_{-\nu-1}^{0(1)}(\xi,\theta) &= \frac{\sin(\nu\pi)}{\pi}\,\mathrm{e}^{i\pi\nu}K_{-\nu-1}^0(\theta)Qs_\nu^0(\xi,\theta). \end{aligned} \tag{2.14}$$

Applying the symmetry relation (A.5) for expansion coefficients, we derive that

$$\begin{aligned} Qs_{-\nu-1}^0(\xi,\theta) &= \sum_{r=-\infty}^{\infty}(-1)^r a_{-\nu-1,r}^0(\theta)Q_{-\nu-1+2r}^0(\xi)\\ &= \sum_{r=-\infty}^{\infty}(-1)^r a_{\nu,r}^0(\theta)Q_{-\nu-1-2r}^0(\xi). \end{aligned}$$

Using the asymptotic formulae (see [5])

$$Q_\nu^0(\xi) = -\frac{1}{2}\log\frac{\xi-1}{2} + \Psi(1) - \Psi(\nu+1) + O((\xi-1)\log(\xi-1)),$$

the series expansion in (A.11) and formulae (2.14), we deduce that, as $\xi \to 1+$,

$$S_\nu^{0(1)}(\xi,\theta) \sim -\frac{\sin(\nu\pi)}{\pi}\,\mathrm{e}^{-i\pi(\nu+1)}K_\nu^0(\theta) \times\left[s_\nu^0(\theta)^{-1}\left(\frac{1}{2}\log\frac{\xi-1}{2}-\Psi(1)+\pi\cot(\nu\pi)\right)+\Psi s_\nu(\theta)\right],$$

$$S_{-\nu-1}^{0(1)}(\xi,\theta) \sim -\frac{\sin(\nu\pi)}{\pi}\,\mathrm{e}^{i\pi\nu}K_{-\nu-1}^0(\theta) \times\left[s_\nu^0(\theta)^{-1}\left(\frac{1}{2}\log\frac{\xi-1}{2}-\Psi(1)\right)+\Psi s_\nu(\theta)\right],$$

where the coefficients $s_n^\mu(\theta)$ are introduced in (A.7),

$$\Psi s_\nu(\theta) := \sum_{r=-\infty}^{\infty}(-1)^r a_{\nu,r}^0(\theta)\Psi(\nu+1+2r),$$

and where we have made use of the following relation for the digamma function: $\Psi(-z)=\Psi(z+1)+\pi\cot(\pi z)$.

We conclude that

$$S_\nu^{0(3)}(\xi,\theta) \sim \alpha\log(\xi-1)+\beta+O\left((\xi-1)\log(\xi-1)\right) \quad \text{as } \xi\to 1+,$$

where

$$\alpha = \frac{i\tan(\nu\pi)}{2\pi s_\nu^0(\theta)}\left(\mathrm{e}^{i\pi\nu}K_{-\nu-1}^0(\theta)-\mathrm{e}^{-i\pi(2\nu+3/2)}K_\nu^0(\theta)\right),$$

$$\beta = \alpha\left(-\log 2-2\Psi(1)+2\Psi s_\nu(\theta)s_\nu^0(\theta)\right)+\mathrm{e}^{-2i\pi\nu}s_\nu^0(\theta)^{-1}K_\nu^0(\theta).$$

The substitution for α, β into (2.13) yields

$$Q(z) = -\frac{1}{4\pi a^2}\left(-\log 2-2\Psi(1)+2\,\Psi s_\nu\left(-\frac{a^4\omega^2}{16}\right)s_\nu^0\left(-\frac{a^4\omega^2}{16}\right)\right) + \frac{1}{2a^2\tan(\nu\pi)}\left(\mathrm{e}^{i\pi(3\nu+3/2)}\frac{K_{-\nu-1}^0(-\frac{a^4\omega^2}{16})}{K_\nu^0(-\frac{a^4\omega^2}{16})}-1\right)^{-1}+\frac{\log(2a^2)}{4\pi a^2} \tag{2.15}$$

where ν is chosen so that

$$\lambda_\nu^0\left(-\frac{a^4\omega^2}{16}\right) = -z-\frac{1}{4}. \tag{2.16}$$

For $\nu=n$ being an integer, we can immediately use the known asymptotic formulae for spheroidal functions (see Section 16.12 in [5]) which yield

$$S_n^{0(3)}(\xi,\theta) = \frac{is_n^0(\theta)}{4\sqrt{\theta}K_n^0(\theta)}\log(\xi-1)-\frac{is_n^0(\theta)\log 2}{4\sqrt{\theta}K_n^0(\theta)} + \frac{is_n^0(\theta)^2}{2\sqrt{\theta}K_n^0(\theta)}\sum_{2r\ge -n}(-1)^r a_{n,r}^0(\theta)h_{n+2r}+\frac{K_n^0(\theta)}{s_n^0(\theta)}+O(\xi-1),$$

as $\xi\to 1+$. Here, $h_0=1, h_k=1/1+1/2+\cdots+1/k$. By (2.13), one can calculate the Q-function in this case, too.

2.5. The spectrum of a quantum dot with impurity

The Green function of the Hamiltonian describing a quantum dot with impurity is given by the Krein resolvent formula

$$\mathcal{G}_z^{H(\chi)}(\xi,\phi;\xi',\phi') = \mathcal{G}_z^H(\xi,\phi;\xi',\phi') - \frac{1}{Q^H(z)-\chi}\,\mathcal{G}_z^H(\xi,0;1,0)\mathcal{G}_z^H(1,0;\xi',0)$$

(recall that, due to the rotational symmetry, $\mathcal{G}_z^H(\xi,\phi;1,0) = \mathcal{G}_z^H(\xi,0;1,0)$). The parameter $\chi := a^2\kappa \in (-\infty,\infty]$ determines the corresponding self-adjoint extension $H(\chi)$ of H. In the physical interpretation, this parameter is related to the strength of the δ interaction. Recall that the value $\chi = \infty$ corresponds to the Friedrichs extension of H representing the case with no impurity. This fact is also apparent from the Krein resolvent formula.

The unperturbed Hamiltonian $H(\infty)$ describes a harmonic oscillator on the Lobachevsky plane. As is well known (see, for example, [14]), for the confinement potential tends to infinity as $\varrho \to \infty$, the resolvent of $H(\infty)$ is compact, and the spectrum of $H(\infty)$ is discrete and semibounded. The eigenvalues of $H(\infty)$ are solutions of a scalar equation whose introduction also relies heavily on the theory of spheroidal functions. We are sceptic about the possibility of deriving an explicit formula for the eigenvalues. But the equation turned out to be convenient enough to allow for numerical solutions. A more detailed discussion jointly with a basic numerical analysis is provided in a separate paper [15].

A similar observation about the basic spectral properties (discreteness and semiboundedness) is also true for the operators $H(\chi)$ for any $\chi \in \mathbb{R}$ since, by the Krein resolvent formula, the resolvents for $H(\chi)$ and $H(\infty)$ differ by a rank one operator. Moreover, the multiplicities of eigenvalues of $H(\chi)$ and $H(\infty)$ may differ at most by ± 1 (see [9, Section 8.3]).

A more detailed and rather general analysis which is given in [1] can be carried over to our case almost literally. Denote by σ the set of poles of the function $Q^H(z)$ depending on the spectral parameter z. Note that σ is a subset of $\operatorname{spec}(H(\infty))$. Consider the equation

$$Q^H(z) = \chi. \tag{2.17}$$

Theorem 2.4. *The spectrum of $H(\chi)$ is discrete and consists of four nonintersecting parts S_1, S_2, S_3, S_4 described as follows:*

1. *S_1 is the set of all solutions to equation (2.17) which do not belong to the spectrum of $H(\infty)$. The multiplicity of all these eigenvalues in the spectrum of $H(\chi)$ equals 1.*
2. *S_2 is the set of all $\lambda \in \sigma$ that are multiple eigenvalues of $H(\infty)$. If the multiplicity of such an eigenvalue λ in $\operatorname{spec}(H(\infty))$ equals k then its multiplicity in the spectrum of $H(\chi)$ equals $k-1$.*
3. *S_3 consists of all $\lambda \in \operatorname{spec}(H(\infty))\setminus\sigma$ that are not solutions to equation (2.17). the multiplicities of such an eigenvalue λ in $\operatorname{spec}(H(\infty))$ and $\operatorname{spec}(H(\chi))$ are equal.*

4. *S_4 consists of all $\lambda \in \operatorname{spec}(H(\infty)) \setminus \sigma$ that are solutions to equation* (2.17). *If the multiplicity of such an eigenvalue λ in* $\operatorname{spec}(H(\infty))$ *equals k then its multiplicity in the spectrum of $H(\chi)$ equals $k+1$.*

Hence the eigenvalues of $H(\chi)$, $\chi \in \mathbb{R}$, different from those of the unperturbed Hamiltonian $H(\infty)$ are solutions to (2.17). As far as we see it, this equation can be solved only numerically. We have postponed a systematic numerical analysis of equation (2.17) to a subsequent work. Note that the Krein Q-function (2.15) is in fact a function of ν, and hence dependence (2.16) of the spectral parameter z on ν is fundamental. In this context, it is quite useful to know for which values of ν the spectral parameter z is real. A partial answer is given by Proposition A.1.

3. Conclusion

We have proposed a Hamiltonian describing a quantum dot in the Lobachevsky plane to which we added an impurity modeled by a δ potential. Formulas for the corresponding Q- and Green functions have been derived. Further analysis of the energy spectrum may be accomplished for some concrete values of the involved parameters (by which we mean the curvature a and the oscillator frequency ω) with the aid of numerical methods.

Appendix: Spheroidal functions

Here we follow the source [5]. Spheroidal functions are solutions to the equation

$$(1-\xi^2)\frac{\partial^2\psi}{\partial\xi^2} - 2\xi\frac{\partial\psi}{\partial\xi} + \left[\lambda + 4\theta(1-\xi^2) - \mu^2(1-\xi^2)^{-1}\right]\psi = 0, \tag{A.1}$$

where all parameters are in general complex numbers. There are two solutions that behave like ξ^ν times a single-valued function and $\xi^{-\nu-1}$ times a single-valued function at ∞. The exponent ν is a function of λ, θ, μ, and is called the characteristic exponent. Usually, it is more convenient to regard λ as a function of ν, μ and θ. We shall write $\lambda = \lambda_\nu^\mu(\theta)$. If ν or μ is an integer we denote it by n or m, respectively. The functions $\lambda_\nu^\mu(\theta)$ obey the symmetry relations

$$\lambda_\nu^\mu(\theta) = \lambda_\nu^{-\mu}(\theta) = \lambda_{-\nu-1}^\mu(\theta) = \lambda_{-\nu-1}^{-\mu}(\theta). \tag{A.2}$$

A first group of solutions (radial spheroidal functions) is obtained as expansions in series of Bessel functions,

$$S_\nu^{\mu(j)}(\xi,\theta) = (1-\xi^{-2})^{-\mu/2} s_\nu^\mu(\theta) \sum_{r=-\infty}^{\infty} a_{\nu,r}^\mu \psi_{\nu+2r}^{(j)}(2\theta^{1/2}\xi), \tag{A.3}$$

$j = 1, 2, 3, 4$, where the factors $s_\nu^\mu(\theta)$ are determined below and

$$\psi_\nu^{(1)}(\zeta) = \sqrt{\frac{\pi}{2\zeta}}\, J_{\nu+1/2}(\zeta), \quad \psi_\nu^{(2)}(\zeta) = \sqrt{\frac{\pi}{2\zeta}}\, Y_{\nu+1/2}(\zeta),$$
$$\psi_\nu^{(3)}(\zeta) = \sqrt{\frac{\pi}{2\zeta}}\, H^{(1)}_{\nu+1/2}(\zeta), \quad \psi_\nu^{(4)}(\zeta) = \sqrt{\frac{\pi}{2\zeta}}\, H^{(2)}_{\nu+1/2}(\zeta).$$

The coefficients $a^\mu_{\nu,r}(\theta)$ (denoted only a_r for the sake of simplicity) satisfy a three term recurrence relation

$$\frac{(\nu+2r-\mu)(\nu+2r-\mu-1)}{(\nu+2r-3/2)(\nu+2r-1/2)}\,\theta a_{r-1} + \frac{(\nu+2r+\mu+2)(\nu+2r+\mu+1)}{(\nu+2r+3/2)(\nu+2r+5/2)}\,\theta a_{r+1}$$
$$+ \left[\lambda_\nu^\mu(\theta) - (\nu+2r)(\nu+2r+1) + \frac{(\nu+2r)(\nu+2r+1)+\mu^2-1}{(\nu+2r-1/2)(\nu+2r+3/2)}\,2\theta\right] a_r = 0. \tag{A.4}$$

Here and in what follows we assume that $\nu+1/2$ is not an integer (to our knowledge, the omitted case is not yet fully investigated).

The coefficients $a^\mu_{\nu,r}(\theta)$ may be chosen such that

$$a^\mu_{\nu,0}(\theta) = a^\mu_{-\nu-1,0}(\theta) = a^{-\mu}_{\nu,0}(\theta),$$

and so (see (A.2))

$$a^\mu_{\nu,r}(\theta) = a^\mu_{-\nu-1,-r}(\theta) = \frac{(\nu-\mu+1)_{2r}}{(\nu+\mu+1)_{2r}}\, a^{-\mu}_{\nu,r}(\theta) \tag{A.5}$$

where $(a)_r := a(a+1)(a+2)\dots(a+r-1) = \Gamma(a+r)/\Gamma(a)$, $(a)_0 := 1$. Equation (A.4) leads to a convergent infinite continued fraction and this way one can prove that

$$\lim_{r\to\infty} \frac{r^2 a_r}{a_{r-1}} = \lim_{r\to-\infty} \frac{r^2 a_r}{a_{r+1}} = \frac{\theta}{4}. \tag{A.6}$$

From (A.6) and the asymptotic formulae for Bessel functions, it follows that (A.3) converges if $|\xi| > 1$.

If we set in (A.3)

$$s_\nu^\mu(\theta) = \left[\sum_{r=-\infty}^{\infty} (-1)^r a^\mu_{\nu,r}(\theta)\right]^{-1} \tag{A.7}$$

then

$$S_\nu^{\mu(j)}(\xi,\theta) \sim \psi_\nu^{(j)}(2\theta^{1/2}\xi), \quad \text{for } |\arg(\theta^{1/2}\xi)| < \pi \quad \text{as } \xi \to \infty.$$

We have the asymptotic forms, valid as $\xi \to \infty$,

$$S_\nu^{\mu(3)}(\xi,\theta) = \frac{1}{2}\,\theta^{-1/2}\xi^{-1}\mathrm{e}^{i(2\theta^{1/2}\xi-\nu\pi/2-\pi/2)}[1+O(|\xi|^{-1})], \quad \text{for } -\pi < \arg(\theta^{1/2}\xi) < 2\pi, \tag{A.8}$$

and

$$S_\nu^{\mu(4)}(\xi,\theta) = \frac{1}{2}\,\theta^{-1/2}\xi^{-1}\mathrm{e}^{-i(2\theta^{1/2}\xi-\nu\pi/2-\pi/2)}[1+O(|\xi|^{-1})], \quad \text{for } -2\pi < \arg(\theta^{1/2}\xi) < \pi. \tag{A.9}$$

The radial spheroidal functions satisfy the relation

$$S_\nu^{\mu(3)} = \frac{1}{i\cos(\nu\pi)}\left(S_{-\nu-1}^{\mu(1)} + i\,\mathrm{e}^{-i\pi\nu}S_\nu^{\mu(1)}\right). \tag{A.10}$$

The radial spheroidal functions are especially useful for large ξ; the larger is the ξ the better is the convergence of the expansion. To obtain solutions useful near ± 1, and even on the segment $(-1,1)$, one uses expansions in series in Legendre functions,

$$\begin{aligned} Ps_\nu^\mu(\xi,\theta) &= \sum_{r=-\infty}^{\infty}(-1)^r a_{\nu,r}^\mu(\theta)P_{\nu+2r}^\mu(\xi),\\ Qs_\nu^\mu(\xi,\theta) &= \sum_{r=-\infty}^{\infty}(-1)^r a_{\nu,r}^\mu(\theta)Q_{\nu+2r}^\mu(\xi). \end{aligned} \tag{A.11}$$

These solutions are called the angular spheroidal functions and are related to the radial spheroidal functions by the following formulae:

$$\begin{aligned} S_\nu^{\mu(1)}(\xi,\theta) &= \pi^{-1}\sin[(\nu-\mu)\pi]\mathrm{e}^{-i\pi(\nu+\mu+1)}K_\nu^\mu(\theta)Qs_{-\nu-1}^\mu(\xi,\theta),\\ S_n^{m(1)}(\xi,\theta) &= K_n^m(\theta)Ps_n^m(\xi,\theta), \end{aligned} \tag{A.12}$$

where $K_\nu^\mu(\theta)$ can be expressed as a series in coefficients $a_{\nu,r}^\mu(\theta)$, and sometimes it is called the joining factor. In more detail, for any $k\in\mathbb{Z}$ it holds true that

$$\begin{aligned} K_\nu^\mu(\theta) = \frac{1}{2}\left(\frac{\theta}{4}\right)^{\nu/2+k} & \Gamma(1+\nu-\mu+2k)\,\mathrm{e}^{(\nu+k)\pi i}s_\nu^\mu(\theta)\\ & \times \frac{\displaystyle\sum_{r=-\infty}^{k}\frac{(-1)^r a_{\nu,r}^\mu(\theta)}{(k-r)!\,\Gamma(\nu+k+r+3/2)}}{\displaystyle\sum_{r=k}^{\infty}\frac{(-1)^r a_{\nu,r}^\mu(\theta)}{(r-k)!\,\Gamma(1/2-\nu-k-r)}}. \end{aligned}$$

Proposition A.1. *Let $\nu,\theta\in\mathbb{R}$ and set $\mu=0$. Then $\lambda_\nu^0(\theta)\in\mathbb{R}$.*

Proof. To simplify the notation we denote, in (A.4),

$$\begin{aligned} \alpha_r^{\mu,\nu}(\theta) &= \frac{(\nu+2r+\mu+2)(\nu+2r+\mu+1)}{(\nu+2r+3/2)(\nu+2r+5/2)}\,\theta,\\ \beta_r^{\mu,\nu}(\theta) &= -(\nu+2r)(\nu+2r+1) + \frac{(\nu+2r)(\nu+2r+1)+\mu^2-1}{(\nu+2r-1/2)(\nu+2r+3/2)}\,2\theta,\\ \gamma_r^{\mu,\nu}(\theta) &= \frac{(\nu+2r-\mu)(\nu+2r-\mu-1)}{(\nu+2r-3/2)(\nu+2r-1/2)}\,\theta. \end{aligned}$$

The resulting formula may be written in the matrix form,

$$\begin{pmatrix} \ddots & & & & & \\ & \gamma_{-1} & \beta_{-1} & \alpha_{-1} & & \\ & & \gamma_0 & \beta_0 & \alpha_0 & \\ & & & \gamma_1 & \beta_1 & \alpha_1 \\ & & & & & \ddots \end{pmatrix} \begin{pmatrix} \vdots \\ a_{-1} \\ a_0 \\ a_1 \\ \vdots \end{pmatrix} = -\lambda \begin{pmatrix} \vdots \\ a_{-1} \\ a_0 \\ a_1 \\ \vdots \end{pmatrix} \tag{A.13}$$

where we have omitted the fixed indices.

As one can see,

$$\gamma_{r+1}^{0,\nu}(\theta) = \frac{\nu + 2r + 5/2}{\nu + 2r + 1/2}\, \alpha_r^{0,\nu}(\theta)$$

and so

$$\frac{\nu + 2r + 1/2}{\nu + 2r - 3/2}\alpha_{r-1}^{0,\nu}(\theta)a_{\nu,r-1}^{0}(\theta) + \beta_r^{0,\nu}(\theta)a_{\nu,r}^{0}(\theta) + \alpha_r^{0,\nu}(\theta)a_{\nu,r+1}^{0}(\theta) = -\lambda_\nu^0(\theta)a_{\nu,r}^0(\theta).$$

Substitution $a_{\nu,r}^0 = L_r(\nu)\tilde{a}_{\nu,r}^0$, where $L_r(\nu)$ are non-zero constants, yields

$$\frac{\nu+2r+1/2}{\nu+2r-3/2}\alpha_{r-1}^{0,\nu}(\theta)\tilde{a}_{\nu,r-1}^{0}(\theta)\frac{L_{r-1}(\nu)}{L_r(\nu)} + \beta_r^{0,\nu}(\theta)\tilde{a}_{\nu,r}^{0}(\theta) + \alpha_r^{0,\nu}(\theta)\tilde{a}_{\nu,r+1}^{0}(\theta)\frac{L_{r+1}(\nu)}{L_r(\nu)}$$
$$= -\lambda_\nu^0(\theta)\tilde{a}_{\nu,r}^0(\theta).$$

We require the matrix in (A.13) to be symmetric in the new coordinates $\{\tilde{a}_r\}$. This implies that

$$\frac{\nu + 2r + 1/2}{\nu + 2r - 3/2}\frac{L_{r-1}(\nu)}{L_r(\nu)} = \frac{L_r(\nu)}{L_{r-1}(\nu)}.$$

For $r \notin (-\nu/2 - 1/4, -\nu/2 + 3/4)$, the solution is $L_r(\nu) = \sqrt{|\nu + 2r + 1/2|}$. For $r_0 \equiv r \in (-\nu/2 - 1/4, -\nu/2 + 3/4)$, there is no real solution and so we set $L_{r_0}(\nu) = \sqrt{|\nu + 2r_0 + 1/2|}$ and make another transformation of coordinates:

$$\tilde{\tilde{a}}_r = \begin{cases} -\tilde{a}_r & \text{for } r = r_0 - (2k-1),\ k \in \mathbb{N}, \\ \tilde{a}_r & \text{for all other } r. \end{cases}$$

Relation (A.13) can be viewed as an eigenvalue equation with a symmetric matrix in the coordinate system $\{\tilde{\tilde{a}}_k\}$, hence $\lambda_\nu^0(\theta)$ must be real. □

Acknowledgments

The authors wish to acknowledge gratefully partial support from the following grants: grant No. 201/05/0857 of the Grant Agency of Czech Republic (P. Š.) and grant No. LC06002 of the Ministry of Education of Czech Republic (M. T.).

References

[1] J. Brüning, V. Geyler, and I. Lobanov, *Spectral Properties of a Short-range Impurity in a Quantum Dot*, J. Math. Phys. **46** (2004), 1267–1290.

[2] A. Comtet, *On the Landau levels on the hyperbolic plane*, Ann. Physics **173** (1987), 185–209.

[3] M. Antoine, A. Comtet and S. Ouvry, *Scattering on a hyperbolic torus in a constant magnetic field*, J. Phys. A: Math. Gen. **23** (1990), 3699–3710.

[4] O. Lisovyy, *Aharonov-Bohm effect on the Poincaré disk*, J. Math. Phys. **48** (2007), 052112.

[5] H. Bateman and A. Erdélyi, *Higher Transcendental Functions III.* McGraw-Hill Book Company, 1955.

[6] J. Meixner and F.V. Schäfke, *Mathieusche Funktionen und Sphäroidfunktionen.* Springer-Verlag, 1954.

[7] M.F. Rañada and M. Santader, *On Harmonic Oscillators on the Two-dimensional Sphere S^2 and the Hyperbolic Plane H^2*, J. Math. Phys. **43** (2002), 431–451.

[8] N. Dunford and J.T. Schwartz, *Linear Operators. Part II: Spectral theory. Self Adjoint Operators in Hilbert Space.* Wiley-Interscience Publication, 1988.

[9] J. Weidmann, *Linear Operators in Hilbert Spaces.* Springer, 1980.

[10] S. Albeverio, F. Gesztesy, R. Høegh-Krohn and H. Holden, *Solvable Models in Quantum Mechanics.* Springer-Verlag, 1988.

[11] J. Brüning, V. Geyler, and K. Pankrashkin, *Spectra of Self-adjoint Extensions and Applications to Solvable Schrödinger Operators*, arXiv:math-ph/0611088 (2007).

[12] M. Reed, and B. Simon, *Methods of Modern Mathematical Physics II.* Academic Press, 1975.

[13] J. Brüning, V. Geyler, and K. Pankrashkin, *On-diagonal Singularities of the Green Function for Schrödinger Operators*, J. Math. Phys. **46** (2005), 113508.

[14] F.A. Berezin, and M.A. Shubin, *The Schrödinger Equation.* Kluwer Academic Publishers, 1991.

[15] P. Šťovíček, and M. Tušek, *On the Harmonic Oscillator on the Lobachevsky Plane*, Russian J. Math. Phys. **14** (2007), 401–405.

V. Geyler
Department of Mathematics
Mordovian State University
Saransk, Russia

P. Šťovíček and M. Tušek
Department of Mathematics
Faculty of Nuclear Sciences
Czech Technical University
Prague, Czech Republic
e-mail: stovicek@kmalpha.fjfi.cvut.cz
e-mail: tusekmat@km1.fjfi.cvut.cz

Operator Theory:
Advances and Applications, Vol. 188, 149–173

One-dimensional Perturbations, Asymptotic Expansions, and Spectral Gaps

Seppo Hassi, Adrian Sandovici, Henk de Snoo and Henrik Winkler

Abstract. Let S be a closed symmetric operator or relation with defect numbers $(1,1)$ and let A be a self-adjoint extension of S. The self-adjoint extensions $A(\tau)$, $\tau \in \mathbb{R} \cup \{\infty\}$, of S, when parametrized by means of Kreĭn's formula, can be seen as one-dimensional (graph) perturbations of A. The spectral properties of the self-adjoint extension $A(\tau)$ of (the completely non-self-adjoint part of) S can be determined via the analytic properties of the Weyl function (Q-function) $Q_\tau(z)$ corresponding to S and $A(\tau)$, and conversely. In order to study the limiting properties of these functions at spectral points, local analogs of the Kac-Donoghue classes of Nevanlinna functions are introduced, giving rise to asymptotic expansions at real points. In the case where the self-adjoint extension A has a (maximal) gap in its spectrum, all the perturbations $A(\tau)$ have the same gap in their spectrum with the possible exception of an isolated eigenvalue $\lambda(\tau)$, $\tau \in \mathbb{R} \cup \{\infty\}$. By means of the Weyl function $Q_\tau(z)$ the (analytic) properties of this eigenvalue are established.

Mathematics Subject Classification (2000). Primary 47A55, 47B25; Secondary 47A11, 47A57.

Keywords. Boundary triplet, Weyl function, one-dimensional perturbation, Nevanlinna function, spectral measure, moment, spectral gap, asymptotic expansion.

1. Introduction

Let S be a closed symmetric operator or relation (multi-valued operator) in a Hilbert space $\mathfrak{H}$ and assume that S has defect numbers $(1,1)$. The self-adjoint extensions $A(\tau)$ of S are in one-to-one correspondence with $\tau \in \mathbb{R} \cup \{\infty\}$ via Kreĭn's resolvent formula:

$$(A(\tau)-z)^{-1} = (A-z)^{-1} - \gamma(z)\,\frac{1}{Q(z)+1/\tau}\,(\cdot,\gamma(\bar{z})), \quad z \in \rho(A)\cap\rho(A(\tau)), \tag{1.1}$$

with the understanding that $A(0) = A$. Here $\gamma(z)$ and $Q(z)$ stand for the γ-field and the Weyl function (or Q-function) corresponding to S and A, which are holomorphic on $\rho(A)$, cf. [10, 11, 8, 14, 27]. Kreĭn's formula shows that the resolvent of the self-adjoint extension $A(\tau)$ can be interpreted as a one-dimensional analytic perturbation of the resolvent of the self-adjoint extension A. Another interpretation is obtained by considering the graphs of the self-adjoint extensions $A(\tau)$ as one-dimensional *perturbations of the graph* of A; see [19]. Recall that if S is densely defined then S is an operator and all self-adjoint extensions of S are densely defined operators. However, if S is a relation with a nontrivial multi-valued part then all self-adjoint extensions of S are relations with a nontrivial multi-valued part. Finally, if S is a nondensely defined operator then all but one of its self-adjoint extensions are operators, and the remaining relation extension is called the generalized Friedrichs extension, see [16, 19, 20]. If S is a nondensely defined operator and A is a self-adjoint operator extension of S, then S is a one-dimensional domain restriction of A, i.e., there exists a nontrivial element $\omega \in \mathfrak{H}$ such that $\operatorname{dom} S = \{\, f \in \operatorname{dom} A : (f, \omega) = 0 \,\}$. In this case one can choose $Q(z) = ((A - z)^{-1}\omega, \omega)$ and for $\tau \in \mathbb{R}$ all self-adjoint extensions $A(\tau)$ in (1.1) are operators which can be seen as rank one perturbations of A:

$$A(\tau) = A + \tau(\cdot, \omega)\omega, \quad \tau \in \mathbb{R}. \tag{1.2}$$

The remaining self-adjoint extension $A(\infty)$ in (1.1) is the generalized Friedrichs extension; it is not an operator, but a relation whose multi-valued part $\operatorname{mul} A(\infty)$ is one-dimensional:

$$A(\infty) = S \,\widehat{+}\, (\{0\} \times \operatorname{span}\{\omega\}), \tag{1.3}$$

where $\widehat{+}$ stands for the componentwise sum in the Cartesian product $\mathfrak{H} \times \mathfrak{H}$.

In the general case, when S is not necessarily densely defined, the spectrum of the extension $A(\tau)$ in (1.1) is related to the spectrum of $A (= A(0))$ and this relation is well understood. In particular, the essential spectrum of $A(\tau)$ does not depend on $\tau \in \mathbb{R} \cup \{\infty\}$ (cf. [12]); this fact is independent of S being densely defined or not. The spectral properties of the self-adjoint extension $A(\tau)$ of (the completely non-self-adjoint part of) S can be determined via the analytic properties of the Weyl function (Q-function) $Q_\tau(z)$ corresponding to S and $A(\tau)$, and conversely. In order to study the limiting properties of these functions at spectral points, local analogs of the Kac-Donoghue classes of Nevanlinna functions are introduced, giving rise to asymptotic expansions at real points. These expansions reflect the smoothness of the Weyl function in terms of nontangential derivatives.

The self-adjoint extension A is said to have a *spectral gap* when there is a maximal interval in $\mathbb{R}$ belonging to the resolvent set $\rho(A)$ of A. It is not difficult to see that such a gap also belongs to the resolvent set of $A(\tau)$, $\tau \in \mathbb{R} \cup \{\infty\}$, with the possible exception of one eigenvalue $\lambda(\tau)$, and this eigenvalue is a real zero of the function $Q(z) + 1/\tau = 0$, cf. (1.1). The holomorphic dependence of the eigenvalue $\lambda(\tau)$, which is lying in the spectral gap, on the perturbation parameter τ is described by means of the Weyl function $Q_\tau(z)$. In particular, it is shown that as a function of the perturbation parameter $\tau \in \mathbb{R} \cup \{\infty\}$ the isolated eigenvalue $\lambda(\tau)$

has a power series expansion inside the interval. These power series expansions are derived directly from the associated Weyl function $Q_\tau(z)$.

The contents of this paper are as follows. In Section 2 some facts concerning self-adjoint extensions of a symmetric relation with defect numbers $(1,1)$ are presented. In particular, the notions of γ-field and Weyl function (or Q-function) are introduced in a form suitable for the interpretations in terms of graph perturbations. Some necessary information about Nevanlinna functions is treated in Section 3. Asymptotic expansions of Nevanlinna functions, either at ∞ or at a finite real point, are considered. This also involves local versions of the so-called Kac-Donoghue classes of Nevanlinna functions. Section 4 contains information about the gaps of Nevanlinna functions. The location of the eigenvalues of $A(\tau)$ in terms of (the boundary values of) the Weyl function $Q(z)$ is given in Section 5. The analytic behavior of the eigenvalues of the self-adjoint extensions $A(\tau)$ of S in spectral gaps as a function of $\tau \in \mathbb{R} \cup \{\infty\}$ is taken up in Section 6. Further results and applications will be given in [18], where the asymptotics of the eigenvalue $\lambda(\tau)$ will be studied near the endpoints of the spectral gap.

2. Self-adjoint extensions of symmetric relations

2.1. Weyl functions and γ-fields

A linear relation T in a Hilbert space $\mathfrak{H}$ is a linear subspace T of the Cartesian product $\mathfrak{H} \times \mathfrak{H}$. The domain $\operatorname{dom} T$, the range $\operatorname{ran} T$, and the kernel $\ker T$, are defined as usual. The multi-valued part $\operatorname{mul} T$ is defined by $\operatorname{mul} T = \{f' : \{0, f'\} \in T\}$. A linear relation T is (the graph of) an operator if and only if $\operatorname{mul} T = \{0\}$. The adjoint T^* of a linear relation T is defined by

$$T^* = \{\, \{f, f'\} \in \mathfrak{H} \times \mathfrak{H} : \langle \{f, f'\}, \{h, h'\} \rangle = 0, \ \{h, h'\} \in T \,\},$$

as a closed linear relation (which is the graph of an operator if and only if T is densely defined). Here and in the following the notation

$$\langle \{f, f'\}, \{h, h'\} \rangle = (f', h) - (f, h'), \quad \{f, f'\}, \{h, h'\} \in \mathfrak{H} \times \mathfrak{H},$$

will be used. Furthermore, the notation $\widehat{f} = \{f, f'\}$ will be used.

Next some basic notions and facts associated with *boundary triplets* are recalled from [10, 11, 14]. Let S be a closed symmetric relation in a Hilbert space and assume that S has defect numbers $(1,1)$. Associated with S is a boundary triplet $\Pi = \{\mathbb{C}, \Gamma_0, \Gamma_1\}$. This means that Γ_0 and Γ_1 are mappings from S^* to $\mathbb{C}$ such that the mapping mapping $\{\Gamma_0, \Gamma_1\} : S^* \to \mathbb{C} \times \mathbb{C}$ is surjective and the following analog of Green's identity holds:

$$\langle \widehat{f}, \widehat{g} \rangle = \Gamma_1 \widehat{f}\, \overline{\Gamma_0 \widehat{g}} - \Gamma_0 \widehat{f}\, \overline{\Gamma_1 \widehat{g}}, \quad \widehat{f} = \{f, f'\}, \quad \widehat{g} = \{g, g'\} \in S^*. \tag{2.1}$$

Then $A_0 = \ker \Gamma_0$ and $A_1 = \ker \Gamma_1$ are self-adjoint extensions of S, and $S = \ker \Gamma_0 \cap \ker \Gamma_1$. The eigenspaces of S^*, also called the *defect subspaces* of S, are denoted by

$$\widehat{\mathfrak{N}}_z(S^*) = \{\, \{f, zf\} : f \in \mathfrak{N}_z(S^*)\,\}, \quad \mathfrak{N}_z(S^*) = \ker\,(S^* - z).$$

Associated with a boundary triplet $\Pi = \{\mathbb{C}, \Gamma_0, \Gamma_1\}$ are two functions: the *γ-field* $\gamma(z)$ defined by

$$\gamma(z) = \{\, \{\Gamma_0 \widehat{f}_z, f_z\} : \widehat{f}_z \in \widehat{\mathfrak{N}}_z(S^*)\,\}, \quad z \in \rho(A_0), \tag{2.2}$$

and the *Weyl function* $Q(z)$, defined by

$$Q(z) = \{\, \{\Gamma_0 \widehat{f}_z, \Gamma_1 \widehat{f}_z\} : \widehat{f}_z \in \widehat{\mathfrak{N}}_z(S^*)\,\}, \quad z \in \rho(A_0). \tag{2.3}$$

Both notions are graphs of operators in $\mathbb{C}$ (by identifying the numbers in $\mathbb{C} \cup \{\infty\}$ with the linear relations in $\mathbb{C} \times \mathbb{C}$):

$$\gamma(z) = \frac{1}{\Gamma_0 \widehat{f}_z} f_z, \quad Q(z) = \frac{\Gamma_1 \widehat{f}_z}{\Gamma_0 \widehat{f}_z}, \quad z \in \rho(A_0).$$

The γ-field satisfies the identity

$$\gamma(z) - \gamma(w) = (z - w)(A_0 - z)^{-1}\gamma(w), \quad z, w \in \rho(A_0), \tag{2.4}$$

and the relation between the Weyl function $Q(z)$ and the γ-field $\gamma(z)$ is given by

$$\frac{Q(z) - Q(w)^*}{z - \bar{w}} = (\gamma(z), \gamma(w))\,, \quad z, w \in \rho(A_0). \tag{2.5}$$

The identity (2.5) implies that $Q(z)$ is the *Q-function* of the pair $\{S, A\}$; cf. [10, 11]. The functions $\gamma(z)$ and $Q(z)$ are holomorphic on $\rho(A_0)$. Moreover, it follows from (2.5) that the function $Q(z)$ belongs to the class $\mathbf{N}$ of *Nevanlinna functions*, i.e., it is holomorphic on $\mathbb{C} \setminus \mathbb{R}$ and satisfies $Q(z)^* = Q(\bar{z})$ and $(\operatorname{Im} z)(\operatorname{Im} Q(z)) \geq 0$. In addition $Q(z)$ is strict, i.e., $\operatorname{Im} Q(z) > 0$ for all $z \in \mathbb{C} \setminus \mathbb{R}$. In general, the Weyl function $Q(z)$ determines, up to unitary isomorphisms, a model for the symmetric operator S and its self-adjoint extension A; cf. [24, 27].

The self-adjoint extensions of S are in one-to-one correspondence with $\mathbb{R} \cup \{\infty\}$ via the formula

$$A(\tau) = \ker\,(\Gamma_0 + \tau\Gamma_1), \quad \tau \in \mathbb{R}, \quad A(\infty) = \ker\,\Gamma_1, \tag{2.6}$$

or, equivalently, via Kreĭn's resolvent formula (1.1). Note that $A(0) = A_0$ and $A(\infty) = A_1$. The boundary triplets associated with the self-adjoint extensions can be expressed via certain J-unitary transforms of the boundary triplet $\Pi = \{\mathbb{C}, \Gamma_0, \Gamma_1\}$; see [11]. The following statement is formulated along the lines of [7, Proposition 5.2].

Proposition 2.1. *Let $\Pi = \{\mathbb{C}, \Gamma_0, \Gamma_1\}$ be a boundary triplet for S^* with the γ-field $\gamma(z)$ and the Weyl function $M(z)$. Then:*

(i) $\Pi_\tau = \{\mathbb{C}, \Gamma_0^\tau, \Gamma_1^\tau\}$, *where*

$$\Gamma_0^\tau = \frac{1}{\sqrt{\tau^2 + 1}}(\Gamma_0 + \tau\Gamma_1), \quad \Gamma_1^\tau = \frac{1}{\sqrt{\tau^2 + 1}}(\Gamma_1 - \tau\Gamma_0) \tag{2.7}$$

for $\tau \in \mathbb{R}$ *and* $\Gamma_0^\infty = \Gamma_1$, $\Gamma_1^\infty = -\Gamma_0$ *for* $\tau = \infty$, *defines a boundary triplet for* S^* *with* $\ker \Gamma_0^\tau = A(\tau)$*;*

(ii) *the corresponding* γ*-field* $\gamma_\tau(z)$ *and the Weyl function* $Q_\tau(z)$ *are given by*

$$\gamma_\tau(z) = \frac{\sqrt{\tau^2+1}}{1+\tau Q(z)}\gamma(z), \quad z \in \mathbb{C}\setminus\mathbb{R}, \quad \tau \in \mathbb{R}\cup\{\infty\}, \tag{2.8}$$

and

$$Q_\tau(z) = \frac{Q(z)-\tau}{1+\tau Q(z)}, \quad z \in \mathbb{C}\setminus\mathbb{R}, \quad \tau \in \mathbb{R}\cup\{\infty\}. \tag{2.9}$$

Proof. (i) Consider the transforms

$$W_\tau = \frac{1}{\sqrt{\tau^2+1}}\begin{pmatrix} 1 & \tau \\ -\tau & 1 \end{pmatrix}, \quad \tau \in \mathbb{R}, \qquad W_\infty = \begin{pmatrix} 0 & 1 \\ -1 & 0 \end{pmatrix}, \quad \tau = \infty.$$

Clearly W_τ, $\tau \in \mathbb{R}\cup\{\infty\}$, is J-unitary with respect to

$$J = \begin{pmatrix} 0 & -i \\ i & 0 \end{pmatrix},$$

which immediately implies that $\Gamma^\tau = W_\tau \Gamma$, where $\Gamma^\tau = \operatorname{col}(\Gamma_0^\tau, \Gamma_1^\tau)$ has the form (2.7), is surjective and satisfies Green's identity (2.1). Thus Γ^τ, $\tau \in \mathbb{R}\cup\{\infty\}$, defines a boundary triplet for S^* with $\ker \Gamma_0^\tau = \ker\,(\Gamma_0 + \tau\Gamma_1) = A(\tau)$.

(ii) It is straightforward to check that for Γ^τ, $\tau \in \mathbb{R}$, the definitions in (2.2) and (2.3) imply

$$\{\Gamma_0^\tau \widehat{f}_z, f_z\} = \{(\Gamma_0 + \tau\Gamma_1)\widehat{f}_z, \sqrt{\tau^2+1}\, f_z\} = \{(1+\tau Q(z))\Gamma_0 \widehat{f}_z, \sqrt{\tau^2+1}\, f_z\},$$

where $\widehat{f}_z \in \widehat{\mathfrak{N}}_z(S^*)$. Similarly one treats the case $\tau = \infty$ and this leads to (2.8). The formula (2.9) is obtained analogously. □

Note that the identity (2.9) can be rewritten as follows:

$$Q_\tau(z) = \frac{1}{\tau} - \frac{1+\tau^2}{\tau^2}\frac{1}{Q(z)+1/\tau} = -\tau + \frac{1+\tau^2}{1+\tau Q(z)}Q(z), \quad \tau \neq 0. \tag{2.10}$$

Clearly, the linear fractional transform $Q_\tau(z)$ satisfies

$$\frac{Q_\tau(z) - \overline{Q_\tau(w)}}{z - \bar{w}} = (\gamma_\tau(z), \gamma_\tau(w)), \quad z, w \in \mathbb{C}\setminus\mathbb{R}. \tag{2.11}$$

The interpretation of (2.8) and (2.9) for $\tau = \infty$ is that $\gamma_\infty(z) = \gamma(z)/Q(z)$ and $Q_\infty(z) = -1/Q(z)$, which agrees with interpretation of (2.11) for $\tau = \infty$. Whereas Kreĭn's formula (1.1) completely describes the self-adjoint extensions $A(\tau)$, the Weyl function $Q_\tau(z)$ in (2.9) only describes the *completely non-self-adjoint part* S_s of S and the corresponding reduction $A(\tau)_s$ of $A(\tau)$ in the Hilbert space $\mathfrak{H}_s = \overline{\operatorname{span}}\,\{\gamma(z) : z \in \mathbb{C}\setminus\mathbb{R}\}$.

2.2. Self-adjoint extensions as one-dimensional graph perturbations

Following [19] a closed symmetric relation S is a one-dimensional restriction of a self-adjoint relation A in a Hilbert space $\mathfrak{H}$ if and only if there exists a one-dimensional subspace Z, $Z \not\subset A$, of the Cartesian product $\mathfrak{H} \times \mathfrak{H}$ such that

$$S = A \cap Z^*. \tag{2.12}$$

This defines Z uniquely modulo A. It follows from the definition (2.12) that

$$S^* = A \,\widehat{+}\, Z, \tag{2.13}$$

which is a direct sum decomposition. There is always a symmetric relation $\widetilde{Z}$ (i.e., $\widetilde{Z} \subset \widetilde{Z}^*$) equal to Z modulo A. Hence, without loss of generality Z will be chosen to be symmetric. Assume that Z is spanned by the pair $\{\varphi, \varphi'\} \in \mathfrak{H} \times \mathfrak{H}$. Then the assumptions $Z \not\subset A$ and $Z \subset Z^*$ are equivalent to $\{\varphi, \varphi'\} \notin A$ and $(\varphi', \varphi) \in \mathbb{R}$, respectively, and one can rewrite (2.12) in the form

$$S = \{\{h, h'\} \in A : \langle\{h, h'\}, \{\varphi, \varphi'\}\rangle = 0\}. \tag{2.14}$$

This leads to some explicit formulas for the associated γ-fields and Weyl functions, which were established in [19]. In the next proposition these results are reformulated using the terminology of boundary triplets.

Proposition 2.2. *Let A be a self-adjoint relation in the Hilbert space $\mathfrak{H}$ and let the restriction S of A be defined by* (2.14) *with a symmetric $Z = \operatorname{span}\{\varphi, \varphi'\} \subset \mathfrak{H} \times \mathfrak{H}$, $\{\varphi, \varphi'\} \notin A$, i.e., with $(\varphi', \varphi) \in \mathbb{R}$. Then:*

(i) *the adjoint S^* of S is given by*

$$S^* = \big\{\{f, f'\} \in \mathfrak{H} \times \mathfrak{H} : \{f, f'\} = \{h, h'\} \,\widehat{+}\, c\{\varphi, \varphi'\},\ \{h, h'\} \in A,\ c \in \mathbb{C}\big\};$$

(ii) *a boundary triplet $\Pi = \{\mathbb{C}, \Gamma_0, \Gamma_1\}$ for S^* is determined by*

$$\Gamma_0\{f, f'\} = c, \quad \Gamma_1\{f, f'\} = \langle\{f, f'\}, \{\varphi, \varphi'\}\rangle; \tag{2.15}$$

(iii) *the corresponding γ-field $\gamma(z)$ and the Weyl function $Q(z)$ are given by*

$$\gamma(z) = (A - z)^{-1}(z\varphi - \varphi') + \varphi, \tag{2.16}$$

$$Q(z) = ((A - z)^{-1}(z\varphi - \varphi') + \varphi, \bar{z}\varphi - \varphi'); \tag{2.17}$$

(iv) *the self-adjoint extensions $A(\tau)$ of S in* (1.1) *and* (2.6) *have the following representations: for $\tau \in \mathbb{R}$ one has*

$$A(\tau) = \{\,\{f, f'\} - \tau\,\langle\{f, f'\}, \{\varphi, \varphi'\}\rangle\,\{\varphi, \varphi'\} : \{f, f'\} \in A\,\}, \tag{2.18}$$

and for $\tau = \infty$ one has

$$A(\infty) = S \,\widehat{+}\, \operatorname{span}\{\varphi, \varphi'\}. \tag{2.19}$$

Proof. (i) It follows from (2.13) that for each $\{f,f'\} \in S^*$ there exist a unique element $\{h,h'\} \in A$ and a unique number $c \in \mathbb{C}$, such that

$$\{f,f'\} = \{h,h'\} \,\widehat{+}\, c\{\varphi,\varphi'\},$$

which gives the formula for S^*.

(ii) The mappings Γ_0 and Γ_1 in (2.15) are well defined on S^*. Moreover, with $\widehat{f} = \{f,f'\} \in S^*$ and $\widehat{g} = \{g,g'\} = \{k,k'\} \,\widehat{+}\, d\{\varphi,\varphi'\} \in S^*$ a simple calculation shows that

$$\begin{aligned}\langle \widehat{f}, \widehat{g} \rangle &= \bar{d}\langle \{h,h'\},\{\varphi,\varphi'\}\rangle - c\overline{\langle \{k,k'\},\{\varphi,\varphi'\}\rangle} \\ &= \bar{d}\langle \{f,f'\},\{\varphi,\varphi'\}\rangle - c\overline{\langle \{g,g'\},\{\varphi,\varphi'\}\rangle},\end{aligned}$$

where the second identity is due to $\langle\{\varphi,\varphi'\},\{\varphi,\varphi'\}\rangle = 0$ by the symmetry of Z. Hence, the Green's identity (2.1) is satisfied. Surjectivity of the mapping $\{\Gamma_0,\Gamma_1\}: S^* \to \mathbb{C}\times\mathbb{C}$ is clear. Thus $\Pi = \{\mathbb{C},\Gamma_0,\Gamma_1\}$ is a boundary triplet for S^*.

(iii) A simple calculation shows that the defect space ker (S^*-z) is spanned by the function

$$f_z = (A-z)^{-1}(z\varphi - \varphi') + \varphi, \quad z \in \rho(A).$$

Moreover,

$$\widehat{f}_z = \{f_z, zf_z\} = \{(A-z)^{-1}(z\varphi-\varphi'), (I + z(A-z)^{-1})(z\varphi - \varphi') + \{\varphi,\varphi'\}.$$

Now, by applying the definitions of the boundary mappings in (2.15) one sees immediately that

$$\Gamma_0 \widehat{f}_z = 1, \quad \Gamma_1 \widehat{f}_z = \langle \{f_z, zf_z\},\{\varphi,\varphi'\}\rangle = z(f_z,\varphi) - (f_z,\varphi') = (f_z, \bar{z}\varphi - \varphi').$$

By applying the definitions of the γ-field and the Weyl function in (2.2) and (2.3) one obtains the formulas for $\gamma(z)$ and $Q(z)$ in (2.16) and (2.17).

(iv) The self-adjoint extensions $A(\tau) = \ker\,(\Gamma_0 + \tau\Gamma_1)$ of S in (2.6) (and in (1.1)) can be expressed in terms of the boundary mappings Γ_0 and Γ_1 in (2.15) as follows: $\{f,f'\} \in A(\tau)$ if and only if $c + \tau\langle\{f,f'\},\{\varphi,\varphi'\}\rangle = 0$ for $\tau \in \mathbb{R}$ and for $\tau = \infty$ one has the condition $\langle\{f,f'\},\{\varphi,\varphi'\}\rangle = 0$. This leads to (2.18) and (2.19). □

Note that $A = A_0 = \ker \Gamma_0$ and that $S = \ker \Gamma_0 \cap \ker \Gamma_1$. Proposition 2.2 shows that each self-adjoint extension of S is a one-dimensional graph perturbation of the self-adjoint extension A; see also [19, Theorem 2.4]. The function $Q_\tau(z)$ in (2.9) can also be expressed in terms of the pair $\{\varphi,\varphi'\}$.

Corollary 2.3. *Let A, $Z = \operatorname{span}\{\varphi,\varphi'\}$, and S be as in Proposition 2.2. Then the function $Q_\tau(z)$, $z \in \mathbb{C}\setminus\mathbb{R}$, in (2.9) with $\tau \in \mathbb{R}$ admits the representation*

$$Q_\tau(z) = -\tau + (\tau^2+1)((A(\tau)-z)^{-1}(z\varphi-\varphi') + \varphi, \bar{z}\varphi - \varphi'), \tag{2.20}$$

while for $\tau = \infty$ one has

$$Q_\infty(z) = \left((A(\infty)-z)^{-1}(z^2\varphi - z\varphi') + \varphi', \gamma(\bar{z})/Q(\bar{z})^2\right). \tag{2.21}$$

Proof. It follows from the Kreĭn's formula (1.1) that for all $\tau \in \mathbb{R} \cup \{\infty\}$

$$
\begin{aligned}
\left((A(\tau)-z)^{-1}(z\varphi-\varphi')+\varphi, \bar{z}\varphi-\varphi'\right) &= Q(z) - \frac{(\gamma(z), \bar{z}\varphi-\varphi')(z\varphi-\varphi', \gamma(\bar{z}))}{Q(z)+1/\tau} \\
&= Q(z) - \frac{Q(z)\overline{Q(\bar{z})}}{Q(z)+1/\tau} \\
&= \frac{Q(z)}{1+\tau Q(z)}.
\end{aligned}
$$

It remains to apply (2.10) to get the formula (2.20) with $\tau \in \mathbb{R}$. For $\tau = \infty$ one has $\{\varphi, \varphi'\} \in A(\infty)$, see (2.19), so that $(A(\infty)-z)^{-1}(z\varphi-\varphi')+\varphi = 0$, which yields

$$
\begin{aligned}
\left((A(\infty)-z)^{-1}(z^2\varphi - z\varphi') + \varphi', \gamma(\bar{z})/Q(\bar{z})^2\right) &= \left(\varphi' - z\varphi, \gamma(\bar{z})\right)/Q(z)^2 \\
&= -1/Q(z).
\end{aligned}
$$

This gives (2.21) since $Q_\infty(z) = -1/Q(z)$. □

Let A be a self-adjoint operator in a Hilbert space $\mathfrak{H}$ and assume that in the pair $\{\varphi, \varphi'\} \in Z$ the element φ belongs to $\operatorname{dom} A$. Then the element $\{0, \varphi' - A\varphi\}$ is equal to $\{\varphi, \varphi'\}$ modulo A. In this case the one-dimensional restriction S of A is given by

$$
S = \{\, \{f, f'\} \in A : \ (f, \omega) = 0 \,\}
$$

with $\omega = A\varphi - \varphi'$. Obviously, the operator S is not densely defined. The γ-field and the Weyl function are now given by

$$
\gamma(z) = (A-z)^{-1}\omega, \quad Q(z) = ((A-z)^{-1}\omega, \omega) - (A\varphi, \varphi) + (\varphi', \varphi),
$$

where $-(A\varphi, \varphi) + (\varphi', \varphi) \in \mathbb{R}$. In this case the formulas (2.18) and (2.19) reduce to (1.2) and (1.3); cf. [19]. In general, the pair of elements $\{\varphi, \varphi'\}$ in the symmetric restriction (2.14) represents an element in the space of generalized elements associated with A, see [21].

3. Nevanlinna functions and their asymptotic expansions

3.1. Integral representations

It is well known that each Nevanlinna function $Q(z) \in \mathbf{N}$ has an integral representation of the form

$$
Q(z) = bz + a + \int_{\mathbb{R}} \left(\frac{1}{t-z} - \frac{t}{t^2+1} \right) d\sigma(t), \quad z \in \mathbb{C} \setminus \mathbb{R}, \tag{3.1}
$$

with constants $a \in \mathbb{R}$ and $b \geq 0$, and a nonnegative measure σ, called spectral measure of $Q(z)$, which satisfies

$$
\int_{\mathbb{R}} \frac{d\sigma(t)}{t^2+1} < \infty,
$$

cf. [13, 23]. Recall for any point $\gamma \in \mathbb{R}$ the identity

$$\lim_{z \widehat{\rightarrow} \gamma} (\gamma - z)Q(z) = \sigma(\{\gamma\}),$$

where the notation $z \widehat{\rightarrow} \gamma$ means that z approaches γ in a sector which is bounded away from the real line $\mathbb{R}$. If $\sigma(\{\gamma\}) > 0$, then this may be interpreted as the pointmass of σ at the point γ. Similarly, one has

$$\lim_{z \widehat{\rightarrow} \infty} \frac{Q(z)}{z} = b, \tag{3.2}$$

which, when $b > 0$, may be interpreted as the pointmass of σ at ∞. Furthermore, recall the Stieltjes inversion formula (for details see [13, p. 24]): if $\sigma(\{\alpha\}) = \sigma(\{\beta\}) = 0$, then

$$\sigma(\beta) - \sigma(\alpha) = \lim_{y \downarrow 0} \pi^{-1} \int_\alpha^\beta \operatorname{Im} Q(t + iy) dt$$

In particular, Fatou's lemma (cf. [13, Theorem IV.I]) states that

$$\sigma'(t) = \pi^{-1} \lim_{y \downarrow 0} \operatorname{Im} Q(t + iy),$$

whenever $\sigma'(t)$ exists (which is the case almost everywhere).

3.2. Global growth conditions

A Nevanlinna function $Q(z)$ is said to belong to the class $\mathbf{N}_r$ with $0 \leq r < 2$, if $b = 0$ in the representation (3.1) and

$$\int_{\mathbb{R}} \frac{1}{|t|^r + 1} d\sigma(t) < \infty.$$

Observe that $\mathbf{N}_s \subset \mathbf{N}_r$ when $s < r$. In particular, if $Q(z)$ belongs to the so-called Kac-Donoghue class $\mathbf{N}_1$, it has an integral representation of the form

$$Q(z) = a_1 + \int_{\mathbb{R}} \frac{d\sigma(t)}{t - z}, \quad \int_{\mathbb{R}} \frac{d\sigma(t)}{|t| + 1} < \infty, \tag{3.3}$$

where

$$a_1 = a - \int_{\mathbb{R}} \frac{t d\sigma(t)}{t^2 + 1} = \lim_{z \widehat{\rightarrow} \infty} Q(z),$$

cf. [23]. Recall that $Q(z) \in \mathbf{N}_1$ if and only if

$$\int_1^\infty \frac{\operatorname{Im} Q(iy)}{y} dy < \infty,$$

and that $Q(z) \in \mathbf{N}_0$ if and only if

$$\sup_{y > 0} y \operatorname{Im} Q(iy) < \infty,$$

cf. [16, 23].

3.3. Global moments

Let $Q(z)$ be a Nevanlinna function with integral representation (3.1). Then the function $Q(z)$ belongs to $\mathbf{N}_{-n}$ if and only if $b = 0$ and

$$\int_{\mathbb{R}} |t|^n \, d\sigma(t) < \infty,$$

see [22, 25]; cf. also [9]. In this case $Q(z)$ (or the spectral measure σ) is said to have $n \geq 0$ *(global) moments*

$$m_k = \int_{\mathbb{R}} t^k \, d\sigma(t), \quad k = 0, \dots, n. \tag{3.4}$$

In particular, $Q(z)$ has an integral representation of the form (3.3). Nevanlinna functions with a finite number of moments give rise to asymptotic expansions at ∞, cf. [1]. The Nevanlinna classes $\mathbf{N}_{-n}$ play an important role in the description of such asymptotic expansions. Recall the following result from [22]; see also [9, 25].

Proposition 3.1. *Assume that $Q(z) \in \mathbf{N}_{-n}$ for some integer $n \geq 0$. Then the moments (3.4) exist and $Q(z)$ has an asymptotic expansion of the form*

$$Q(z) = a_1 - \sum_{k=0}^{n} \frac{m_k}{z^{k+1}} + o\left(\frac{1}{z^{n+1}}\right), \quad z \widehat{\rightarrow} \infty. \tag{3.5}$$

Conversely, let $Q(z) \in \mathbf{N}$ and assume that for some $a_1, m_0, \dots, m_n \in \mathbb{R}$ the function $Q(z)$ has the asymptotic expansion (3.5). Then

$$q(z) := z^{n+1}\left(Q(z) - a_1 + \sum_{k=0}^{n} \frac{m_k}{z^{k+1}}\right) \in \mathbf{N}. \tag{3.6}$$

If n is even, then $Q(z) \in \mathbf{N}_{-n}$ and m_k, $0 \leq k \leq n$, are the moments given by the identity (3.4). If n is odd, then $Q(z) \in \mathbf{N}_{-n+1}$ and m_k, $0 \leq k \leq n-1$, are the moments given by the identity (3.4). Moreover, if n is odd and additionally $q(z)$ in (3.6) belongs to $\mathbf{N}_1$, then $Q(z) \in \mathbf{N}_{-n}$ and m_n is the nth moment of $Q(z)$.

3.4. Local growth conditions

For any $\gamma \in \mathbb{R}$ the mapping $z \to -1/(z-\gamma)$ is a Nevanlinna function. If $Q(z)$ is a Nevanlinna function, then $-Q(-z)$ is a Nevanlinna function and also the composition function $Q(z;\gamma)$ defined by

$$Q(z;\gamma) = -Q\left(\frac{1}{z-\gamma}\right), \quad z \in \mathbb{C} \setminus \mathbb{R}, \tag{3.7}$$

is a Nevanlinna function. Introduce the measure $\sigma(t;\gamma)$ by

$$d\sigma(t;\gamma) = (t-\gamma)^2 d\sigma\left(\frac{1}{t-\gamma}\right), \quad t \in \mathcal{I}_\gamma, \quad \sigma(\{\gamma\};\gamma) = b, \tag{3.8}$$

where $\mathcal{I}_\gamma = (-\infty, \gamma) \cup (\gamma, \infty)$, and the numbers $\tilde{b}$ and $\tilde{a}$ by

$$\tilde{b} = \sigma(\{0\}), \tag{3.9}$$

and

$$\tilde{a} = -a - \gamma\sigma(\{0\}) + \int_{\mathcal{I}_\gamma} \frac{\gamma(1+\gamma t - t^2)}{(t^2+1)((t-\gamma)^2+1)} d\sigma(t;\gamma). \tag{3.10}$$

Proposition 3.2. *Let $Q(z)$ be a Nevanlinna function with an integral representation of the form (3.1) and let $\gamma \in \mathbb{R}$. Then the Nevanlinna function $Q(z;\gamma)$, $z \in \mathbb{C} \setminus \mathbb{R}$, defined by (3.7), has the integral representation*

$$Q(z;\gamma) = \tilde{b}z + \tilde{a} + \int_{\mathbb{R}} \left(\frac{1}{t-z} - \frac{t}{t^2+1} \right) d\sigma(t;\gamma), \quad z \in \mathbb{C} \setminus \mathbb{R}, \tag{3.11}$$

where $\tilde{b} \geq 0$ and $\tilde{a} \in \mathbb{R}$ are given by (3.9) and (3.10) and where $\sigma(t;\gamma)$ is a monotonically nondecreasing function given by (3.8) with

$$\int_{\mathbb{R}} \frac{d\sigma(t;\gamma)}{t^2+1} < \infty.$$

Proof. It follows from the identity (3.1) that

$$-Q\left(\frac{1}{z-\gamma}\right) = -a + \frac{b}{\gamma - z} + (z-\gamma)\sigma(\{0\}) + \int_{\mathcal{I}_0} \left(\frac{-1}{t - \frac{1}{z-\gamma}} + \frac{t}{t^2+1} \right) d\sigma(t).$$

Replace in the last integral t by $1/(t-\gamma)$ and use the identity

$$\frac{-1}{\frac{1}{t-\gamma} - \frac{1}{z-\gamma}} + \frac{\frac{1}{t-\gamma}}{\left(\frac{1}{t-\gamma}\right)^2 + 1} = \frac{(t-\gamma)^2}{t-z} - \frac{(t-\gamma)^3}{(t-\gamma)^2+1}, \quad t \in \mathcal{I}_\gamma.$$

Then it follows that for $z \in \mathbb{C} \setminus \mathbb{R}$,

$$\int_{\mathcal{I}_0} \left(\frac{-1}{t + \frac{1}{z-\gamma}} + \frac{t}{t^2+1} \right) d\sigma(t) = \int_{\mathcal{I}_\gamma} \left(\frac{1}{t-z} - \frac{t-\gamma}{(t-\gamma)^2+1} \right) d\sigma(t;\gamma),$$

which results in the identity (3.11). □

Let $Q(z)$ be a Nevanlinna function with the integral representation (3.1) and let $\gamma \in \mathbb{R}$. Then the function $Q(z)$ is said to belong to $\mathbf{N}_{\gamma,r}$, $r \in (-\infty, 2)$, if

$$\int_{\gamma-\varepsilon}^{\gamma+\varepsilon} |t-\gamma|^{r-2} d\sigma(t) < \infty, \tag{3.12}$$

for some $\varepsilon > 0$. The class $\mathbf{N}_{\gamma,r}$ can be seen as the local analog of the Kac-Donoghue class $\mathbf{N}_r$. Observe that $\mathbf{N}_{\gamma,s} \subset \mathbf{N}_{\gamma,r}$ when $s < r$. Local growth conditions can be related to global growth conditions via the transformation in (3.7).

Proposition 3.3. *Let $Q(z)$ be a Nevanlinna function with the integral representation (3.1), let $\gamma \in \mathbb{R}$, and let $r \in (-\infty, 2)$. Then $Q(z) \in \mathbf{N}_r$ if and only if $Q(z;\gamma) \in \mathbf{N}_{\gamma,r}$.*

Proof. The statement follows from the identity

$$\begin{aligned}\int_{\mathbb{R}\setminus[-1,1]} |t|^{-r} d\sigma(t) &= \int_{(\gamma-1,\gamma+1)} |t-\gamma|^{r} d\sigma((t-\gamma)^{-1}) \\ &= \int_{(\gamma-1,\gamma+1)} |t-\gamma|^{r-2} d\sigma(t;\gamma)\end{aligned}$$

by using (3.12) and the fact that the class $\mathbf{N}_r$ is defined by the corresponding growth condition at ∞. □

It is useful to consider the inverse operation of (3.7), i.e., for any Nevanlinna function $Q(z)$ and $\gamma \in \mathbb{R}$ define

$$R(z) = -Q(\gamma + 1/z), \quad z \in \mathbb{C} \setminus \mathbb{R}. \tag{3.13}$$

Then $R(z)$ is also a Nevanlinna function and it satisfies $R(z;\gamma) = Q(z)$.

Corollary 3.4. *Let $Q(z)$ be a Nevanlinna function with the integral representation (3.1), let $\gamma \in \mathbb{R}$, and let $r \in (-\infty, 2)$. Then the Nevanlinna function $R(z)$ defined in (3.13) belongs to $\mathbf{N}_r$ if and only if $Q(z)$ belongs to $\mathbf{N}_{\gamma,r}$.*

3.5. Local moments

The local analog $\mathbf{N}_{\gamma,1}$ of the Kac-Donoghue subclass $\mathbf{N}_1$ of Nevanlinna functions can be characterized as follows; cf. [17, 15].

Lemma 3.5. *Let $Q(z)$ be a Nevanlinna function with the integral representation (3.1). Then $Q(z)$ belongs to $\mathbf{N}_{\gamma,1}$ if and only if*

$$\int_0^1 \frac{\operatorname{Im} Q(\gamma + iy)}{y}\, dy < \infty. \tag{3.14}$$

In this case $Q(\gamma)$ exists as a nontangential real limit and one has

$$Q(z) = bz + \hat{a} + \int_{\mathbb{R}} \frac{z-\gamma}{(t-z)(t-\gamma)}\, d\sigma(t), \tag{3.15}$$

in particular, $Q(\gamma) = b\gamma + \hat{a}$, where

$$\hat{a} = a + \int_{\mathbb{R}} \frac{1+t\gamma}{(t-\gamma)(t^2+1)}\, d\sigma(t).$$

Proof. The criterion (3.14) and the existence of the limit value $Q(\gamma)$ are proved in [17, Theorem 5.1] and in the special case where $\gamma = 0$ also in [15, Proposition 3.1]. Hence, it remains to rewrite the integral representation (3.1) in the form (3.15) by means of the condition $\int_{\gamma-\varepsilon}^{\gamma+\varepsilon} \frac{d\sigma(t)}{|t-\gamma|} < \infty$. □

A Nevanlinna function $Q(z)$ belongs to $\mathbf{N}_{\gamma,0}$ if and only if $Q'(\gamma)$ exists as a nontangential limit. It is clear from (3.15) that in this case

$$Q'(\gamma) = b + \int_{\mathbb{R}} \frac{d\sigma(t)}{(t-\gamma)^2}\ (\geq 0). \tag{3.16}$$

Further smoothness properties are guaranteed by the classes $\mathbf{N}_{\gamma,-n}$. In general, an induction argument shows that a Nevanlinna function $Q(z)$ belongs to $\mathbf{N}_{\gamma,-n}$, i.e.,

$$\int_{\mathbb{R}} \frac{d\sigma(t)}{|t-\gamma|^{n+2}} < \infty,$$

if and only if $Q^{(n+1)}(\gamma)$ exists as a nontangential limit, and then

$$\frac{Q^{(n+1)}(\gamma)}{(n+1)!} = \int_{\mathbb{R}} \frac{d\sigma(t)}{(t-\gamma)^{n+2}}.$$

In this case the numbers

$$A_k = \int_{\mathbb{R}} \frac{d\sigma(t)}{(t-\gamma)^{k+2}}, \quad k = 0, \dots, n, \tag{3.17}$$

are called the *local moments* of the spectral measure σ, and shortly, also local moments of $Q(z)$ at γ.

Proposition 3.6. *Assume that $Q(z) \in \mathbf{N}_{\gamma,-n}$ for some integer $n \geq 0$. Then $Q(z)$ has an asymptotic expansion of the form*

$$Q(z) = bz + \hat{a} + \sum_{k=0}^{n} A_k (z-\gamma)^{k+1} + o\left((z-\gamma)^{n+1}\right), \quad z \widehat{\to} \gamma, \tag{3.18}$$

where A_k, $0 \leq k \leq n$, are the local moments of $Q(z)$ at γ given by the identity (3.17). Assume that for some numbers b, $\hat{a}$, $A_0, \dots, A_n \in \mathbb{R}$ the function $Q(z) \in \mathbf{N}$ has an asymptotic expansion of the form (3.18). Then

$$q_\gamma(z) := (z-\gamma)^{-n-1} \left(Q(z) - bz - \hat{a} - \sum_{k=0}^{n} A_k (z-\gamma)^{k+1} \right) \in \mathbf{N}. \tag{3.19}$$

If n is even, then $Q(z) \in \mathbf{N}_{\gamma,-n}$ and A_k, $0 \leq k \leq n$, are the local moments given by the identity (3.17). If n is odd, then $Q(z) \in \mathbf{N}_{\gamma,-n+1}$ and A_k, $0 \leq k \leq n-1$, are the moments given by the identity (3.17). Moreover, if n is odd and additionally $q_\gamma(z)$ in (3.19) belongs to $\mathbf{N}_{\gamma,1}$, then $Q(z) \in \mathbf{N}_{-n}$ and A_n is the nth moment of $Q(z)$ at γ.

Proof. Let $Q(z) \in \mathbf{N}_{\gamma,-n}$ and define $R(z)$ as in (3.13), so that $R(z) \in \mathbf{N}_{-n}$, cf. Corollary 3.4. By Proposition 3.1, $R(z)$ has an asymptotic expansion

$$R(z) = a_1 - \sum_{k=0}^{n} \frac{m_k}{z^{k+1}} + o\left(\frac{1}{z^{n+1}}\right), \quad z \widehat{\to} \infty,$$

where m_k stands for the kth moment of $R(z)$. If the spectral measure of $R(z)$ is denoted by ρ and the spectral measure of $Q(z)$ is denoted by σ, then the definition $Q(z) = -R(1/(z-\gamma))$ and Proposition 3.2 imply that

$$d\sigma(t) = (t-\gamma)^2 d\rho\left(\frac{1}{t-\gamma}\right), \quad t \in (-\infty, \gamma) \cup (\gamma, \infty).$$

Therefore,

$$m_k = \int_{\mathbb{R}} s^k \, d\rho(s) = \int_{\mathbb{R}} \frac{1}{(t-\gamma)^{k+2}} \, d\sigma(t) = A_k, \quad k = 0, \dots, n,$$

and the asymptotic expansion (3.18) follows. The other statements follow from Proposition 3.1 and Proposition 3.3 by the fact that $q_\gamma(z) = q(z;\gamma)$, cf. (3.7). □

4. Nevanlinna functions with spectral gaps

An interval (α, β) with $-\infty \le \alpha < \beta \le \infty$ is called a *spectral gap* of a Nevanlinna function $Q(z)$ if it is a maximal interval on which $Q(z)$ is holomorphic, see [10, 26]. The following lemma is well known (cf. [13]) and follows immediately, e.g., from (3.16).

Lemma 4.1. *Let $Q(z) \in \mathbf{N}$ have a spectral gap (α, β). Then the real function $Q(x)$, $x \in (\alpha, \beta)$, is nondecreasing.*

Hence, on a spectral gap (α, β) the function $Q(x)$, $x \in (\alpha, \beta)$, has limits in the endpoints which may be proper or improper. The following propositions are refinements of earlier characterizations, now for the endpoints of a gap.

Proposition 4.2. *Let $Q(z) \in \mathbf{N}$ have a spectral gap on (α, β) with $-\infty \le \alpha < \beta \le \infty$. If $\alpha \in \mathbb{R}$, then*

$$\lim_{y \downarrow \alpha} Q(y) > -\infty \quad \textit{if and only if} \quad Q(z) \in \mathbf{N}_{\alpha,1},$$

and if $\beta \in \mathbb{R}$, then

$$\lim_{y \uparrow \beta} Q(y) < \infty \quad \textit{if and only if} \quad Q(z) \in \mathbf{N}_{\beta,1}.$$

If $\alpha = -\infty$, then

$$\lim_{y \downarrow -\infty} Q(y) > -\infty \quad \textit{if and only if} \quad Q(z) \in \mathbf{N}_1,$$

and if $\beta = \infty$, then

$$\lim_{y \uparrow \infty} Q(y) < \infty \quad \textit{if and only if} \quad Q(z) \in \mathbf{N}_1.$$

Proof. It suffices to prove the cases for the left endpoint $\alpha \in \mathbb{R}$ and $\alpha = -\infty$, respectively.

Assume that $\lim_{y \downarrow \alpha} Q(y) > -\infty$. Then

$$\int_{[\alpha-\epsilon,\alpha]} (t-y)^{-1} d\sigma(t)$$

is bounded for all $y > \alpha$. The monotone convergence theorem implies that

$$\int_{[\alpha-\epsilon,\alpha]} (t-\alpha)^{-1} d\sigma(t) < \infty.$$

The proof of the converse statement is clear.

Now assume that $\lim_{y\downarrow-\infty} Q(y) = a_1 > -\infty$. The identity (3.2) shows that $b = 0$. The assumption $\operatorname{supp}\sigma \subset [\beta,\infty)$ implies that $Q(z)$ has a spectral gap $(-\infty,\beta)$. Hence, it follows from Lemma 4.1 that $Q(y) \geq a_1$ for $y \in (-\infty,\beta)$. Therefore,

$$a_1 \leq a + \int_\beta^\infty \left(\frac{1}{t-y} - \frac{t}{t^2+1}\right) d\sigma(t), \quad y < \beta.$$

By the monotone convergence theorem

$$a_1 = a - \int_\beta^\infty \frac{t d\sigma(t)}{t^2+1},$$

which implies that $Q(z) \in \mathbf{N}_1$. Conversely, assume that $Q(z) \in \mathbf{N}_1$. The assumption $\operatorname{supp}\sigma \subset [\beta,\infty)$ and the representation formula (3.3) show that

$$Q(z) = a_1 + \int_\beta^\infty \frac{d\sigma(t)}{t-z}, \quad \int_\beta^\infty \frac{d\sigma(t)}{|t|+1} < \infty.$$

Again by the monotone convergence theorem $\lim_{y\downarrow-\infty} Q(y) = a_1$. □

In Proposition 4.2 it has been shown for a spectral gap of $Q(z)$ that the $\mathbf{N}_1$ behaviour of $Q(z)$ implies that the limit of $Q(z)$ exists at the endpoints of the gap. Now observe that $0 < r < 1$ implies $\mathbf{N}_r \subset \mathbf{N}_1$ and similarly $\mathbf{N}_{\gamma,r} \subset \mathbf{N}_{\gamma,1}$. Note that for $0 < r < 1$ there are no moments (global or local). However, the next proposition presents some asymptotic results near the endpoints of the spectral gap as a further specification of Proposition 4.2.

Proposition 4.3. *Let $Q(z) \in \mathbf{N}$ have a spectral gap (α,β) with $-\infty \leq \alpha < \beta \leq \infty$. Let $\alpha \in \mathbb{R}$. If $Q(z) \in \mathbf{N}_{\alpha,r}$ for some $r \in (0,1)$ and $Q(\alpha) = 0$ then*

$$\lim_{y\downarrow\alpha} (y-\alpha)^{r-1} Q(y) = 0.$$

Let $\beta \in \mathbb{R}$. If $Q(z) \in \mathbf{N}_{\beta,r}$ for some $r \in (0,1)$ and $Q(\beta) = 0$ then

$$\lim_{y\downarrow\alpha} (\beta-y)^{r-1} Q(y) = 0.$$

Let $\alpha = -\infty$. If $Q(z) \in \mathbf{N}_r$ for some $r \in (0,1)$ and $\lim_{y\downarrow-\infty} Q(y) = 0$ then

$$\lim_{y\downarrow-\infty} (-y)^{1-r} Q(y) = 0.$$

Let $\beta = \infty$. If $Q(z) \in \mathbf{N}_r$ for some $r \in (0,1)$ and $\lim_{y\uparrow\infty} Q(y) = 0$ then

$$\lim_{y\uparrow\infty} y^{1-r} Q(y) = 0.$$

Proof. Let $\alpha \in \mathbb{R}$ and assume that $Q(z) \in \mathbf{N}_{\alpha,r}$ for some $r \in (0,1)$ with $Q(\alpha) = 0$. Then the identity (3.15) implies that $Q(z)$ has a representation of the form

$$Q(z) = \int_{\mathbb{R}} \frac{z-\alpha}{(t-z)(t-\alpha)} d\sigma(t).$$

If $t < \alpha$ and $y > \alpha$ then $y - t \geq y - \alpha > 0$ and $y - t \geq \alpha - t > 0$, and hence

$$\frac{(y-\alpha)^r}{|t-y||t-\alpha|} \leq \frac{1}{|t-y|^{1-r}|t-\alpha|} \leq \frac{1}{|t-\alpha|^{2-r}}.$$

It follows that the function $(z-\alpha)^r(t-z)^{-1}(t-\alpha)^{-1}$ has a $d\sigma$-integrable majorant in a neighborhood of α, and the result follows from Lebesgue's theorem.

Now let $\alpha = -\infty$ and assume that $Q(z) \in \mathbf{N}_r$ for some $r \in (0,1)$ with $Q(-\infty) = 0$. Since $\mathbf{N}_r \subset \mathbf{N}_1$ the function $Q(z)$ has the integral representation (3.3) with $a_1 = 0$. On each compact interval $[c,d]$ the limit

$$\lim_{y\to\infty} \int_c^d \frac{y^{1-r}}{t+y}\, d\sigma(t) = 0$$

is clear by dominated convergence, since the integrand is $O(1/y^r)$. Furthermore, the limit

$$\lim_{y\to\infty} \int_1^\infty \frac{y^{1-r}}{t+y}\, d\sigma(t) = 0$$

follows from the definition of $\mathbf{N}_r$ and the dominated convergence theorem, since

$$\frac{y^{1-r}}{t+y} \leq t^{-r},$$

due to $x^r \leq x+1$, $x \geq 0$.

The remaining cases can be treated in a similar way. □

5. Eigenvalues of self-adjoint extensions

5.1. Eigenvalues and linear fractional transforms

Let S be a closed symmetric operator in a Hilbert space $\mathfrak{H}$ and assume that the defect numbers of S are $(1,1)$. Let A be a fixed self-adjoint extension of S and recall that all self-adjoint extensions $A(\tau)$ of S are in one-to-one correspondence with $\tau \in \mathbb{R}\cup\{\infty\}$ via (1.1), and that the Weyl functions $Q_\tau(z)$ of $A(\tau)$ and S are expressed in terms of the Weyl function $Q(z)$ of A and S via (2.9) in Proposition 2.1. A classical result states that the essential spectrum of $A(\tau)$ is independent of $\tau \in \mathbb{R}\cup\{\infty\}$, see [12]; cf. [5]. The following result gives an analytic description of the point spectrum of $A(\tau)$.

Proposition 5.1. *Let $Q(z) \in \mathbf{N}$ be a Nevanlinna function and let $\gamma \in \mathbb{R}$. Then the function $Q_\tau(z)$, $\tau \neq 0$, defined by the transform (2.9) has a pointmass at γ if and only if $Q(z) \in \mathbf{N}_{\gamma,0}$ and $Q(\gamma) = -1/\tau$, in which case the spectral measure σ_τ of $Q_\tau(z)$ satisfies*

$$\sigma_\tau(\{\gamma\}) = \frac{1+\tau^2}{\tau^2(b+A_0)}, \tag{5.1}$$

where A_0 is given by (3.17). Moreover, the function $Q_\tau(z)$, $\tau \neq 0$, has a pointmass at ∞, that is $b_\tau > 0$, if and only if $Q(z) \in \mathbf{N}_0$ and $\lim_{z \hat{\to} \infty} Q(z) = -1/\tau$, in which case

$$b_\tau = \frac{1+\tau^2}{\tau^2 m_0}. \tag{5.2}$$

Proof. Recall that the identity (2.9) is equivalent to (2.10), i.e.,

$$Q_\tau(z) = \frac{1}{\tau} - \frac{1+\tau^2}{\tau^2} \frac{1}{Q(z)+1/\tau}, \quad \tau \neq 0. \tag{5.3}$$

It follows from this equality that $\lim_{z \hat{\to} \gamma} (\gamma - z) Q_\tau(z) \neq 0$ if and only if

$$\lim_{z \hat{\to} \gamma} \frac{Q(z)+1/\tau}{z-\gamma} < \infty,$$

that is, $Q(\gamma) = -\tau^{-1}$ and $Q'(\gamma)$ exists. The identity (5.1) is now obtained from (3.16) and (3.17). Moreover, note that by the identities (3.2) and (5.3)

$$b_\tau = -\lim_{z \hat{\to} \infty} \frac{1+\tau^2}{\tau^2} \frac{1}{z(Q(z)+1/\tau)}.$$

The equality $\lim_{z \hat{\to} \infty} Q(z) = -1/\tau$ implies with the identities (3.3) and (3.4) that $\lim_{z \hat{\to} \infty} z(Q(z)+1/\tau) = -m_0$. □

Equivalently, the position of the eigenvalues of the self-adjoint extension $A(\tau)$ can be described in terms of the Weyl function $Q(z)$ of S corresponding to the self-adjoint extension A, provided that the symmetric operator S is completely non-self-adjoint.

Corollary 5.2. *Let S be a completely non-self-adjoint closed symmetric operator in a Hilbert space $\mathfrak{H}$ with defect numbers $(1,1)$ and let A be a self-adjoint extension of S. Let $Q(z)$ be the Weyl function corresponding to S and A. Then $A(\tau)$, $\tau \neq 0$, has an eigenvalue at γ if and only if $Q(z) \in \mathbf{N}_{\gamma,0}$ and $Q(\gamma) = -1/\tau$. Moreover, $A(\tau)$, $\tau \neq 0$, has an eigenvalue at ∞ if and only if $Q(z) \in \mathbf{N}_0$ and $\lim_{z \hat{\to} \infty} Q(z) = -1/\tau$.*

5.2. The existence of eigenvalues in a spectral gap

Let S be a completely non-self-adjoint closed symmetric operator in $\mathfrak{H}$ with defect numbers $(1,1)$ and let A be a self-adjoint extension of S. Let $Q(z)$ be the Weyl function of A and S and assume that $Q(z)$ has a spectral gap (α, β), where $-\infty \leq \alpha < \beta \leq \infty$. A self-adjoint extension $A(\tau)$ defined by (2.18), (2.19), or by (1.1), has an eigenvalue in (α, β) if and only if its Weyl function $Q_\tau(z)$ defined in (2.9) has a pole in (α, β).

Note that $z \in (\alpha, \beta)$ is a pole of $Q_\tau(z)$ if and only if $1 + \tau Q(z) = 0$ if $\tau \in \mathbb{R}$, or $Q(z) = 0$ if $\tau = \infty$, in which case the (unique) solution is denoted by $\lambda(\tau)$. Note that

$$Q(\lambda(\tau)) = -\frac{1}{\tau}, \quad \tau \in \mathbb{R} \cup \{\infty\}. \tag{5.4}$$

In terms of the notations $A = Q(\alpha+)$ and $B = Q(\beta-)$, so that $-\infty \le A < B \le \infty$, two cases can therefore be distinguished:

- $A \ge 0$ or $B \le 0$. The function $Q(z)$ has no zero in (α, β) and there is a solution $\lambda(\tau)$ precisely when $-A^{-1} < \tau < -B^{-1}$.
- $A < 0$ and $B > 0$. The function $Q(z)$ has a unique zero in (α, β) and there is a solution $\lambda(\tau)$ precisely when $-\infty < \tau < -B^{-1}$, $-A^{-1} < \tau < \infty$, or $\tau = \infty$.

Denote the solution set in each of these cases by $I_{A,B}$. The following proposition is now clear.

Proposition 5.3. *For each $\tau \in I_{A,B}$ there is a unique solution $\lambda(\tau)$ of the equation (5.4) and $\alpha < \lambda(\tau) < \beta$. Moreover, the function $\lambda(\tau)$ is differentiable. In fact, $\lambda(\tau)$ is monotonically increasing in τ and*

$$\lambda'(\tau) = \frac{1}{\tau^2 Q'(\lambda(\tau))} \qquad (\tau \ne 0).$$

The dependence of $\lambda(\tau)$ on the parameter $\tau \in \mathbb{R} \cup \{\infty\}$ will be studied in the rest of this paper and in [18].

5.3. Meromorphic Weyl functions

Let S be a completely non-self-adjoint symmetric operator with defect numbers $(1, 1)$. Let A be a self-adjoint extension of S and assume that the corresponding Weyl fuction $Q(z)$ is meromorphic. Then $Q(z)$ has a sequence of poles (λ_i) ordered by $\lambda_i < \lambda_{i+1}$, so that there are gaps of the form $(\lambda_i, \lambda_{i+1})$. Here $i \in \mathbb{Z}$ when A is not semibounded, and $i \in \mathbb{N} \cup \{0\}$ when A is semibounded from below. It follows from Proposition 5.3 that for each $\tau \in \mathbb{R} \cup \{\infty\}$ there exists a unique eigenvalue $\lambda_i(\tau)$ of $A(\tau)$ with

$$\lambda_i < \lambda_i(\tau) < \lambda_{i+1}.$$

Note that $\lambda_i(\tau)$ is monotonically increasing, running through the above interval as τ runs through $(0+, \infty]$ and then, identifying ∞ and $-\infty$, through $[-\infty, 0-)$. Recall that the mapping $\gamma(z)$ in (2.2) (with $A_0 = A$) is defined for all $z \in \rho(A)$. Hence, it follows that for each $\tau \in \mathbb{R} \cup \{\infty\}$ the element $\gamma(\lambda_i(\tau))$ is well defined. The following result identifies the eigenspace corresponding to $\lambda_i(\tau)$ in terms of the γ-field $\gamma(z)$.

Proposition 5.4. *Let S be a completely non-self-adjoint symmetric operator with defect numbers $(1, 1)$. Let A be a self-adjoint extension of S and assume that the corresponding Weyl fuction $Q(z)$ is meromorphic. Then for each $\tau \in \mathbb{R} \cup \{\infty\}$ with the possible exception of ∞, the elements*

$$\frac{\gamma(\lambda_i(\tau))}{\sqrt{Q'(\lambda_i(\tau))}} \tag{5.5}$$

form a complete orthonormal system of eigenvectors of $A(\tau)$ in $\mathfrak{H}$. In the exceptional case, which occurs precisely when S is not densely defined, the elements in (5.5) form a complete orthonormal system in $\overline{\operatorname{dom} S}$.

Proof. Apply Kreĭn's formula (1.1) with $z \in \rho(A) \cap \rho(A(\tau))$ to the element $\gamma(\lambda_i(\tau))$:

$$(A(\tau) - z)^{-1}\gamma(\lambda_i(\tau)) = (A - z)^{-1}\gamma(\lambda_i(\tau)) - \gamma(z)\,\frac{1}{Q(z) + 1/\tau}\,(\gamma(\lambda_i(\tau)), \gamma(\bar{z})). \tag{5.6}$$

By (2.4) and (2.5) (with $A_0 = A$ and $w \in \rho(A)$) and Corollary 5.2, the last element in the right-hand side of (5.6) is equal to

$$\frac{1}{\lambda_i(\tau) - z}\,\gamma(z) = \frac{1}{\lambda_i(\tau) - z}\,\gamma(\lambda_i(\tau)) - (A - z)^{-1}\,\gamma(\lambda_i(\tau)), \quad z \in \rho(A) \cap \rho(A(\tau)).$$

Therefore, it follows from (5.6) that

$$(A(\tau) - z)^{-1}\gamma(\lambda_i(\tau)) = \frac{1}{\lambda_i(\tau) - z}\,\gamma(\lambda_i(\tau)), \quad z \in \rho(A) \cap \rho(A(\tau)),$$

in other words $\{\gamma(\lambda_i(\tau)), \lambda_i(\tau)\,\gamma(\lambda_i(\tau))\} \in A(\tau)$, so that $\gamma(\lambda_i(\tau))$ is an eigenvector of $A(\tau)$ corresponding to the eigenvalue $\lambda_i(\tau)$.

The normalization of the eigenelement $\gamma(\lambda_i(\tau))$ in (5.5) is a direct consequence of (2.5).

Since the spectrum of $A(\tau)$ is discrete, the completeness properties follow from the spectral representation of self-adjoint relations. □

If the operator S is not densely defined, so that $A(\infty)$ is the generalized Friedrichs extension of S, then the elements

$$\frac{\gamma(\lambda_i(\infty))}{\sqrt{Q'(\lambda_i(\infty))}}$$

give an orthonormal basis for the Hilbert space $\overline{\operatorname{dom} S} = (\operatorname{mul} A(\infty))^{\perp}$ corresponding to the generalized Friedrichs extension $A(\infty)$. In other words, this system gives a partial isometric expansion for the whole Hilbert space $\mathfrak{H}$.

6. Taylor expansions for the eigenvalues in spectral gaps

6.1. Holomorphic properties of the eigenvalue in a spectral gap

Let $\tau \in I_{A,B} \cap \mathbb{R}$, then the function $Q_\tau(z)$ has a simple pole at $\lambda(\tau)$. If this pole is isolated, it leads to a representation of $Q_\tau(z)$ of the form

$$Q_\tau(z) = (1 + \tau^2)\left(\frac{c(\tau)}{\lambda(\tau) - z} - h_\tau(z)\right) - \tau, \tag{6.1}$$

where $h_\tau(z)$ is holomorphic around $\lambda(\tau)$ and $c(\tau)$ is a positive constant which can be calculated in the following way:

$$\begin{aligned} c(\tau) &= \frac{1}{1+\tau^2} \lim_{z\to\lambda(\tau)} (\lambda(\tau)-z)Q_\tau(z) \\ &= \frac{-1}{\tau^2}\frac{\lambda(\tau)-z}{Q(z)+1/\tau} \\ &= \frac{1}{\tau^2}\frac{1}{Q'(\lambda(\tau))} \\ &= \lambda'(\tau), \end{aligned}$$

where the identity (2.10) has been used. The function $h_\tau(z)$ in the representation (6.1) gives rise to a sequence of coefficients $(\gamma_n(\tau))_{n=1}^\infty$ as follows:

$$\gamma_n(\tau) = \begin{cases} c(\tau), & n=1, \\ \sum_{k=1}^{n-1} c(\tau)^{n-k}\binom{n}{k}\binom{n-1}{k}k!\, \frac{d^{n-1-k}}{dz^{n-1-k}}\, h_\tau(z)^k\Big|_{z=\lambda(\tau)}, & n>1. \end{cases} \tag{6.2}$$

Theorem 6.1. *Let the Nevanlinna function $Q(z)$ have a spectral gap (α,β) and let $\tau_0 \in I_{A,B} \cap \mathbb{R}$. Then the solution $\lambda(\tau)$ of $1+\tau Q(z)=0$ in an appropriate neighborhood of τ_0 is a holomorphic function in τ with the expansion*

$$\lambda(\tau) = \lambda(\tau_0) + \sum_{n=1}^\infty \frac{\gamma_n(\tau_0)}{n!}(\tau-\tau_0)^n,$$

where the sequence $(\gamma_n(\tau_0))_{n=1}^\infty$ is given in (6.2).

Proof. On $I_{A,B}$ the function $\lambda(\tau)$, $\tau\in\mathbb{R}$, satisfies $Q_\infty(\lambda(\tau))=\tau$. As $Q_\infty(z)$ is holomorphic in a neighborhood of τ_0 with positive derivative, the holomorphic dependence of $\lambda(\tau)$ on τ in an appropriate neighborhood of τ_0 is guaranteed by the implicit function theorem. The expansion follows from the Lagrange formula for the inversion of a power series, cf. [6, 29]. Let

$$Q_\infty(z) = \frac{z-\lambda(\tau_0)}{\varphi(z)} + \tau_0,$$

so that

$$\varphi(\lambda(\tau)) = \frac{\lambda(\tau)-\lambda(\tau_0)}{\tau-\tau_0}.$$

Then the coefficients γ_n in the expansion of $\lambda(\tau)$ in the proposition are given by

$$\gamma_n = \frac{d^{n-1}}{dz^{n-1}}\varphi(z)^n\Big|_{z=\lambda(\tau_0)}$$

cf. [6, p. 125]. Now observe that the function $\varphi(z)$ can be written in terms of the function $h(z) := h_{\tau_0}(z)$:

$$\varphi(z) = (z-\lambda(\tau_0))h(z) + c(\tau_0),$$

which implies for the coefficients $\gamma_n(\tau_0)$ that

$$\gamma_n(\tau_0) = \frac{d^{n-1}}{dz^{n-1}} \left[(z - \lambda(\tau_0))h(z) + c(\tau_0)\right]^n \bigg|_{z=\lambda(\tau_0)} .$$

Hence the sequence $(\gamma_n(\tau_0))$ in (6.2) above with $\tau = \tau_0$ is obtained. This completes the proof. □

The coefficients $\gamma_1(\tau_0), \gamma_2(\tau_0), \gamma_3(\tau_0), \ldots$ can be expressed in terms of the function $Q_\infty(z)$; in particular, one gets

$$\gamma_1(\tau_0) = \frac{1}{Q'_\infty(\lambda(\tau_0))}, \quad \gamma_2(\tau_0) = -\frac{Q''_\infty(\lambda(\tau_0))}{Q'_\infty(\lambda(\tau_0))^3},$$

$$\gamma_3(\tau_0) = 3\frac{Q''_\infty(\lambda(\tau_0))^2}{Q'_\infty(\lambda(\tau_0))^5} - \frac{Q'''_\infty(\lambda(\tau_0))}{Q'_\infty(\lambda(\tau_0))^4}.$$

There exists an analog of Theorem 6.1 when the function $Q(z)$ has a zero in the spectral gap (α, β), which is then given by $\lambda(\infty)$. Since $Q(z)$ is holomorphic around its zero $\lambda(\infty)$, the asymptotic behaviour of the eigenvalue $\lambda(\tau)$ as $\tau \to \infty$ can be given up to any order. To express its series expansion around $\tau = \infty$, let $c(\infty) = 1/Q'(\lambda(\infty))$, see (3.16), (5.1), and define the function $l(z)$ in an appropriate neighborhood of $\lambda(\infty)$ by

$$Q_\infty(z) = \frac{c(\infty)}{\lambda(\infty) - z} - l(z).$$

Then $l(z)$ is holomorphic around $z = \lambda(\infty)$. The function $l(z)$ gives rise to a sequence of coefficients $(\delta_n)_{n=1}^\infty$ as follows:

$$\delta_n = \begin{cases} c(\infty), & n = 1, \\ \sum_{k=1}^{n-1} c(\infty)^{n-k} \binom{n}{k} \binom{n-1}{k} k! \, \frac{d^{n-1-k}}{dz^{n-1-k}} \, l(z)^k \Big|_{z=\lambda(\infty)}, & n > 1. \end{cases} \tag{6.3}$$

Theorem 6.2. *Let the Nevanlinna function $Q(z)$ have a spectral gap (α, β) and assume that $Q(z)$ has a zero $\lambda(\infty)$ on (α, β). Then the solution $\lambda(\tau)$ of $Q(z) + 1/\tau = 0$ has the expansion*

$$\lambda(\tau) = \lambda(\infty) + \sum_{n=1}^{\infty} \frac{(-1)^n \delta_n}{n!} \frac{1}{\tau^n}, \quad \tau \to \pm\infty,$$

where the sequence $(\delta_n)_{n=1}^\infty$ is given in (6.3).

Proof. The function $\lambda(\tau)$ satisfies $Q(\lambda(\tau)) = -1/\tau$, $\tau \neq 0$. Around the point $\lambda(\infty)$ the function $Q(z)$ is holomorphic, hence the expansion for $\tau \to \infty$ follows from the Lagrange formula for the inversion of power series. In fact, this formula gives the expansion of $\lambda(\tau)$ in the proposition when the coefficients γ_n are given by

$$\delta_n = \frac{d^{n-1}}{dz^{n-1}} \psi(z)^n \bigg|_{z=\lambda(\infty)}, \quad \text{where} \quad Q(z) = \frac{z - \lambda(\infty)}{\psi(z)}.$$

Now observe that the function $\psi(z)$ can be written in terms of the function $l(z)$:

$$\psi(z) = (\lambda(\infty) - z)Q_\infty(z) = (z - \lambda(\infty))l(z) + c(\infty),$$

which implies for the coefficients δ_n that

$$\delta_n = \frac{d^{n-1}}{dz^{n-1}} \left[(z - \lambda(\infty))l(z) + c(\infty)\right]^n\Big|_{z=\lambda(\infty)}.$$

Hence the sequence (δ_n) in (6.3) is obtained. This completes the proof. □

It follows from (6.3) that the coefficients δ_1, δ_2, and δ_3 can be expressed in terms of the function $Q(z)$ by

$$\delta_1 = \frac{1}{Q'(\lambda(\infty))}, \quad \delta_2 = -\frac{Q''(\lambda(\infty))}{Q'(\lambda(\infty))^3}, \quad \delta_3 = 3\frac{Q''(\lambda(\infty))^2}{Q'(\lambda(\infty))^5} - \frac{Q'''(\lambda(\infty))}{Q'(\lambda(\infty))^4}.$$

The result in Theorem 6.2 contains in particular the case of eigenvalues of self-adjoint operators in a finite-dimensional space $\mathfrak{H}$, and a special case of this theorem in a finite-dimensional space can be found from [2, Proposition 6.1].

6.2. A first-order differential operator

Let $\mathfrak{H} = L^2(0,1)$ and let A be the self-adjoint operator in $\mathfrak{H}$ generated by iD and the self-adjoint boundary condition $y(0) = y(1)$. As illustration several one-dimensional restrictions of A will be discussed, cf. [2]. The interpretation of the corresponding self-adjoint extensions as graph perturbations (cf. Section 2) can be found in [3], [4], [28].

Usually, the operator A is viewed as a self-adjoint extension of the operator iD, restricted by the boundary conditions $y(0) = y(1) = 0$. This restriction is a densely defined, symmetric operator with defect numbers $(1,1)$, which is clearly completely non-self-adjoint. The corresponding functions $\gamma(z)$ and $Q(z)$ are given by

$$\gamma(z) = \frac{e^{-izx}}{1 - e^{-iz}}, \quad Q(z) = -\frac{1}{2}\cot\frac{z}{2}, \quad z \in \rho(A).$$

Since $Q(z)$ is a meromorphic Nevanlinna function, its Mittag-Leffler expansion is given by

$$Q(z) = -\frac{1}{z} + \sum_{k \neq 0} \left(\frac{1}{2k\pi - z} - \frac{1}{2k\pi} \right),$$

which follows from the series expansion of $\frac{z}{2}\cot\frac{z}{2}$, cf. [29, p. 125]. The spectral gaps are of the form $(2k\pi, 2(k+1)\pi)$, $k \in \mathbb{Z}$. The eigenvalue $\lambda_k(\tau)$ of the self-adjoint extension $A(\tau)$ on such a gap is explicitly given by

$$\lambda_k(\tau) = (2k+1)\pi - 2\arctan\frac{2}{\tau}, \quad \tau \in (0+, \infty] \cup [-\infty, 0-),$$

giving an expansion as in Theorem 6.2. The corresponding normalized eigenfunctions $\gamma(\lambda_k(\tau))$ as in (5.5) form a complete orthonormal system in $\mathfrak{H} = L^2(0,1)$.

A domain restriction of A is obtained by the pair $\{\varphi, \omega\} = \{0, \omega\}$, cf. Proposition 2.2. When $\omega(x) = x$, the function $\gamma(z) = (A - z)^{-1}\omega$ and the Weyl function $Q(z) = [(A - z)^{-1}\omega, \omega]$ are given by

$$\gamma(z) = \frac{e^{-izx}}{z(e^{-iz} - 1)} - \frac{x}{z} - \frac{i}{z^2}, \quad Q(z) = -\frac{1}{2z^2} \cot\frac{z}{2} + \frac{1}{z^3} - \frac{1}{3z}.$$

The function $Q(z)$ is a meromorphic function with poles at $z = 2k\pi$, $k \in \mathbb{Z}$. The Mittag-Leffler expansion is now

$$Q(z) = -\frac{1}{4z} + \sum_{k \in \mathbb{Z} \setminus \{0\}} \frac{1}{(2k\pi)^2} \left(\frac{1}{2k\pi - z} - \frac{1}{2k\pi} \right).$$

Clearly, the minimal operator has no eigenvectors and, hence, it is completely non-self-adjoint. Note that $A(\infty)$ is multi-valued and that the appearance of the eigenvalue ∞ is discontinuous as the eigenvalues $\lambda_k(\tau)$ of the self-adjoint extensions $A(\tau)$ are all trapped in the intervals $(2k\pi, 2(k+1)\pi)$, $k \in \mathbb{Z}$. When $\omega(x) = 1$, the functions $\gamma(z)$ and $Q(z)$ are given by

$$\gamma(z) = \frac{1}{z}, \quad Q(z) = -\frac{1}{z}.$$

In this case the minimal operator is clearly not completely non-self-adjoint. In fact the completely non-self-adjoint part of S is the trivial relation in $\mathbb{C}$ (i.e., the subspace $\{0, 0\}$ in $\mathbb{C} \times \mathbb{C}$) and in this Hilbert space the self-adjoint extensions $A(\tau)$, $\tau \in \mathbb{R} \cup \{\infty\}$, have the form

$$A(\tau) = \{\, \{h, \tau h\} : h \in \mathbb{C} \,\}, \quad \tau \in \mathbb{R}, \quad A(\infty) = \{\, \{0, h\} : h \in \mathbb{C} \,\}.$$

The eigenvalue ∞ of $A(\infty)$ can now be seen as a limit of the eigenvalues τ of $A(\tau)$.

References

[1] N.I. Achieser, *The classical moment problem*, Oliver & Boyd, Edinburgh, 1965.

[2] Yu.M. Arlinskiĭ, S. Hassi, H.S.V. de Snoo, and E.R. Tsekanovskiĭ, "One-dimensional perturbations of selfadjoint operators with finite or discrete spectrum", Contemporary Mathematics, 323 (2003), 419–433.

[3] E.A. Catchpole, "A Cauchy problem for an ordinary integro-differential equation", Proc. Roy. Soc. Edinburgh (A), 72 (1972–73), 39–55.

[4] E.A. Catchpole, "An integro-differential operator", J. London Math. Soc., 6 (1973), 513–523.

[5] E.A. Coddington and N. Levinson, "On the nature of the spectrum of singular second order linear differential equations", Canadian J. Math., 3 (1951), 335–338.

[6] E.T. Copson, *An introduction to the theory of functions of a complex variable*, Oxford University Press, Oxford, 1935.

[7] V.A. Derkach, S. Hassi, M.M. Malamud, and H.S.V. de Snoo, "Generalized resolvents of symmetric operators and admissibility", Methods of Functional Analysis and Topology, 6, No. 3, 2000, 24–55.

[8] V.A. Derkach, S. Hassi, M.M. Malamud, and H.S.V. de Snoo, "Boundary relations and their Weyl families", Trans. Amer. Math. Soc., 358, No. 12 (2006), 5351–5400.

[9] V.A. Derkach, S. Hassi, and H.S.V. de Snoo, "Asymptotic expansions of generalized Nevanlinna functions and their spectral properties", Oper. Theory Adv. Appl., 175 (2007), 51–88.

[10] V.A. Derkach and M.M. Malamud, "Generalized resolvents and the boundary value problems for Hermitian operators with gaps", J. Funct. Anal., 95 (1991) 1–95.

[11] V.A. Derkach and M.M. Malamud, "The extension theory of hermitian operators and the moment problem", J. Math. Sciences, 73 (1995), 141–242.

[12] W.F. Donoghue, "On the perturbation of spectra", Comm. Pure Appl. Math., 18 (1965), 559–579.

[13] W.F. Donoghue, *Monotone matrix functions and analytic continuation*, Springer-Verlag, Berlin-Heidelberg-New York, 1974.

[14] V.I. Gorbachuk and M.L. Gorbachuk, *Boundary value problems for operator differential equations*, Mathematics and its Applications (Soviet Series), 48, Kluwer Academic Publishers, Dordrecht, 1991.

[15] S. Hassi, M. Kaltenbäck, and H.S.V. de Snoo, "Generalized Kreĭn-von Neumann extensions and associated operator models", Acta Sci. Math. (Szeged), 64 (1998), 627–655.

[16] S. Hassi, H. Langer, and H.S.V. de Snoo, "Selfadjoint extensions for a class of symmetric operators with defect numbers $(1,1)$", 15th OT Conference Proceedings, (1995), 115–145.

[17] S. Hassi, C. Remling, and H.S.V. de Snoo, "Subordinate solutions and spectral measures of canonical systems", Integral Equations Operator Theory, 37 (2000), 48–63.

[18] S. Hassi, A. Sandovici, H.S.V. de Snoo, and H.Winkler, "Limit properties of eigenvalues in spectral gaps and the spectrum at the endpoints of gaps", in preparation.

[19] S. Hassi and H.S.V. de Snoo, "One-dimensional graph perturbations of selfadjoint relations", Ann. Acad. Sci. Fenn. A.I. Math., 22 (1997), 123–164.

[20] S. Hassi and H.S.V. de Snoo, "On rank one perturbations of selfadjoint operators", Integral Equations Operator Theory, 29 (1997), 288–300.

[21] S. Hassi and H.S.V. de Snoo, "Nevanlinna functions, perturbation formulas and triplets of Hilbert spaces", Math. Nachr., 195 (1998), 115–138.

[22] S. Hassi, H.S.V. de Snoo, and A.D.I. Willemsma, "Smooth rank one perturbations of selfadjoint operators", Proc. Amer. Math. Soc., 126 (1998), 2663–2675.

[23] I.S. Kac and M.G. Kreĭn, "R-functions – analytic functions mapping the upper halfplane into itself", Supplement to the Russian edition of F.V. Atkinson, *Discrete and continuous boundary problems*, Mir, Moscow, 1968 (Russian) (English translation: Amer. Math. Soc. Transl. Ser. 2, 103 (1974), 1–18).

[24] M.G. Kreĭn and H. Langer, "Über die Q-Funktion eines π-hermiteschen Operators im Raume Π_κ", Acta Sci. Math. (Szeged), 34 (1973), 191–230.

[25] M.G. Kreĭn and H. Langer, "Über einige Fortsetzungsprobleme, die eng mit der Theorie hermitescher Operatoren im Raume Π_κ zusammenhängen. 1. Einige Funktionenklassen und ihre Darstellungen", Math. Nachr., 77 (1977), 187–236.

[26] M.G. Kreĭn and A. Nudelman, *The Markov moment problem and extremal problems*, Transl. Math. Monographs, 51, A.M.S., 1977.

[27] H. Langer and B. Textorius, "On generalized resolvents and Q-functions of symmetric linear relations", Pacific J. Math., 72 (1977), 135–165.

[28] I. Stakgold, *Green's functions and boundary value problems*, John Wiley and Sons, New York, 1979.

[29] E.T. Whittaker and G.N. Watson, *A course of modern analysis*, Fourth edition, Cambridge University Press, 1963.

Seppo Hassi
Department of Mathematics and Statistics
University of Vaasa
P.O. Box 700
65101 Vaasa, Finland
e-mail: `sha@uwasa.fi`

Adrian Sandovici
Colegiul Naţional "Petru Rareş"
Str. Ştefan Cel Mare, Nr. 4
RO-610101 Piatra Neamt, Romania
e-mail: `adrian.sandovici@allstudio.ro`

Henk de Snoo
Department of Mathematics and Computing Science
University of Groningen
P.O. Box 407
9700 AK Groningen, Nederland
e-mail: `desnoo@math.rug.nl`

Henrik Winkler
Institut für Mathematik, MA 6-4
Technische Universität Berlin
Strasse des 17. Juni 136
D-10623 Berlin, Deutschland
e-mail: `winkler@math.tu-berlin.de`

Operator Theory:
Advances and Applications, Vol. 188, 175–195

Abstract Kinetic Equations with Positive Collision Operators

I.M. Karabash

Abstract. We consider "forward-backward" parabolic equations in the abstract form $Jd\psi/dx + L\psi = 0$, $0 < x < \tau \leq \infty$, where J and L are operators in a Hilbert space H such that $J = J^* = J^{-1}$, $L = L^* \geq 0$, and $\ker L = \{0\}$. The following theorem is proved: if the operator $B = JL$ is similar to a self-adjoint operator, then associated half-range boundary problems have unique solutions. We apply this theorem to corresponding nonhomogeneous equations, to the time-independent Fokker-Plank equation $\mu \frac{\partial \psi}{\partial x}(x,\mu) = b(\mu) \frac{\partial^2 \psi}{\partial \mu^2}(x,\mu)$, $0 < x < \tau$, $\mu \in \mathbb{R}$, as well as to other parabolic equations of the "forward-backward" type. The abstract kinetic equation $Td\psi/dx = -A\psi(x) + f(x)$, where $T = T^*$ is injective and A satisfies a certain positivity assumption, is also considered. The method is based on the Krein space theory.

Mathematics Subject Classification (2000). Primary 47N55, 35K70, 47B50; Secondary 35M10, 35K90.

Keywords. Forward-backward parabolic equations, kinetic equations, transport equations, equations of mixed type, J-self-adjoint operator, similarity, regular and singular critical points.

1. Introduction

Consider the equation

$$w(\mu)\frac{\partial \psi}{\partial x}(x,\mu) = \frac{\partial^2 \psi}{\partial \mu^2}(x,\mu) - q(\mu)\psi(x,\mu) \qquad (0 < x < \tau < \infty,\ \mu \in \mathbb{R}), \tag{1.1}$$

and the associated boundary value problem

$$\psi(0,\mu) = \varphi_+(\mu) \quad \text{if} \quad \mu > 0, \qquad \psi(\tau,\mu) = \varphi_-(\mu) \quad \text{if} \quad \mu < 0. \tag{1.2}$$

This work was partly supported by a Postdoctoral Fellowship at the University of Calgary.

Here w and q are locally summable on $\mathbb{R}$ and $\mu w(\mu) > 0$, $\mu \in \mathbb{R}$. So the weight function w changes its sign at 0. We assume also that the Sturm-Liouville operator

$$L : y \mapsto \frac{1}{|w|}(-y'' + qy) \tag{1.3}$$

defined on the maximal domain in the Hilbert space $L^2(\mathbb{R}, |w(x)|dx)$ is self-adjoint. Boundary value problems of this (forward-backward) type arise as various kinetic equations (e.g., [6, 22, 19, 55]; for other applications see [42, 56, 54] and references therein).

In this paper we will consider the following abstract version of equation (1.1):

$$\frac{d\psi}{dx} = -JL\psi(x) \qquad (0 < x < \tau < \infty). \tag{1.4}$$

Here L and J are operators in an abstract Hilbert space H such that L is a self-adjoint (bounded or unbounded) operator and J is a *signature operator* in H, that is $J = J^* = J^{-1}$.

By $P_\pm$ we denote the orthogonal projections onto $H_\pm := \ker(J \mp I)$. Clearly,

$$H = H_+ \oplus H_- \quad \text{and} \quad J = P_+ \oplus P_-.$$

The aim is to find *strong solutions* of the associated boundary problem, i.e., to find continuous functions $\psi : [0, \tau] \to H$ ($\psi \in C([0, \tau]; H)$) which are strongly continuously differentiable on $(0, \tau)$ ($\psi \in C^1((0, \tau); H)$) and satisfy Eq. (1.4) with the following boundary conditions

$$P_+\psi(0) = \varphi_+, \qquad P_-\psi(\tau) = \varphi_-, \tag{1.5}$$

where $\varphi_+ \in H_+$ and $\varphi_- \in H_-$ are given vectors. If we define L by (1.3) and put

$$H = L^2(\mathbb{R}, |w(\mu)|d\mu), \qquad (Jf)(\mu) := (\operatorname{sgn} \mu) f(\mu), \tag{1.6}$$

we get problem (1.1)–(1.2).

For the case when L is nonnegative and has discrete spectrum, problem (1.4)–(1.5) has been described in great detail (see [3, 48, 6, 52, 5, 19, 35, 12, 53] and references therein). For methods used in these papers, the assumption $\sigma(L) = \sigma_{\text{disc}}(L)$ or the weaker assumption $\inf \sigma_{\text{ess}}(L) > 0$ is essential. Generally, the latter assumption is not fulfilled for Eq. (1.1). The simplest example is the equation

$$(\operatorname{sgn} \mu)|\mu|^\alpha \frac{\partial\psi}{\partial x}(x, \mu) = \frac{\partial^2\psi}{\partial\mu^2}(x, \mu) \qquad (0 < x < \tau \leq \infty,\ \mu \in \mathbb{R}), \quad \alpha > -1, \tag{1.7}$$

which arises in kinetic theory and in the theory of stochastic processes (see [47, 19, 55] and references in [46]). Indeed, for this equation the operator $L = -|\mu|^{-\alpha}\frac{d^2}{d\mu^2}$ is self-adjoint in $L^2(\mathbb{R}, |\mu|^\alpha d\mu)$ and $\sigma(L) = \sigma_{\text{ess}}(L) = [0, +\infty)$.

In the case $\tau < \infty$, problem (1.7), (1.2) was studied in [56] with $\alpha = 0$. In the case $\tau = \infty$ (the half-space problem), one has a boundary condition of the type (1.2) at $x = 0$ and, in addition, a growth condition on $\psi(x, \mu)$ for large x. The half-space problem for Eq. (1.7) was considered in [46, 47, 18] in connection with stationary equations of Brownian motion. Note that the methods of [46, 47, 18, 56] use the special form of the weight w and corresponding integral transforms. The results achieved in [47] for the sample case $\alpha = 1$ was used in [48], where a wider class of problems was considered under the hypotheses that

$$\text{the weight} \quad w \quad \text{is bounded} \tag{1.8}$$

and $w(\mu) = \mu + o(\mu)$ as $\mu \to 0$. However, all these results were obtained under additional assumptions on the boundary data. In particular, it was supposed that $\varphi_\pm$ are continuous.

The case when L may be unbounded and may have a continuous spectrum was considered in [17, Section 4], where the half-space problem was studied (in an abstract setting) under the assumptions (1.8) and $L > \delta > 0$. This assumption was changed to $L > 0$ in [7], where also more tricky case $\tau < \infty$ was studied. However, it is difficult to apply the results of [7] to equation (1.7) since an additional assumption on the boundary values $\varphi_\pm$ appears. This assumption is close to assumption of [46, 47, 48]. The method of [17, 7] is based on the spectral theory of self-adjoint operators in Krein spaces.

The aim of this paper is to prove that problem (1.4)–(1.5) has a unique solution for arbitrary $\varphi_\pm \in H_\pm$. In particular, it will be shown that the problem (1.7), (1.2) has a unique solution for arbitrary $\varphi_\pm \in L^2(\mathbb{R}_\pm, |\mu|^\alpha d\mu)$. More general equations of the Fokker-Plank type will also be considered.

Recall that two closed operators T_1 and T_2 in a Hilbert space H are called similar if there exists a bounded and boundedly invertible operator S in H such that $S\,\mathrm{dom}(T_1) = \mathrm{dom}(T_2)$ and $T_2 = ST_1S^{-1}$.

The central result is the following theorem.

Theorem 1.1. *Assume that the operator L is self-adjoint and positive (i.e., $L \geq 0$ and $\ker L = \{0\}$). Assume that the operator JL is similar to a self-adjoint operator in the Hilbert space H. Then for each pair $\{\varphi_+, \varphi_-\}$, $\varphi_\pm \in H_\pm$, there is a unique strong solution ψ of problem (1.4)–(1.5). This solution is given by (2.19).*

The proof is given in Subsection 2.2. It is based on the Krein space approach of [17, 7]. The half-space problem ($\tau = \infty$) is considered in Subsection 2.3. In Section 3 we consider well-posedness and nonhomogeneous equations.

The formal similarity between Eqs. (1.1)–(1.2) and certain problems of neutron transport, radiative transfer, and rarefied gas dynamics has given rise to the emergence of abstract kinetic equation

$$T\frac{d\psi}{dx} = -A\psi(x) \qquad (0 < x < \tau \leq \infty); \tag{1.9}$$

see [20, 4, 43, 22, 19, 17] and references therein. When Eq. (1.9) is considered in a Hilbert space H, the operator T is self-adjoint and injective. The operator A is called *a collision* operator, usually it satisfies certain positivity assumptions (see, e.g., [19]). For unbounded collision operators, equation (1.9) is usually considered in the space $\mathcal{H}_T$ that is a completion of $\mathrm{dom}(T)$ with respect to (w.r.t.) the scalar product $\langle\cdot,\cdot\rangle_T := (|T|\cdot,\cdot)_H$. The interplay between $\mathrm{dom}(T)$ and $\mathrm{dom}(A)$ may be various. This leads to additional assumptions on the operators T and A. It is assumed in [17, Section 4] that T is bounded and $A > \delta > 0$; in [7], $\mathrm{ran}(T) \subset \mathrm{ran}(A)$. Note that equation (1.7) can not be included in these settings.

The second goal of the present paper is to remove the assumptions mentioned above. We will show that the following condition is natural for the case when A is unbounded: *the operator A is a positive self-adjoint operator from $\mathcal{H}_T$ to the space $\mathcal{H}'_T$* that is a completion of $\mathrm{dom}(T^{-1})$ w.r.t. the scalar product $(|T|^{-1}\cdot, \cdot)_H$, see Section 4 for details. It is weaker than the assumptions mentioned above. On the other hand, it characterizes the case when equation (1.9) may be reduced to equation (1.4).

Theorem 1.1 leads to the similarity problem for J-positive differential operators. In Section 5, we use recent results concerning the similarity [8, 16, 31, 37, 32, 38, 39, 30] (see also [24, 25, 13, 33, 28, 29] and references in [32, 30]) to prove uniqueness and existence theorems for various equations of type (1.1).

Note also that abstract kinetic equations with nonsymmetric collision operators may be found in [22, 19, 17, 7, 44, 54]. From another point of view, equation (1.1) belongs to the class of second-order equations with nonnegative characteristic form. Boundary problems for this class of equations were considered by various authors (see [36, 45] and references therein). But some restrictions imposed in this theory makes it inapplicable to Eq. (1.1) (see a discussion in [48]). The case when w is dependent of μ or the operator L is dependent of x was considered, e.g., in [1, 51].

The main results of this paper were announced in the short communications [27, 26].

Notation. Let $\mathcal{A}$ be a linear operator from a Banach space H_1 to a Banach space H_2. In what follows, $\mathrm{dom}(\mathcal{A})$, $\ker \mathcal{A}$, $\mathrm{ran}\, \mathcal{A}$, $\|\mathcal{A}\|_{H_1 \to H_2}$ are the domain, kernel, range, and norm of $\mathcal{A}$, respectively. If M is a subset of H_1, then $\mathcal{A}M := \{\mathcal{A}h : h \in M\}$. In the case $H_1 = H_2$, $\sigma(\mathcal{A})$ and $\rho(\mathcal{A})$ denote the spectrum and the resolvent set of $\mathcal{A}$, respectively. As usual, $\sigma_{\mathrm{disc}}(\mathcal{A})$ denotes the discrete spectrum of $\mathcal{A}$, that is, the set of isolated eigenvalues of finite algebraic multiplicity; the essential spectrum is $\sigma_{\mathrm{ess}}(\mathcal{A}) := \sigma(\mathcal{A}) \setminus \sigma_{\mathrm{disc}}(\mathcal{A})$. By $E^{\mathcal{A}}(\cdot)$ we denote the spectral function of a self-adjoint (or J-self-adjoint) operator $\mathcal{A}$. We write $f \in AC_{\mathrm{loc}}(\mathbb{R})$ if the function f is absolutely continuous on each bounded interval in $\mathbb{R}$. Put $\mathbb{R}_+ := (0, +\infty)$, $\mathbb{R}_- := (-\infty, 0)$, and $\overline{\mathbb{R}} := \mathbb{R} \cup \infty$.

2. Existence and uniqueness of solutions

2.1. Preliminaries

In this section basic facts from the theory of operators in Krein spaces are collected. The reader can find more details in [2, 41].

Consider a complex Hilbert space H with a scalar product $(\cdot, \cdot)_H$ and the norm $\|\cdot\|_H := (\cdot, \cdot)^{1/2}$. An operator $\mathcal{A}$ in H is called *positive* (*negative*) if we have $(\mathcal{A}h, h)_H > 0$ (resp., < 0) for all nonzero $h \in H$. We write $\mathcal{A} > 0$ if $\mathcal{A}$ is positive.

Suppose that $H = H_+ \oplus H_-$, where H_+ and H_- are (closed) subspaces of H. Denote by $P_\pm$ the orthogonal projections from H onto $H_\pm$. Let $J = P_+ - P_-$ and $[\cdot,\cdot] := (J\cdot,\cdot)_H$. Then the pair $\mathcal{K} = (\mathcal{H}, [\cdot,\cdot])$ is called a *Krein space* (see [2, 41] for the original definition). The form $[\cdot,\cdot]$ is called *an inner product* in the Krein space $\mathcal{K}$ and the operator J is called *a fundamental symmetry* (or *a signature operator*) in the Krein space $\mathcal{K}$. Evidently, $(\mathcal{H}, [\cdot,\cdot])$ is a Hilbert space if and only if $H_- = 0$.

A subspace $H_1 \subset H$ is called *non-negative* (*non-positive*) if $[h,h] \geq 0$ (≤ 0, resp.) for all $h \in H_1$. A non-negative (non-positive) subspace H_1 is *uniformly positive (uniformly negative, resp.)* if there is a constant $\alpha > 0$ such that $\alpha\|h\|_H^2 \leq [h,h]$ ($\leq -[h,h]$, resp.) for $h \in H_1$. A non-negative (non-positive) subspace H_1 is called maximal non-negative (non-positive) if for any non-negative (non-positive) subspace $H_2 \supset H_1$ we have $H_2 = H_1$. Subspaces H_1 and H_2 are called J-orthogonal if $[h_1,h_2] = 0$ for all $h_1 \in \mathcal{H}_1$, $h_2 \in \mathcal{H}_2$. We write $H_1 = H_2[\dot{+}]H_3$ if H_1 admits a decomposition into the direct sum of two J-orthogonal subspaces H_2 and H_3.

Suppose that subspaces $\mathcal{H}_\pm$ possess the properties

(i) $\mathcal{H}_+$ is non-negative, $\mathcal{H}_-$ is non-positive,

(ii) $(\mathcal{H}_+, [\cdot,\cdot])$ and $(\mathcal{H}_-, -[\cdot,\cdot])$ are Hilbert spaces;

and suppose that $H = \mathcal{H}_+[\dot{+}]\mathcal{H}_-$; then this decomposition is called a *canonical decomposition* of the Krein space $\mathcal{K}$.

Evidently, $H = H_+ \oplus H_-$ is a canonical decomposition. Note that there exist other canonical decompositions (see Proposition 2.2).

Let $H = \mathcal{H}_+ \oplus \mathcal{H}_-$ be a canonical decomposition. Then the norm $\|\cdot\|_{\mathcal{H}_\pm} := \sqrt{\pm[\cdot,\cdot]}$ in the Hilbert space $(\mathcal{H}_\pm, \pm[\cdot,\cdot])$ is called *an intrinsic norm.* It is easy to prove (see [2, Theorem I.7.19]) that the norms $\|\cdot\|_{\mathcal{H}_\pm}$ and $\|\cdot\|_H$ are equivalent on $\mathcal{H}_\pm$; moreover,

$$\gamma\|h_\pm\|_H \leq \|h_\pm\|_{\mathcal{H}_\pm} \leq \|h_\pm\|_H, \quad h_\pm \in \mathcal{H}_\pm, \tag{2.1}$$

where $\gamma \in (0,1]$ is a constant.

Statement (i) of the following proposition is due to Ginzburg (see [2, Theorems I.4.1 and I.4.5]), Statement (ii) is due to Phillips and Ginzburg (see [2, Lemma I.8.1]).

Proposition 2.1 (e.g., [2]). *Let $H = \mathcal{H}_+ \oplus \mathcal{H}_-$ be a canonical decomposition and let $\mathcal{P}_+$ and $\mathcal{P}_-$ be corresponding mutually complementary projections on $\mathcal{H}_+$ and $\mathcal{H}_-$, respectively.*

(i) *If H_1 is a maximal non-negative subspace in H, then the restriction $\mathcal{P}_+ \restriction H_1 : H_1 \to \mathcal{H}_+$ is a homeomorphism, that is, it is bijective, continuous, and the inverse mapping $(\mathcal{P}_+ \restriction H_1)^{-1} : \mathcal{H}_+ \to H_1$ is also continuous.*

(ii) *If H_1 is a uniformly positive subspace in H, then there is a constant $\beta \in (0,1)$ such that $|[\mathcal{P}_-h, \mathcal{P}_-h]| \leq \beta[\mathcal{P}_+h, \mathcal{P}_+h]$.*

Let $\mathcal{A}$ be a densely defined operator in H. Its J-*adjoint operator of* $\mathcal{A}^{[*]}$ is defined by the relation

$$[\mathcal{A}f, g] = [f, \mathcal{A}^{[*]}g]\ , \qquad f \in \mathrm{dom}(\mathcal{A}),$$

on the set of all $g \in H$ such that the mapping $f \mapsto [\mathcal{A}f, g]$ is a continuous linear functional on $\operatorname{dom}(\mathcal{A})$. The operator $\mathcal{A}$ is called *J-self-adjoint* if $\mathcal{A} = \mathcal{A}^{[*]}$. It is easy to see that $\mathcal{A}^{[*]} = J\mathcal{A}^*J$ and the operator $\mathcal{A}$ is J-self-adjoint if and only if $J\mathcal{A}$ is self-adjoint. Note that $J = J^* = J^{-1} = J^{[*]}$. A closed operator $\mathcal{A}$ is called *J-positive* if $[\mathcal{A}f, f] > 0$ for $f \in \operatorname{dom}(\mathcal{A}) \setminus \{0\}$ (this is equivalent to $J\mathcal{A} > 0$).

Let $\mathfrak{S}$ be the semiring consisting of all bounded intervals with endpoints different from 0 and their complements in $\overline{\mathbb{R}} := \mathbb{R} \cup \infty$.

Let $\mathcal{A}$ be a J-positive J-self-adjoint operator in H with a nonempty resolvent set, $\rho(\mathcal{A}) \neq \emptyset$. Then $\mathcal{A}$ admits a spectral function $E(\Delta)$. Namely,

(i) the spectrum of $\mathcal{A}$ is real, $\sigma(\mathcal{A}) \subset \mathbb{R}$;
(ii) there exists a mapping $\Delta \to E(\Delta)$ from $\mathfrak{S}$ into the set of bounded linear operators in H with the following properties ($\Delta, \Delta' \in \mathfrak{S}$):
   (E1) $E(\Delta \cap \Delta') = E(\Delta)E(\Delta')$, $\quad E(\emptyset) = 0$, $\quad E(\overline{\mathbb{R}}) = I$, $\quad E(\Delta) = E(\Delta)^{[*]}$;
   (E2) $E(\Delta \cup \Delta') = E(\Delta) + E(\Delta')$ if $\Delta \cap \Delta' = \emptyset$;
   (E3) the form $\pm[\cdot, \cdot]$ is positive definite on $E(\Delta)H$ if $\Delta \subset \mathbb{R}_\pm$;
   (E4) $E(\Delta)$ is in the double commutant of the resolvent $R_{\mathcal{A}}(\lambda) = (\mathcal{A} - \lambda)^{-1}$ and $\sigma(\mathcal{A} \restriction E(\Delta)H) \subset \overline{\Delta}$;
   (E5) if Δ is bounded, then $E(\Delta)H \subset \operatorname{dom}(\mathcal{A})$ and $\mathcal{A} \restriction E(\Delta)H$ is a bounded operator.

According to [41, Proposition II.4.2], a number $s \in \{0, \infty\}$ is called *a critical point* of $\mathcal{A}$, if for each $\Delta \in \mathfrak{S}$ such that $s \in \Delta$, the form $[\cdot, \cdot]$ is indefinite on $E(\Delta)\mathfrak{H}$ (i.e., there exist $h_\pm \in E(\Delta)\mathfrak{H}$ such that $[h_+, h_+] < 0$ and $[h_-, h_-] > 0$). The set of critical points is denoted by $c(\mathcal{A})$.

If $\alpha \notin c(\mathcal{A})$, then for arbitrary $\lambda_0, \lambda_1 \in \mathbb{R} \setminus c(\mathcal{A})$, $\lambda_0 < \alpha$, $\lambda_1 > \alpha$, the limits

$$\lim_{\lambda \uparrow \alpha} E([\lambda_0, \lambda]), \qquad \lim_{\lambda \downarrow \alpha} E([\lambda, \lambda_1])$$

exist in the strong operator topology. If $\alpha \in c(\mathcal{A})$ and the above limits do still exist, then the critical point α is called *regular*, otherwise it is called *singular*. Here we agree that, if $\alpha = \infty$, then $\lambda > \alpha$, $\lambda < \alpha$, $\lambda \downarrow \alpha$, and $\lambda \uparrow \alpha$ mean $\lambda > -\infty$, $\lambda < +\infty$, $\lambda \downarrow -\infty$, and $\lambda \uparrow +\infty$, respectively.

The mapping $\Delta \to E(\Delta)$ can be extended to the semiring generated by those intervals whose endpoints are not singular critical points of $\mathcal{A}$. For this extension Properties (E1)–(E5) are preserved.

2.2. Proof of Theorem 1.1.

Let L be a positive operator in a Hilbert space H and let J be a signature operator in H. Put

$$B := JL.$$

Then B is a J-self-adjoint and J-positive operator in the Krein space $\mathcal{K} := (H, [\cdot, \cdot])$, where $[\cdot, \cdot] := (J\cdot, \cdot)_H$. The following proposition is well known.

Proposition 2.2. *Let B be a J-positive J-self-adjoint operator in H such that $\rho(B) \neq \emptyset$. Then the following assertions are equivalent:*

(i) *B is similar to a self-adjoint operator.*
(ii) *0 and ∞ are not singular critical points of B.*
(iii) *There exists a fundamental decomposition $H = H_+^B[\dot{+}]H_-^B$ of the Krein space $\mathcal{K}$ such that $B = B^+\dot{+}B^-$, where $B^\pm := B \restriction \operatorname{dom}(B) \cap H_\pm^B$, and $\sigma(B^\pm) = \overline{\sigma(B) \cap \mathbb{R}_\pm}$.*

If these assertions hold and P_+^B and P_-^B are corresponding mutually complementary projections on H_+^B and H_-^B, respectively, then $P_\pm^B = E^B(\mathbb{R}_\pm)$, where $E^B(\cdot)$ is the spectral function of B defined in Subsection 2.1.

Assume that

$$B \quad \text{is similar to a self-adjoint operator in} H. \tag{2.2}$$

Then $\rho(B) \subset \mathbb{R}$ and assertions (i)–(iii) of Proposition 2.2 hold true. Note that Property (E3) implies that the form $\pm[\cdot,\cdot]$ is non-negative on $H_\pm^B$. Evidently,

$$H_+^B \quad (H_-^B) \quad \text{is a maximal non-negative (non-positive) subspace in} \quad H. \tag{2.3}$$

Moreover, it follows from (2.1) that H_+^B (H_-^B) is uniformly positive (resp., negative). In the sequel, we consider $H_\pm^B$ as Hilbert spaces with the inner products $\pm[\cdot,\cdot]$ and the (intrinsic) norms $\|\cdot\|_{H_\pm^B} := \sqrt{\pm[h,h]}$.

Proposition 2.3. *Let B be similar to a self-adjoint operator. Then the operator B^+ (B^-) defined in Proposition 2.2 is a positive (negative, resp.) self-adjoint operator in the Hilbert space H_+^B (resp., H_-^B).*

Proof. Positivity of B^+ follows immediately from J-positivity of B. B^+ is symmetric since it is positive. It follows from (2.2) that the spectrum of B is real, $\sigma(B) \subset \mathbb{R}$. Thus, $\sigma(B^+) \subset \mathbb{R}$ and therefore B^+ is self-adjoint in H_+^B. Similar arguments hold for B^-. □

Now we write equation (1.4) in the form

$$\frac{d\psi}{dx} = -B\psi(x) \qquad (0 < x < \tau < \infty), \tag{2.4}$$

and suppose that ψ is a solution of (2.4), (1.5).

We put $\psi_\pm(x) := P_\pm^B\psi(x)$ and $E_t^B := E^B((-\infty,t])$. Note that, by Proposition 2.2, E_t^B is well defined for all $t \in \mathbb{R}$ and $E^B(\cdot) \restriction H_\pm^B$ is the spectral function of the self-adjoint operator $B^\pm$. Eq. (2.4) implies that for $x \in [0,\tau]$,

$$\psi_+(x) = \int_{+0}^{+\infty} e^{-xt}\, dE_t^B\, \psi_+(0) = e^{-xB^+}\psi_+(0), \tag{2.5}$$

$$\psi_-(x) = \int_{-\infty}^{-0} e^{(\tau-x)t}\, dE_t^B\, \psi_-(\tau) = e^{(\tau-x)B^-}\psi_-(\tau). \tag{2.6}$$

The integrals converge in the norm topologies of $H_\pm^B$ as well as in the norm topology of H. Recall that, by (2.1), these topologies are equivalent.

It follows immediately from Proposition 2.3 that for all $h \in H_\pm^B$,

$$\|e^{\mp xB^\pm} h\|_{H_\pm^B} \le \|h\|_{H_\pm^B} \qquad \text{if} \quad x \ge 0. \tag{2.7}$$

So for each pair $\widetilde{\psi}_+ \in H_+^B$, $\widetilde{\psi}_- \in H_-^B$, there is a unique solution of (2.4) such that $\psi_+(0) = \widetilde{\psi}_+$, $\psi_-(\tau) = \widetilde{\psi}_-$.

The boundary conditions (1.5) become

$$P_+[\psi_+(0) + e^{\tau B^-}\psi_-(\tau)] = \varphi_+, \quad P_-[\psi_-(\tau) + e^{-\tau B^+}\psi_+(0)] = \varphi_-. \tag{2.8}$$

It follows from (2.3) and Proposition 2.1 (i) that there exist operators

$$R_\pm := (P_\pm \restriction H_\pm^B)^{-1} \ : \ H_\pm \to H_\pm^B; \qquad \text{moreover,} \quad R_\pm \quad \text{are homeomorphisms.} \tag{2.9}$$

Let us introduce operators $G_\pm : H_\pm^B \to H_\mp^B$ by

$$G_+ h_+ := R_- P_- e^{-\tau B^+} h_+, \quad G_- h_- := R_+ P_+ e^{\tau B^-} h_-, \quad h_\pm \in H_\pm^B.$$

Using the operators $R_\pm$ and $G_\pm$, we write (2.8) in the form

$$\psi_+(0) + G_-\psi_-(\tau) = R_+\varphi_+, \quad \psi_-(\tau) + G_+\psi_+(0) = R_-\varphi_-. \tag{2.10}$$

Combining these equations, one gets

$$(I - G_-G_+)\psi_+(0) = R_+\varphi_+ - G_-R_-\varphi_-, \tag{2.11}$$

$$(I - G_+G_-)\psi_-(\tau) = R_-\varphi_- - G_+R_+\varphi_+. \tag{2.12}$$

Lemma 2.4. $\|G^+\|_{H_+^B \to H_-^B} < 1$ *and* $\|G^-\|_{H_-^B \to H_+^B} < 1$.

Proof. Let us prove that for $h_\pm \in H_\pm^B$,

$$\|R_\mp P_\mp h_\pm\|_{H_\mp^B} \le \beta_\pm \|h_\pm\|_{H_\pm^B} \qquad \text{with certain constants} \quad \beta_\pm < 1. \tag{2.13}$$

Indeed, since $R_-P_-h_+ \in H_-^B$, we have $h_+ = P_+^B(h_+ - R_-P_-h_+)$ and therefore

$$\|h_+\|_{H_+^B} = \|P_+^B(h_+ - R_-P_-h_+)\|_{H_+^B}. \tag{2.14}$$

Note that $H = H_+^B[\dot{+}]H_-^B$ is a fundamental decomposition of the Krein space $\mathcal{K}$ and H_+ is a uniformly positive subspace in $\mathcal{K}$. Hence Proposition 2.1 (ii) implies

$$\beta_+\|P_+^B g_+\|_{H_+^B} \ge \|P_-^B g_+\|_{H_-^B} \qquad \text{for all} \quad g_+ \in H_+ \tag{2.15}$$

with a certain $\beta_+ < 1$. Further, $P_-(h_+ - R_-P_-h_+) = 0$, that is, $(h_+ - R_-P_-h_+) \in H_+$. Therefore (2.15) yields

$$\beta_+\|P_+^B(h_+ - R_-P_-h_+)\|_{H_+^B} \ge \|P_-^B(h_+ - R_-P_-h_+)\|_{H_-^B} = \|R_-P_-h_+\|_{H_-^B}.$$

From this and (2.14), we get (2.13) for $h_+ \in H_+^B$. The proof of (2.13) for $h_- \in H_-^B$ is similar. To conclude the proof, it remains to combine (2.13) and (2.7). $\square$

Lemma 2.4 implies that

$$I - G_{\mp}G_{\pm} \qquad \text{are linear homeomorphisms in} \quad H_{\pm}^{B}. \tag{2.16}$$

Let us consider Eqs. (2.8) as a system for the unknowns $\psi_+(0)$ and $\psi_-(\tau)$.

Lemma 2.5 ([7]). *System* (2.8) *has a unique solution*

$$\psi_+(0) = (I - G_-G_+)^{-1}(R_+\varphi_+ - G_-R_-\varphi_-), \tag{2.17}$$

$$\psi_-(\tau) = (I - G_+G_-)^{-1}(R_-\varphi_- - G_+R_+\varphi_+). \tag{2.18}$$

Remark 2.6. Lemma 2.5 was obtained in another form in [7] (see Lemma 2.2 and the end of the proof of Theorem 3.4 there). Earlier, it was proved under the additional condition $\sigma(B) = \sigma_{\text{disc}}(B)$ in [6], see also [52, 19]. We give another proof, which is based on Lemma 2.4 and is an improvement of treatments from [52].

Proof. First note that system (2.8) is equivalent to system (2.10). Further, (2.10) implies (2.11), (2.12). On the other hand, (2.11), (2.12) is equivalent to (2.17)–(2.18) due to (2.16). Thus we should only prove that (2.11), (2.12) implies (2.10).

Let us write Eqs. (2.11), (2.12) as

$$R_+\varphi_+ = (I - G_-G_+)\psi_+(0) + G_-R_-\varphi_-,$$
$$R_-\varphi_- = (I - G_+G_-)\psi_-(\tau) + G_+R_+\varphi_+.$$

Combining these two equalities, we get

$$(I - G_-G_+)R_+\varphi_+ = (I - G_-G_+)\psi_+(0) + G_-(I - G_+G_-)\psi_-(\tau),$$
$$(I - G_+G_-)R_-\varphi_- = (I - G_+G_-)\psi_-(\tau) + G_+(I - G_-G_+)\psi_+(0).$$

Note that $G_{\pm}(I - G_{\mp}G_{\pm}) = (I - G_{\pm}G_{\mp})G_{\pm}$ and therefore, using (2.16), one obtains (2.10). □

This lemma shows that the function

$$\psi(x) = e^{-tB^+}(I - G_-G_+)^{-1}(R_+\varphi_+ - G_-R_-\varphi_-) + \\ + e^{(\tau-t)B_-}(I - G_+G_-)^{-1}(R_-\varphi_- - G_+R_+\varphi_+) \tag{2.19}$$

is a unique solution of problem (1.4)–(1.5).

One can check that the functions $\psi_{\pm}$ are continuous on $[0,\tau]$, the strong derivatives $d\psi_{\pm}/dx$ exist and are (strongly) continuous on $(0,\tau)$ (see Subsection 3). This completes the proof of Theorem 1.1.

2.3. Half-space problems

Under the same assumptions, let us consider the equation

$$\frac{d\psi}{dx} = -JL\psi(x) \qquad (0 < x < \infty), \tag{2.20}$$

on the infinite interval $(0, +\infty)$. The boundary conditions

$$P_+\psi(0) = \varphi_+, \tag{2.21}$$

$$\|\psi(x)\|_H = O(1) \qquad (x \to +\infty) \tag{2.22}$$

correspond to this feature of the problem (see, e.g., [19]). As above, $\varphi_+ \in H_+$ is a given vector.

Theorem 2.7. *Assume that $L = L^* > 0$ and that the operator $B := JL$ is similar to a self-adjoint operator in the Hilbert space H. Then for each $\varphi_+ \in H_+$ there is a unique solution ψ of (2.20)–(2.22). This solution is given by*

$$\psi(x) = \int_{+0}^{+\infty} e^{-xt}\, dE_t^B R_+\varphi_+ \quad (= e^{-xB^+} R_+\varphi_+), \tag{2.23}$$

where E_t^B, R_+, and B^+ are the operators defined in Subsection 2.2.

The proof is simpler than the proof of Theorem 1.1. It is similar to the treatments from [17, Section 4], where equation (1.9) with bounded T and uniformly positive A was considered. We give a sketch here.

Proof. Let ψ be a solution of problem (2.20)–(2.22) and $\psi_\pm(\cdot) := P_\pm^B \psi(\cdot)$. It follows from (2.6) that for all $x \in (0, +\infty)$,

$$\psi_-(0) \in \operatorname{ran}(e^{xB^-}) = \operatorname{dom}(e^{-xB^-}) \qquad \text{and} \quad \psi_-(x) = e^{-xB^-}\psi_-(0).$$

Since $B^- < 0$, we see that

$$\lim_{x\to+\infty} \|\psi_-(x)\|_{H_-^B} = \infty \qquad \text{whenever} \quad \psi_-(0) \neq 0.$$

Taking into account (2.1) and (2.22), we get $\psi_-(0) = \psi_-(x) = 0$ for all $x \in (0, \infty)$. Hence, $\psi(x) = \psi_+(x)$, $x \geq 0$, and therefore (2.21) yields $P_+\psi_+(0) = \varphi_+$. Combining this with (2.21) and (2.9), one can see that $\psi(x) = \psi_+(x) = e^{-xB^+}R_+\varphi_+$ is an only function that satisfies (2.20) and (2.21). Finally, note that (2.22) follows from (2.7) and (2.1). □

Remark 2.8. Clearly, Theorem 2.7 is valid with the condition $\|\psi(x)\|_H = o(1)$ as $x \to +\infty$ instead of (2.22).

3. Well-posedness and nonhomogeneous problems

Let the assumptions of Subsection 2.2 be fulfilled.

Since B^+ is a positive self-adjoint operator in H_+^B, we see that $U_+(z) := e^{-zB^+}$ is a bounded holomorphic semigroup in the sector $|\arg z| < \pi/2$ (see, e.g., [34, Subsection IX.1.6]). The same is true of the function $U_-(z) := e^{zB^-}$. In particular, this implies that for any $\widetilde{\psi}_\pm \in H_\pm^B$ the problems

$$d\psi_+(x)/dx = -B^+\psi_+(x), \quad x > 0, \qquad \psi_+(0) = \widetilde{\psi}_+, \tag{3.1}$$

$$d\psi_-(x)/dx = -B^-\psi_-(x), \quad x < \tau, \qquad \psi_-(\tau) = \widetilde{\psi}_-, \tag{3.2}$$

have unique solutions $\psi_+(x) = U_+(x)\widetilde{\psi}_+$ and $\psi_-(x) = U_-(\tau - x)\widetilde{\psi}_-$, respectively. Besides, these solutions are infinitely differentiable on $(0, \tau)$ and problems (3.1),

(3.2) are uniformly well posed (see, e.g., [40, Subsection I.2]). The latter means that the mappings $\widetilde{\psi}_\pm \to \psi_\pm(\cdot)$ from $H_\pm^B$ to $C([0,\tau];H_\pm^B)$ are continuous.

Lemma 2.5 and (2.5)–(2.6) show that the solution ψ of problem (1.4)–(1.5) possesses the same properties:

(i) ψ is infinitely differentiable on $(0,\tau)$,
(ii) the mapping $\{\varphi_+,\varphi_-\} \to \psi(\cdot)$ from $H_+ \oplus H_-$ to $C([0,\tau];H)$ is continuous.

One can obtain similar statements for the solution of problem (2.20)–(2.22).

Now consider the nonhomogeneous equation

$$\frac{d\psi}{dx} = -JL\psi(x) + f(x) \qquad (0 < x < \tau \le \infty), \tag{3.3}$$

where f is an H-valued function.

We assume that f is Hölder continuous on all finite intervals $[0,x_1]$, i.e.,

$$\text{for each} \quad x_1 \in \mathbb{R}_+ \quad \text{there are numbers} \quad K = K(x_1) > 0, \quad k = k(x_1) \in (0,1]$$
$$\text{such that} \quad \|f(x) - f(y)\|_H \le K|x-y|^k \qquad \text{for} \quad 0 \le x,y \le x_1. \tag{3.4}$$

Evidently, the functions $f_\pm(x) := P_\pm^B f(x)$ possess the same property.

Let us start with the case $\tau < \infty$ and boundary conditions (1.5).

The fact that $U_\pm(z)$ are bounded holomorphic semigroups enables us to apply [34, Theorem IX.1.27]. This theorem yields that the functions

$$\psi_1^+(x) := \int_0^x U_+(x-y)f_+(y)dy \quad \text{and} \quad \psi_1^-(x) := -\int_x^\tau U_-(y-x)f_-(y)dy \tag{3.5}$$

are continuous for $x \in [0,\tau]$, continuously differentiable for $x \in (0,\tau)$ and $d\psi_1^\pm/dx = -B^\pm\psi_1^\pm + f_\pm$. By Theorem 1.1, there exists a unique solution ψ_0 of homogeneous equation (2.4) satisfying the boundary conditions

$$P_+\psi_0(0) = \varphi_+ - P_+\psi_1^-(0), \qquad P_-\psi_0(\tau) = \varphi_- - P_-\psi_1^+(\tau).$$

The representation of ψ_0 may be obtained from (2.19). Thus we prove the following statement.

Proposition 3.1. *Let assumptions* (2.2) *and* (3.4) *be fulfilled. Then problem* (3.3), (1.5) *has a unique solution ψ given by $\psi = \psi_1^+ + \psi_1^- + \psi_0$.*

In the case $\tau = \infty$, we assume additionally that

$$\int_0^{+\infty} \|f(x)\|_H dx < \infty. \tag{3.6}$$

Let us define $\psi_1^\pm(\cdot)$ by (3.5) (with $\tau = +\infty$). Assumption (3.6) and inequality (2.7) yield that $\psi_1^-(\cdot)$ is well defined on $[0,\infty)$.

Proposition 3.2. *Let $\tau = \infty$, let assumptions* (2.2), (3.4), *and* (3.6) *be fulfilled. Then problem* (3.3), (2.21), (2.22) *has a unique solution ψ given by $\psi = \psi_1^+ + \psi_1^- + \psi_0$, where*

$$\psi_0(x) = e^{-xB^+}R_+(\varphi_+ - P_+\psi_1^-(0)).$$

Proof. For $0 < x < X$, we write $\psi_1^-(x)$ in the form $v_1(x) + v_2(x)$, where

$$v_1(x) = -\int_x^X U_-(y-x)f_-(y)dy, \quad v_2(x) = -\int_X^{+\infty} U_-(y-x)f_-(y)dy.$$

Since $v_2(x) = -U_-(X-x)\int_X^{+\infty} U_-(y-X)f_-(y)dy$, we see that $dv_2(x)/dx = -B^- v_2(x)$ for $0 < x < X$. On the other hand, [34, Theorem IX.1.27] yields that v_1 is a solution of nonhomogeneous equation $d\psi^-(x)/dx = -B^-\psi^-(x) + f_-(x)$ for $0 < x < X$ and, therefore, so is ψ_1^-. The latter holds for all $x > 0$ since X is arbitrary. Thus $\psi_1^+ + \psi_1^-$ is a solution of nonhomogeneous equation (3.3).

It follows from (3.6) and (2.7) that $\|\psi_1^+(x)\| = O(1)$ and $\|\psi_1^-(x)\| = o(1)$ as $x \to \infty$. Using Theorem 2.7, one can conclude the proof. □

If $L \geq \lambda_0 > 0$, then condition (3.6) can be relaxed. One can change it to $\|f(x)\| = O(1)$ as $x \to \infty$ (cf. [19, Section II]).

4. Abstract kinetic equations

Let $\mathcal{H}$ be a complex Hilbert space with scalar product $\langle\cdot,\cdot\rangle$ and norm $\|\cdot\|_{\mathcal{H}}$.

Assume that T is a (bounded or unbounded) self-adjoint operator in $\mathcal{H}$ and that T is injective (i.e., $\ker T = \{0\}$).

Let $Q_+ := E^T(\mathbb{R}_+)$ ($Q_- := E^T(\mathbb{R}_-)$) be the orthogonal projection of $\mathcal{H}$ onto the maximal T-invariant subspace on which T is positive (negative). Then $|T| := (Q_+ - Q_-)T$ is a positive self-adjoint operator. Note that

$$Q_\pm T = TQ_\pm \qquad \text{and} \qquad Q_\pm T^{-1} = T^{-1}Q_\pm. \tag{4.1}$$

Following [4], let us introduce the scalar product $\langle h,g\rangle_T = \langle |T|h, g\rangle$ for $h,g \in \operatorname{dom}(T)$ with corresponding norm $\|\cdot\|_T$ and denote by $\mathcal{H}_T$ the completion of $\operatorname{dom}(T)$ with respect to (w.r.t.) this norm. Clearly, $\mathcal{H} \cap \mathcal{H}_T = \operatorname{dom}(|T|^{1/2})$ and $\|h\|_T = \|\,|T|^{1/2}h\,\|_{\mathcal{H}}$ for $h \in \operatorname{dom}(|T|^{1/2})$.

Similarly, we may introduce another scalar product $\langle h,g\rangle_T' = \langle |T|^{-1}h, g\rangle_T$ on $\operatorname{dom}(T^{-1})$ with associated norm $\|\cdot\|_T'$ and consider the completion $\mathcal{H}_T'$ of $\operatorname{dom}(T^{-1})$ w.r.t. the norm $\|\cdot\|_T'$. As before, $\mathcal{H} \cap \mathcal{H}_T' = \operatorname{dom}(|T|^{-1/2})$.

It is easy to see that for each $g \in \operatorname{ran}(|T|) = \operatorname{dom}(T^{-1})$, the linear functional $\langle\cdot, g\rangle$ is continuous on $\operatorname{dom}(T)$ w.r.t. the norm $\|\cdot\|_T$. Besides, its norm is equal to

$$\sup_{\substack{h\in\operatorname{dom}(T)\\ h\neq 0}} \frac{|\langle h,g\rangle|}{\|\,|T|^{1/2}h\,\|_{\mathcal{H}}} = \sup_{\substack{h'\in\operatorname{dom}(|T|^{1/2})\\ h'\neq 0}} \frac{|\langle |T|^{-1/2}h', g\rangle|}{\|\,h'\,\|_{\mathcal{H}}} = \|\,|T|^{-1/2}g\,\|_{\mathcal{H}} = \|g\|_T'.$$

So one can use the $\mathcal{H}$-scalar product as a pairing to identify $\mathcal{H}_T'$ with the dual $\mathcal{H}_T^*$ of $\mathcal{H}_T$. The operator $|T|$ (the operator $|T|^{-1}$) has a natural extension to an isometric operator from $\mathcal{H}_T$ onto $\mathcal{H}_T'$ (from $\mathcal{H}_T'$ onto $\mathcal{H}_T$). We use the same notation for the extensions.

By (4.1), we may extend the orthogonal projections $Q_\pm$ onto $\mathcal{H}_T$ and $\mathcal{H}_T'$. Put $Th := (Q_+ - Q_-)|T|h$ for $h \in \mathcal{H}_T$ and $T^{-1}h := (Q_+ - Q_-)|T|^{-1}h$ for $h \in \mathcal{H}_T'$.

Now let A be a linear operator in $\mathcal{H}+\mathcal{H}_T+\mathcal{H}'_T$, let φ be a vector in $\mathcal{H}_T$, and let $f(\cdot)$ be a function with values in $\mathcal{H}'_T$. Consider an abstract kinetic equation

$$T\frac{d\psi}{dx} = -A\psi(x) + f(x) \qquad (0 < x < \tau \leq \infty), \tag{4.2}$$

supplemented by "half-range" boundary conditions in the form

$$Q_+\psi(0) = Q_+\varphi, \tag{4.3}$$

$$Q_-\psi(\tau) = Q_-\varphi \ \text{ if } \tau < \infty, \quad \text{or } \|\psi(x)\|_{\mathcal{H}_T} = O(1) \quad (x \to +\infty) \ \text{ if } \tau = \infty. \tag{4.4}$$

In the abstract kinetic theory the operator A is called *a collision operator*. It may have various properties (see [22, 19]).

Here we consider the case when A *is a positive self-adjoint operator from* $\mathcal{H}_T$ *to* $\mathcal{H}'_T$. That is,

(A1) $\operatorname{dom}(A) \subset \mathcal{H}_T$, and $Ah \in \mathcal{H}'_T$ for all $h \in \operatorname{dom}(A)$,
(A2) $\langle Ah, h\rangle > 0$ for all $h \in \operatorname{dom}(A) \setminus \{0\}$,
(A3) $A = A^*$, i.e., $\operatorname{dom}(A)$ coincides with the set of all $g \in \mathcal{H}_T$ such that the mapping $h \mapsto \langle Ah, g\rangle$ is a continuous linear functional w.r.t. the $\mathcal{H}_T$-norm.

We seek $\mathcal{H}_T$*-strong solutions* (week solutions in terms of [19, Section 2]) of problem (4.2)–(4.4). That is, it is supposed that d/dx is the strong derivative in $\mathcal{H}_T$ and that

(i) $\psi(\cdot) \in C([0,\tau];\mathcal{H}_T)$ (or $\psi(\cdot) \in C([0,+\infty);\mathcal{H}_T)$ if $\tau = \infty$),
(ii) $\psi(\cdot) \in C^1((0,\tau);\mathcal{H}_T)$ and $\psi(x) \in \operatorname{dom}(A)$ for $x \in (0,\tau)$.

Theorem 4.1. *Let B be the operator in $\mathcal{H}_T$ defined by $B := T^{-1}A$, $\operatorname{dom}(B) = \operatorname{dom}(A)$. Assume that the function f is Hölder continuous on all finite intervals $[0, x_1]$, $x_1 > 0$, w.r.t. the $\mathcal{H}'_T$-norm (cf. (3.4)), and that*

$$B \quad \textit{is similar to a self-adjoint operator in} \quad \mathcal{H}_T. \tag{4.5}$$

Then problem (4.2)–(4.4) has a unique $\mathcal{H}_T$-strong solution for each $\varphi \in \mathcal{H}_T$.

Proof. Put $P_\pm := Q_\pm \restriction \mathcal{H}_T$. Then P_+ and P_- are mutually complementary orthogonal projections in $\mathcal{H}_T$ and $J = P_+ - P_-$ is a signature operator.

Note that $L := JB$ is a positive self-adjoint operator in $\mathcal{H}_T$ (that is, B is J-positive and J-self-adjoint). Indeed, since $L = (Q_+ - Q_-)T^{-1}A$ and $(Q_+ - Q_-)T^{\pm 1} = |T|^{\pm 1}$, we get

$$\langle Lh, g\rangle_T = \langle |T|(Q_+ - Q_-)T^{-1}Ah, g\rangle = \langle Ah, g\rangle, \qquad h \in \operatorname{dom}(A),\ g \in \mathcal{H}_T. \tag{4.6}$$

Hence (A1)–(A3) implies that $L = L^* > 0$.

So problems (4.2)–(4.4) are reduced to the corresponding problems for equation (3.3). Thus Theorem 4.1 follows from Propositions 3.1 and 3.2. □

The form of the solutions is described by Propositions 3.1, 3.2 and formula (2.19).

Remark 4.2. Actually, assumption (A3) is equivalent to the self-adjointness of the operator $L = (Q_+ - Q_-)B$ in $\mathcal{H}_T$. Note also that (A3) follows from (A1),(A2) and (4.5).

Indeed, (4.5) implies that L is closed and $\mathrm{dom}(L)(= \mathrm{dom}(B))$ is dense in $\mathcal{H}_T$. Assumption (A2) implies that $L > 0$. Assume that $L \neq L^*$. Then L admits a self-adjoint extension $\widetilde{L} \geq 0$, and $\widetilde{L} \supsetneqq L$. Further, the operator $\widetilde{B} := J\widetilde{L}$ is a J-non-negative J-self-adjoint extension of B. Therefore [2, Theorem II.3.25] implies $\sigma_p(\widetilde{B}) \subset \mathbb{R}$. But (4.5) yields that $\mathrm{ran}(B - \lambda I) = \mathcal{H}_T$ for all $\lambda \in \mathbb{C} \setminus \mathbb{R}$. From this and $\widetilde{B} \supsetneqq B$, one gets $\mathbb{C} \setminus \mathbb{R} \subset \sigma_p(\widetilde{B})$. This contradiction concludes the proof.

5. Examples

If the spectrum $\sigma(B)$ is real and discrete, then assumption (2.2) is equivalent to the Riesz basis property for eigenfunctions of B. For ordinary and partial differential operators with indefinite weights, the Riesz basis property was studied in great detail (see [23, 52, 5, 8, 15, 53, 50] and references therein). Below we consider several classes of differential equations with $B(= JL)$ such that $\sigma(B) \neq \overline{\sigma_{\mathrm{disc}}(B)}$. The theorems obtained in the previous sections are combined with known similarity results for Sturm-Liouville operators with an indefinite weight. First, we consider in details a nonhomogeneous version of equation (1.7). Other applications will be indicated briefly (for homogeneous equations and the case $\tau < \infty$ only). Using Propositions 3.1 and 3.2, one can extend these treatments to the half-space problems and the nonhomogeneous case.

5.1. The equation $(\operatorname{sgn} x)|\mu|^\alpha \psi_x = \psi_{\mu\mu} + f$

Let us consider the equation

$$(\operatorname{sgn} \mu)|\mu|^\alpha \frac{\partial \psi}{\partial x}(x, \mu) = \frac{\partial^2 \psi}{\partial \mu^2}(x, \mu) + f(x, \mu) \qquad (0 < x < \tau \leq \infty, \ \mu \in \mathbb{R}), \quad (5.1)$$

where $\alpha > -1$ is a constant. In the case $\tau < \infty$, the associated boundary conditions take the form

$$\psi(0, \mu) = \varphi(\mu) \quad \text{if} \quad \mu > 0, \qquad \psi(\tau, \mu) = \varphi(\mu) \quad \text{if} \quad \mu < 0. \quad (5.2)$$

If $\tau = \infty$, we should change them to

$$\psi(0, \mu) = \varphi(\mu) \quad \text{if} \quad \mu > 0, \qquad \int_{\mathbb{R}} |\psi(x, \mu)|^2 |\mu|^\alpha d\mu = O(1) \quad \text{as } x \to +\infty. \quad (5.3)$$

To write (5.1) in the form (4.2), one can put

$$\mathcal{H} = L^2(\mathbb{R}), \quad (Ty)(\mu) = (\operatorname{sgn} \mu)|\mu|^\alpha y(\mu) \quad \text{and} \quad A : y \mapsto -y''.$$

Then,

$$\mathcal{H}_T = L^2(\mathbb{R}, |\mu|^\alpha d\mu), \qquad \mathcal{H}_T' = L^2(\mathbb{R}, |\mu|^{-\alpha} d\mu).$$

It is assumed that A is an operator from $\mathcal{H}_T$ to $\mathcal{H}'_T$ and that it is defined on the natural domain

$$\operatorname{dom}(A) = \{y \in \mathcal{H}_T : y, y' \in AC_{\mathrm{loc}}(\mathbb{R}) \quad \text{and} \quad y'' \in \mathcal{H}'_T\}.$$

One can find the operators $Q_\pm$ (see Section 4) and check that $J := (Q_+ - Q_-) \upharpoonright \mathcal{H}_T$ coincides with J defined by (1.6). Consider the operators $B := T^{-1}A$ and $L := JB$. Both operators are defined on $\operatorname{dom}(A)$ by the following differential expressions

$$B : y(\mu) \mapsto -\frac{(\operatorname{sgn}\mu)}{|\mu|^\alpha} y''(\mu), \qquad L : y(\mu) \mapsto -\frac{1}{|\mu|^\alpha} y''(\mu).$$

Clearly, $By, Ly \in \mathcal{H}_T$ for all $y \in \operatorname{dom}(A)$. So B and L are operators in $\mathcal{H}_T$. It is easy to check (see, e.g., [16]) that

$$L \quad \text{is a positive self-adjoint operator in} \quad \mathcal{H}_T. \tag{5.4}$$

It follows from (5.4), Remark 4.2 and (4.6) that conditions (A1)–(A3) from Section 4 are fulfilled for the operator A. It was proved in [16] (see also [10, 24, 37]) that the operator B is similar to a self-adjoint operator in the Hilbert space $L^2(\mathbb{R}, |\mu|^\alpha d\mu)$.

By Theorem 4.1, we obtain the following result.

Assume that $\int_{\mathbb{R}} |f(x,\mu)|^2 |\mu|^{-\alpha} d\mu < \infty$, $x \in (0,\tau)$, *and that* $f(x)$ *is Hölder continuous on all finite intervals* $[0, x_1]$ *(see* (3.4)*) as a function with values in* $L^2(\mathbb{R}, |\mu|^{-\alpha} d\mu)$*; then there is a unique solution of problem* (5.1), (5.2) *(problem* (5.1), (5.3), *in the case* $\tau = +\infty$*) for every* $\varphi \in L^2(\mathbb{R}, |\mu|^\alpha d\mu)$.

Remark 5.1. In the case $\tau < \infty$ and $\alpha = 0$, problem (5.1)–(5.2) was considered in [56] under additional assumptions that φ belongs to a certain Hölder class. The half-space problem ($\tau = \infty$) was studied in [46, 18] (see also remarks in [48, Appendix II]). More precisely, in [46], the homogeneous equation was considered for all $\alpha > -1$ under the assumption $\int_{\mathbb{R}_+} (|\varphi|^2 |\mu|^\alpha + |\varphi'|^2) d\mu < \infty$. In [18], the non-homogeneous case was considered for $\alpha = 1$ and $\varphi(\cdot)$, $f(\cdot)$ from certain classes of continuous functions. Explicit integral representations for solutions were obtained in [46, 18, 56]. Note also that Eq. (5.1) was studied in [3] for μ in a finite interval $[-a, a]$, however, the latter makes the spectrum of B discrete.

5.2. The case when L is uniformly positive

Consider equation (1.4) under the assumption $L = L^* > \delta > 0$, i.e., the operator L is uniformly positive in the Hilbert space H. As before, put $B = JL$, where J is a signature operator in H. In this case, B is similar to a self-adjoint operator iff ∞ is not a singular critical point of B (see Proposition 2.2). For ordinary differential operators with indefinite weights, the regularity of the critical point ∞ is well studied even in the case of a finite number of turning points (i.e., the points where the weight w changes sign). We will use one result that follows from [8].

Let the functions w, p, q be such that

$$w, q \in L^1_{\mathrm{loc}}(\mathbb{R}), \qquad p(\mu) > 0 \quad \text{a.e. on} \quad \mathbb{R}, \qquad \text{and} \quad p^{-1}, p \in L^\infty_{\mathrm{loc}}(\mathbb{R}). \tag{5.5}$$

Assume that the maximal operator $L : y \mapsto \frac{1}{|w|}(-(py')' + qy)$ is self-adjoint in the Hilbert space $L^2(\mathbb{R}, |w(\mu)|d\mu)$, i.e., it is in the limit point case both at $+\infty$ and $-\infty$. Assume also that the sets

$$\mathcal{I}_+ := \{\mu \in \mathbb{R} : w(\mu) > 0\} \quad \text{and} \quad \mathcal{I}_- := \{\mu \in \mathbb{R} : w(\mu) < 0\}$$

are both of positive Lebesgue measure.

The elements of the set $\overline{\mathcal{I}_+} \cap \overline{\mathcal{I}_-}$ are called *turning points of* w.

Put $(Jf)(\mu) := (\operatorname{sgn} w(\mu))f(\mu)$ for $f \in L^2(\mathbb{R}, |w(\mu)|d\mu)$, and $B = JL$. Then

$$B : y \mapsto \frac{1}{w}(-(py')' + qy)$$

is a J-self-adjoint operator in $L^2(\mathbb{R}, |w(\mu)|d\mu)$.

The following definition is an improved version of Beals' condition [5].

Definition 5.2 ([8]). A function w is said to be *simple from the right at* μ_0 if there exists $\delta > 0$ such that w is nonnegative (nonpositive) on $[\mu_0, \mu_0 + \delta]$ and

$$w(\mu) = (\mu - \mu_0)^\beta \rho(\mu) \qquad (w(\mu) = -(\mu - \mu_0)^\beta \rho(\mu), \text{ respectively}) \tag{5.6}$$

holds a.e. on $[\mu_0, \mu_0+\delta]$ with some $\beta > -1$, $\rho \in C^1[\mu_0, \mu_0+\delta]$, $\rho(\mu_0) > 0$. A function w is said to be *simple from the left at* μ_0 if the function $\mu \mapsto w(-(\mu - \mu_0) + \mu_0)$ is simple from the right at μ_0. A function w is said to be *simple at* μ_0 if it is simple from the right and simple from the left at μ (with, possibly, different numbers β).

It follows from the results of [8] that ∞ is a regular critical point of B if the following assumptions are fulfilled:

(i) the function w has a finite number of turning points at which it is simple,
(ii) $L = L^* \geq 0$ and $\rho(B) \neq \emptyset$.

For the proof, we note that the set $\mathcal{D}[JB] = \mathcal{D}[L]$ is separated (in terms of [8, Subsection 3.2]) since L is in the limit point case at $+\infty$ and $-\infty$. So the case (i) of [8, Theorem 3.6] holds for B.

If, additionally, $L > \delta > 0$ then $0 \in \rho(L)$ and, consequently, $0 \in \rho(B)$ since J is a unitary operator. Besides, 0 is not a critical point of B. Hence one can obtain the following statement from Proposition 2.2 and [8, Theorem 3.6].

Proposition 5.3. *Assume that the function w has a finite number of turning points at which it is simple, and that $L = L^* > \delta > 0$. Then the operator B is similar to a self-adjoint operator in $L^2(\mathbb{R}, |w(\mu)|d\mu)$.*

Thus, if the conditions of Proposition 5.3 are satisfied, Theorem 1.1 implies that there is a unique strong solution of the corresponding boundary problem (1.4)–(1.5) (cf. [17, Section 4]).

Note that the condition $L > \delta > 0$ is fulfilled whenever

$$\frac{q(\mu)}{|w(\mu)|} > C > 0, \quad \text{where } C \text{ is a constant (cf. [48, Appendix I]).} \tag{5.7}$$

Example (cf. [49]). Consider the equation

$$(\operatorname{sgn}\mu)|\mu|^{\alpha}\frac{\partial\psi}{\partial x} = \frac{\partial^2\psi}{\partial\mu^2} - k\psi \qquad (0 < x < \tau < \infty,\ \mu \in \mathbb{R}), \tag{5.8}$$

where $k > 0$ and $\alpha \in (-1, 0]$ are constants. Evidently, the weight function $w(\mu) = (\operatorname{sgn}\mu)|\mu|^{\alpha}$ is simple at its only turning point 0. Besides, condition (5.7) is satisfied since $\alpha \in (-1, 0]$. So problem (5.8), (5.2) has a unique solution for every $\varphi \in L^2(\mathbb{R}, |\mu|^{\alpha}d\mu)$.

Remark 5.4. The half-space problem for equation (5.8) with $\alpha = 1$ and $k > 0$ was studied in [47]. Note that if $\alpha > 0$, the operator L associated with Eq. (5.8) is not uniformly positive. To extend the method of the present paper to the case $\alpha > 0$, one should prove that 0 is a regular critical point of the operator $B : y \mapsto (\operatorname{sgn}\mu)|\mu|^{-\alpha}(-y'' + ky)$ (cf. Subsection 5.3).

Improvements of condition (5.6) may be found in [8] (the end of Subsection 3.1) and [15]. Note also that [8, Theorem 3.6] is valid for higher order ordinary differential operators.

5.3. The Fokker-Plank equation

In the case when $\inf \sigma_{\mathrm{ess}}(L) = 0$, the similarity problem for the operator $B : y \mapsto \frac{1}{w}(-(py')' + qy)$ is more difficult. This question was considered in [10, 16, 24, 13, 31, 37, 32, 38, 28, 29] (see also references therein). A general method was developed in [31, 32], where the operator B with $w(\mu) = \operatorname{sgn}\mu$ and $p(\mu) \equiv 1$ was studied. However, all results contained in [32, Sections 3–6] are valid without changes in proofs for the general Sturm-Liouville operator B with one turning point. The approach of [32] was applied to the case $q \equiv 0$ in [38, 39, 30], where the following theorem was proved.

Theorem 5.5 ([38, 39, 30]). *Assume that $w \in L^1_{\mathrm{loc}}(\mathbb{R})$, $\mu w(\mu) > 0$ for a.a. $\mu \in \mathbb{R}$, and the function w is simple at its only turning point 0. Assume also that w has the form $w(\mu) = \pm r(\mu)|\mu|^{\alpha_\pm}$ for $\mu \in \mathbb{R}_\pm$, where the function r satisfies the following conditions*

$$\int_1^{+\infty} \mu^{\alpha_+/2}|r(\mu) - c_+|d\mu < \infty, \qquad \int_{-\infty}^{-1} |\mu|^{\alpha_-/2}|r(\mu) - c_-|d\mu < \infty, \tag{5.9}$$

and $\alpha_\pm > -1$, $c_\pm > 0$ are constants. Then the operator $B : y \mapsto -\frac{1}{w}y''$ defined on its maximal domain in $L^2(\mathbb{R}, |w(\mu)|d\mu)$ is similar to a self-adjoint operator.

Evidently, the operator $L : y \mapsto -\frac{1}{|w|}y''$ is nonnegative in $L^2(\mathbb{R}, |w(\mu)|d\mu)$. Moreover, if condition (5.9) is fulfilled, L is positive. Indeed, any solution of the equation $Ly_0 = 0$ has the form $y_0(\mu) = c_1 + c_2\mu$, $\mu \in \mathbb{R}$, where c_1 and c_2 are constants. But it follows from (5.9) that $y_0 \in L^2(\mathbb{R}, |w(\mu)|d\mu)$ only if $c_1 = c_2 = 0$.

Applying Theorem 1.1, we obtain the existence and uniqueness theorem for the time-independent Fokker-Plank equation of the simplest kind (see, e.g., [55] and also references in [6])

$$\mu \frac{\partial \psi}{\partial x}(x,\mu) = b(\mu)\frac{\partial^2 \psi}{\partial \mu^2}(x,\mu) \qquad (0 < x < \tau < \infty, \ \mu \in \mathbb{R}). \tag{5.10}$$

Namely, *if the assumptions of Theorem* 5.5 *are satisfied for the function* $w(\mu) = b(\mu)\mu^{-1}$, *then the boundary value problem* (5.10), (5.2) *has a unique solution for every* $\varphi \in L^2(\mathbb{R}, |w(\mu)|d\mu)$.

Remark 5.6. The case when B is an ordinary differential operator of higher order and $\inf \sigma_{\mathrm{ess}}(L) = 0$ was considered in [10, 24, 25]. For partial differential operators, see [9, 14, 11, 21].

Acknowledgment

The author expresses his gratitude to V.A. Derkach, who drew author's attention to the papers [5, 35], to M.M. Malamud, V.A. Marchenko, and H. Stephan for stimulating discussions about this circle of problems, and to the referee for carefully reading the article. The author wishes to thank for the hospitality the Technische Universität Berlin, the University of Zurich, the Johann Radon Institute for Computational and Applied Mathematics, and the University of Calgary, where various parts of this paper were written.

References

[1] J. Aarão, *A transport equation of mixed type.* J. Differential Equations **150** (1998), no. 1, 188–202.

[2] T.Ya. Azizov, I.S. Iokhvidov, *Linear operators in spaces with an indefinite metric.* John Wiley and Sons, 1989.

[3] M.S. Baouendi, P. Grisvard, *Sur une equation d'evolution changeant de type.* J. Funct. Anal. **2** (1968), 352–367.

[4] R. Beals, *An abstract treatment of some forward-backward problems of transport and scattering.* J. Funct. Anal. **34** (1979), 1–20.

[5] R. Beals, *Indefinite Sturm-Liouville problems and Half-range completeness.* J. Differential Equations **56** (1985), 391–407.

[6] R. Beals, V. Protopopescu, *Half-Range completeness for the Fokker-Plank equation.* J. Stat. Phys. **32** (1983), 565–584.

[7] B. Ćurgus, *Boundary value problems in Kreĭn spaces.* Glas. Mat. Ser. III **35(55)** (2000), no. 1, 45–58.

[8] B. Ćurgus, H. Langer, *A Krein space approach to symmetric ordinary differential operators with an indefinite weight function.* J. Differential Equations **79** (1989), 31–61.

[9] B. Ćurgus, B. Najman, *A Krein space approach to elliptic eigenvalue problems with indefinite weights.* Differential Integral Equations **7** (1994), no. 5–6, 1241–1252.

[10] B. Ćurgus, B. Najman, *Positive differential operators in Krein space* $L^2(\mathbb{R})$. Recent development in operator theory and its applications (Winnipeg, MB, 1994), Oper. Theory Adv. Appl. Vol. 87, Birkhäuser, 1996, 95–104.

[11] B. Ćurgus, B. Najman, *Positive differential operators in the Krein space* $L^2(\mathbb{R}^n)$, Oper. Theory Adv. Appl. Vol. 106, Birkhäuser, 1998, 113–130.

[12] I.E. Egorov, S.G. Pyatkov, S.V. Popov, *Nonclassical operator-differential equations.* Novosibirsk, Nauka, 2000 (Russian).

[13] M.M. Faddeev, R.G. Shterenberg, *On the similarity of some differential operators to selfadjoint operators.* (Russian) Mat. Zametki **72** (2002), no. 2, 292–302; translation in Math. Notes **72** (2002), no. 1–2, 261–270.

[14] M. Faierman, H. Langer, *Elliptic problems involving an indefinite weight function.* Recent developments in operator theory and its applications (Winnipeg, MB, 1994), 105–124, Oper. Theory Adv. Appl. Vol. 87, Birkhäuser, 1996.

[15] A. Fleige, *Spectral theory of indefinite Krein-Feller differential operators.* Mathematical Research 98, Berlin, Akademie Verlag 1996.

[16] A. Fleige, B. Najman, *Nonsingularity of critical points of some differential and difference operators.* Oper. Theory Adv. Appl. Vol. 106, Birkhäuser, 1998, 147–155.

[17] A. Ganchev, W. Greenberg, C.V.M. van der Mee, *A class of linear kinetic equations in a Krein space setting.* Integral Equations Operator Theory **11** (1988), no. 4, 518–535.

[18] Ju.P. Gor'kov, *A formula for the solution of a boundary value problem for the stationary equation of Brownian motion.* Dokl. Akad. Nauk SSSR **223** (1975) (Russian).

[19] W. Greenberg, C.V.M. van der Mee, V. Protopopescu, *Boundary value problems in abstract kinetic theory.* Operator theory, Vol. 23, Birkhäuser, 1987.

[20] R.J. Hangelbroek, *Linear analysis and solution of neutron transport problem.* Transport. Theory Statist. Phys. **5** (1976), 1–85.

[21] S. Hassi, I.M. Karabash, *On the similarity between a* J*-selfadjoint Sturm-Liouville operator with operator potential and a selfadjoint operator.* Mat. Zametki **78** (2005), no. 4, 625–628 (Russian); translation in Math. Notes **78** (2005), no. 3–4, 581–585.

[22] H.G. Kaper, C.G. Lekkerkerker, J. Hejtmanek, *Spectral methods in linear transport theory.* Oper. Theory Adv. Appl. Vol. 5, Birkhäuser, 1982.

[23] H.G. Kaper, C.G. Lekkerkerker, M.K. Kwong, A. Zettl, *Full- and partial-range expansions for Sturm-Liouville problems with indefinite weights.* Proc. Roy. Soc. Edinburgh Sect. A **98** (1984), 69–88.

[24] I.M. Karabash, *J-selfadjoint ordinary differential operators similar to selfadjoint operators.* Methods of Functional Analysis and Topology **6** (2000), no. 2, 22–49.

[25] I.M. Karabash, *On J-selfadjoint differential operators similar to selfadjoint operators.* Math. Notes **68** (2000), no. 6, 798–799.

[26] I.M. Karabash, *Stationary transport equations; the case when the spectrum of collision operators has a negative part.* Proc. of the XVI Crimean Autumn Math. School–Symposium, Simferopol, Spectral and evolution problems **16** (2006), 149–153.

[27] I.M. Karabash, *Existence and uniqueness of solutions of stationary transport equations.* Proc. Appl. Math. Mech. **6** (2006), 635–636.

[28] I.M. Karabash, A.S. Kostenko, *Indefinite Sturm-Liouville operators with the singular critical point zero.* Preprint, arXiv:math.SP/0612173 (submitted to Proc. Roy. Soc. Edinburgh Sect. A).

[29] I.M. Karabash, A.S. Kostenko, *On similarity of a J-nonegative Sturm-Liouville operator to a self-adjoint operator*, to appear in Funct. Anal. Appl.

[30] I.M. Karabash, A.S. Kostenko, M.M. Malamud, *The similarity problem for J-nonnegative Sturm-Liouville operators.* Preprint ESI 1987, Vienna, www.esi.ac.at, 2007.

[31] I.M. Karabash, M.M. Malamud, *The similarity of a J-self-adjoint Sturm-Liouville operator with finite-gap potential to a self-adjoint operator.* Doklady Mathematics **69** (2004), no. 2, 195–199.

[32] I.M. Karabash, M.M. Malamud, *Indefinite Sturm-Liouville operators* $(\operatorname{sgn} x)\,(-\frac{d^2}{dx^2}+q)$ *with finite-zone potentials.* Operators and Matrices **1** (2007), no. 3, 301–368.

[33] I.M. Karabash, C. Trunk, *Spectral properties of singular Sturm-Liouville operators with indefinite weight sgn x.* Preprint, arXiv:0707.0865v1 [math.SP] (submitted to Proc. Roy. Soc. Edinburgh Sect. A).

[34] T. Kato, *Perturbation theory for linear operators.* Springer-Verlag, 1966.

[35] M. Klaus, C.V.M. van der Mee, V. Protopopescu, *Half-range solutions of indefinite Sturm-Liouville problems.* J. Funct. Anal. **70** (1987), no. 2, 254–288.

[36] J.J. Kohn, L. Nirenberg, *Degenerate elliptic-parabolic equations of second order.* Comm. Pure Appl. Math. **20** (1967), 797–872.

[37] A.S. Kostenko, *A spectral analysis of some indefinite differential operators.* Methods Funct. Anal. Topology **12** (2006), no. 2, 157–169.

[38] A.S. Kostenko, *The similarity of some J-nonnegative operators to a selfadjoint operator.* Mat. Zametki **80** (2006), no. 1, 135–138 (Russian); translation in Math. Notes **80** (2006) no. 1, 131–135.

[39] A.S. Kostenko, *Spectral analysis of singularly perturbed differential operators of the second order.* Candidate thesis, The Institute of Applied Mathemtics and Mechanics of NASU, Donetsk, 2006 (Russian).

[40] S.G. Krein, *Linear differential equations in Banach space.* AMS, 1971.

[41] H. Langer, *Spectral functions of definitizable operators in Krein space.* Lecture Notes in Mathematics **948** (1982), 1–46.

[42] I.A. Lar'kin, V.A. Novikov, N.N. Yanenko, *Nonlinear equations of mixed type.* Novosibirsk, Nauka 1983 (Russian).

[43] C.V.M. van der Mee, *Semigroup and Factorization methods in transport theory.* Math. Centre Tract. No. 146, Amsterdam, 1981.

[44] C.V.M. van der Mee, A.C.M. Ran, L. Rodman, *Stability of stationary transport equation with accretive collision operators.* J. Funct. Anal. **174** (2000), 478–512.

[45] O.A. Oleĭnik, E.V. Radkevič, *Second order equations with nonnegative characteristic form.* Plenum Press, 1973.

[46] C.D. Pagani, *On the parabolic equation* $(\operatorname{sgn} x)|x|^p u_y - u_{xx} = 0$ *and a related one.* Ann. Mat. Pura Appl. **99** (1974), no. 4, 333–399.

[47] C.D. Pagani, *On an initial-boundary value problem for the equation* $w_t = w_{xx} - xw_y$. Ann. Scuola Norm. Sup. Pisa, **2** (1975), 219–263.

[48] C.D. Pagani, *On forward-backward parabolic equations in bounded domains.* Bollettino U.M.I. (5) **13-B** (1976), 336–354.

[49] C.D. Pagani, G. Talenti, *On a forward-backward parabolic equation.* Ann. Mat. Pura Appl. **90** (1971), no. 4, 1–57.

[50] A.I. Parfyonov, *On an embedding criterion for interpolation spaces and its applications to indefinite spectral problems.* Sibirsk. Mat. Zh. **44** (2003), no. 4, 810-819; translation in Siberian Math. J. **44** (2003), no. 4, 638–644.

[51] F. Paronetto, *Existence results for a class of evolution equations of mixed type.* J. Funct. Anal. **212** (2004), no. 2, 324–356.

[52] S.G. Pyatkov, *Properties of eigenfunctions of a spectral problem and their applications.* Well-posed boundary value problems for nonclassical equations of mathematical physics, Akad. Nauk SSSR Sibirsk. Otdel., Inst. Mat., Novosibirsk, 1984, 115–130 (Russian).

[53] S.G. Pyatkov, *Operator Theory. Nonclassical Problems.* Utrecht, VSP 2002.

[54] S.G. Pyatkov, N.L. Abasheeva, *Solvability of boundary value problems for operator-differential equations: the degenerate case.* Sibirsk. Mat. Zh. **43** (2002), no. 3, 678–693 (Russian); translation in Siberian Math. J. **43** (2002), no. 3, 549–561.

[55] H. Stephan, *Nichtgleichgewichtsprozesse: Direkte und inverse Probleme.* Aachen: Shaker 1996 (German).

[56] S.A. Tersenov, *Parabolic equations changing theirs time direction.* Novosibirsk, Nauka 1985 (Russian).

I.M. Karabash
Department of Mathematics and Statistics
University of Calgary
2500 University Drive NW
Calgary T2N 1N4
Alberta, Canada

and

Department of PDE
Institute of Applied Mathematics and Mechanics
R. Luxemburg str. 74
Donetsk 83114, Ukraine
e-mail: `karabashi@yahoo.com`, `karabashi@mail.ru`

Operator Theory:
Advances and Applications, Vol. 188, 197–205

On the Structure of Semigroups of Operators Acting in Spaces with Indefinite Metric

V.A. Khatskevich and V.A. Senderov

Abstract. The first part of this paper concludes the cycle of studies in the structure of continuous one-parameter semigroups of operators originated in 2001 in the journal "Nonlinear Analysis" and continued in several other publications. In particular, the following theorem is proved:

If $\mathfrak{H}$ is an (indefinite or definite) complex Krein space and $\mathfrak{J}$ is the K-semigroup of plus-operators acting in $\mathfrak{H}$, then any plus-operator $\mathcal{F}(t) \in \mathfrak{J}$, where $t \geq 0$, is a bistrict operator.

This theorem permits removing several restrictions imposed on the sets of plus-operators in the preceding papers and thus reinforces the results of these papers.

In the second part of the present paper, we consider the heredity problem in discrete one-parametric semigroups. Namely, we study the problem of finding what indefinite properties the generating plus-operator of a semigroup and all its positive integer powers can have only simultaneously and what indefinite properties they have not necessarily simultaneously. In conclusion, we consider several applications to dynamical systems with continuous and discrete time.

Mathematics Subject Classification (2000). Primary 47B50, 47A52.

Keywords. Krein space, linear operator, indefinite metric, continuous one-parameter semigroup.

In recent years, interest in the semigroups mentioned in the title of this paper was created by the study of the so-called KE-problem.

The problem of Koenigs embedding (the KE-problem) was first used by G. Koenigs, P. Lévy, and J. Hadamard to solve various applied problems has a more than century-long history. The general statement of the problem is the following. Let D be a domain in the complex Banach space, $f \in \mathrm{Hol}(D)$. Does there exist a family $\{F(t)\}_{t \geq 0} \subset \mathrm{Hol}(D)$ continuously (in the topology of locally uniform convergence over D) depending on t and satisfying the conditions $F(0) = I$, $F(1) = f$, $F(s+t) = F(s) \circ F(t)$ for all $s, t \geq 0$?

If the family $\{F(t)\}_{t \geq 0}$ exists, then f is said to have the KE-property.

In recent years, new works concerning the KE-problem and its applications have appeared. So the case in which D is the unit open ball of the space $\mathfrak{L}(\mathfrak{H}_1, \mathfrak{H}_2)$, where $\mathfrak{H}_1$ and $\mathfrak{H}_2$ are Hilbert spaces and f is a transformation of D generated by the (plus-)operator A according to the formula

$$K'_+ = \mathcal{F}_A(K_+) = (A_{21} + A_{22}K_+)(A_{11} + A_{12}K_+)^{-1}, \tag{1}$$

with K_+, $K'_+ \in \mathfrak{L}(\mathfrak{H}_1, \mathfrak{H}_2)$ and $A_{ij} \in \mathfrak{L}(\mathfrak{H}_j, \mathfrak{H}_i)$ for i, $j = 1$, 2, was considered in [1–4]. (Here A is an operator on the space $\mathfrak{H}_1 \oplus \mathfrak{H}_2$ with block matrix $\|A_{ij}\|$, where i, $j = 1$, 2.) In particular, in [3, 4], it was shown that the KE-property is inherent in a wide class of mappings with a fixed point in a sufficiently small neighborhood of zero.

In [5], similar results are generalized to the case of Banach spaces.

Thus, all these papers, in fact, deal with **sufficient** conditions, i.e., under these conditions, a given plus-operator can be included in the continuous semigroup of the class under study. The first part of the present paper deals with the inverse problem: what operators (of what indefinite classes) can be elements of the semigroups under study?

In the second part of the paper, we consider one-parametric discrete semigroups generated by plus-operators. Clearly, the generator of such a semigroup can have any indefinite properties. In the second part of the paper, we also investigate the heredity of such properties, namely, which of these properties can be lost or acquired in the course of the "flow."

In what follows, we consider the plus-operators in a Krein space (which has the canonical form $\mathfrak{H} = \mathfrak{H}_1 \oplus \mathfrak{H}_2$, $\mathcal{J} = P_1 - P_2$, where P_1 and P_2 are mutually complementary ($P_1 + P_2 = I$) orthogonal projections onto $\mathfrak{H}_1$ and $\mathfrak{H}_2$, respectively), and in a more general $\mathcal{J}_\nu$-space (i.e., a Banach space $\mathcal{B}$ decomposed into the topological direct sum of subspaces $\mathcal{B}_1$ and $\mathcal{B}_2$,

$$\mathcal{B} = \mathcal{B}_1 \dot{+} \mathcal{B}_2, \tag{2}$$

with the corresponding (bounded) mutually complementary projections P_1 and P_2: $P_1\mathcal{B} = \mathcal{B}_1$ and $P_2\mathcal{B} = \mathcal{B}_2$, endowed with $\mathcal{J}_\nu$-metric of the form

$$\mathcal{J}_\nu(x) = \|P_1 x\|^\nu - \|P_2 x\|^\nu, \quad x \in \mathcal{B}, \tag{3}$$

where $\nu > 0$, see, e.g., [6–12]).

In what follows, we use the definitions and notation introduced in [6–12]. A linear operator T defined everywhere on $\mathcal{B}$ is called a *plus-operator* if $J_\nu(x) \geq 0$ implies that $J_\nu(Tx) \geq 0$ for all $x \in \mathcal{B}$. A plus-operator T is said to be *semistrict* if $J_\nu(x) > 0$ implies that $J_\nu(Tx) > 0$ for all $x \in \mathcal{B}$. A plus-operator T is said to be *strict* if $\mu(T) = \inf_{J_\nu(x)=1} J_\nu(Tx) > 0$. A plus-operator T is said to be *bistrict* if both T and T^* are strict plus-operators.

We note that, for $\nu > 1$, a $J_{\nu'}$-metric on $\mathcal{B}^*$ (where $\frac{1}{\nu} + \frac{1}{\nu'} = 1$) is defined in the same way as a J_ν-metric on $\mathcal{B}$ by using the canonical projections P_1^* and P_2^*, that is,

$$J_{\nu'}(x^*) = \|P_1^* x^*\|^{\nu'} - \|P_2^* x^*\|^{\nu'}, \qquad x^* \in \mathcal{B}^*.$$

We see that $\nu = 2$ for the special case in which $\mathcal{B} = H$ is a Krein space and $\mu(T) = \mu(T^*)$ for a bistrict plus-operator T.

Definition 1. If a family $\{\mathcal{F}(t)\}_{t\geq 0}$ of plus-operators acting in the $\mathcal{J}_\nu$-space is given with the operation of composition and a uniform operator topology and this family continuously depends on t and satisfies the conditions $\mathcal{F}(0) = I$ and $\mathcal{F}(s+t) = \mathcal{F}(s)\mathcal{F}(t)$ for all $s, t \geq 0$, then we shall say that a continuous (i.e., continuously depending on the parameter t) *K-semigroup* is given.

Proposition 1. *Let $\mathcal{B}$ be a $\mathcal{J}_\nu$-space, and let a K-semigroup $\mathfrak{J}$ consist of strict operators acting in $\mathcal{B}$. Then, for any $t \geq 0$, $\mathcal{F}_{11}(t) \equiv \mathcal{P}_1\mathcal{F}(t)\mathcal{P}_1|_{\mathcal{B}_1}$ is a bijective operator.*

Proof. Obviously, $\operatorname{Ker}\mathcal{F}_{11}(t) = \{0\}$ for any $t \geq 0$; we can also easily show that the operator $\mathcal{F}_{11}(t)$ is a Φ_+-operator [13] for any $t \geq 0$. Therefore, the function $\operatorname{Ind}\mathcal{F}_{11}(t)$ is defined on the entire negative semiaxis [13]. As is known (see, e.g., the review [13]), each Φ_+-operator B has a neighborhood (in the uniform operator topology) in which any C is a Φ_+-operator and $\operatorname{Ind} C = \operatorname{Ind} B$ for any such operator C. Further, the function $\mathcal{F}_{11}(t)$ is continuous. Hence any point t_0 in the negative semiaxis has a neighborhood $U(t_0)$ such that $\operatorname{Ind}\mathcal{F}_{11}(t) = \operatorname{Ind}\mathcal{F}_{11}(t_0)$ for any point t in this neighborhood (of course, it is assumed that the neighborhood $U(0)$ is right-sided).

Thus, using the Borel lemma, we easily derive that $\operatorname{Ind}\mathcal{F}_{11}(t) = 0$ for any $t > 0$. In view of $\operatorname{Ker}\mathcal{F}_{11}(t) = \{0\}$, this implies the desired statement. □

Remark 1. We note that we do not use the algebraic properties of $\mathfrak{J}$ in this proof.

Proposition 2. *Suppose that, under the conditions of Proposition 1, $\mathcal{B}_1$ and $\mathcal{B}_2$ are complex Hilbert spaces. Then, for any $t \geq 0$, the operator $\mathcal{F}(t)$ is bistrict.*

Prior to proving this proposition, we present the following lemma.

Lemma 1 (Theorem 1.1 [14]). *Let A be a strict plus-operator for some $\nu > 1$. Then the operator A is strict for any $\mu > 0$.*

Proof of Proposition 2. By Lemma 1, any operator $\mathcal{F}(t)$ is strict for $\mu = 2$, i.e., in the Krein space. In view of Proposition 1, this implies that, for $\mu = 2$, any operator $\mathcal{F}^*(t)$ is also strict [15]. We again apply Lemma 1 and see that any such operator is also strict for the original ν. □

In the case of a Krein space, the statements proved above can be generalized and reinforced.

Proposition 3. *Let $\mathcal{B} = \mathfrak{H}$ be an indefinite complex Krein space*

$$(\min\{\dim \mathfrak{H}_1, \dim \mathfrak{H}_2\} > 0),$$

and let $\mathfrak{J}$ be the K-semigroup of plus-operators acting in $\mathfrak{H}$. Then any plus-operator $\mathcal{F}(t) \in \mathfrak{J}$, where $t \geq 0$, is strict.

Proof. Assume the contrary. Assume also that $\mathcal{F}(t_0)$ is a nonstrict operator. Then all the operators $\mathcal{F}\left(\frac{t_0}{2^n}\right)$, where $n \in \mathbb{N}$, are also nonstrict. It follows from this fact that, for any positive integer n, the lineal $\mathfrak{L}_n = \mathcal{F}\left(\frac{t_0}{2^n}\right)\mathfrak{H}$ is nonnegative [15, 16]. Obviously, $\mathfrak{L}_n \subseteq \mathfrak{L}_{n+1}$ for any $n \in \mathbb{N}$; therefore, the lineal $\mathfrak{L} = \underset{n\in\mathbb{N}}{\mathfrak{Lin}}\{\mathfrak{L}_n\}$ is nonnegative. We consider a vector x such that $(x,\ \mathfrak{L}) = 0$, $x \neq 0$. We obtain

$$0 = \left(\mathcal{F}\left(\frac{t_0}{2^n}\right)x,\ x\right) \to (\mathcal{F}(0)x,\ x) = (x,\ x)$$

as $n \to \infty$. But $(x,\ x) > 0$. This is a contradiction. □

Remark 2. In this proof, we use the fact that the function $\mathcal{F}(t)$ is continuous only in the weak (rather than in the uniform) operator topology. In this case, using the Hahn–Banach–Sukhomlinov theorem, we can generalize our argument to the Banach spaces of several classes considered in [14].

Remark 3. In the case of a definite Krein space, it is already impossible to replace the uniform operator topology in the conditions of Proposition 3 by a weak or even strong topology. This fact is illustrated by the following example.

Let $\mathfrak{H} = \mathfrak{H}_1 = \mathfrak{L}^2(0,\ 1)$. For any $t \geq 0$, for $\mathcal{F}(t)$ we take the operator of multiplication by x^t. Clearly, we thus obtain a semigroup satisfying all the conditions of the K-semigroup except to the condition that $\mathcal{F}(t)$ is continuous in the uniform operator topology; but the continuity in the strong topology also holds. At the same time, for any $t > 0$, the point $\lambda = 0$ belongs to the limit spectrum of the operator $\mathcal{F}(t)$.

It is easy to prove the following assertion.

Proposition 4. *Any K-semigroup of plus-operators acting in an arbitrary (indefinite or definite) $\mathcal{J}_\nu$-space consists of homeomorphic operators.*

Theorem 1. *Let $\mathfrak{H}$ be an (indefinite or definite) complex Krein space, and let $\mathfrak{J}$ be the K-semigroup of plus-operators acting in $\mathfrak{H}$. Then any plus-operator $\mathcal{F}(t) \in \mathfrak{J}$, where $t \geq 0$ is bistrict.*

Proof. The proof in the indefinite case follows from Propositions 2 and 3, and that in the definite case follows from Proposition 4. □

Now we consider the case of discrete one-parameter semigroups. We begin with the following statement about the structure of such semigroups of bistrict plus-operators.

Proposition 5. *Suppose that A is a plus-operator in a Krein space. Suppose also that there exists a positive integer n_0 such that A^{n_0} is bistrict. Then A is a strict plus-operator.*

Proof. We assume that A is nonstrict. Then $\mathfrak{L} \equiv A\mathfrak{H} \subset \mathfrak{p}_+$ [15, 16]. Since $A^{n_0}\mathfrak{H}_1 \subset \mathfrak{L}$ and $A^{n_0}\mathfrak{H}_1 \in \mathfrak{M}_+$ [15], we have $\mathfrak{L} = A^{n_0}\mathfrak{H}_1$. Thus, $\mathfrak{L} = A\mathfrak{H} = A^{n_0}\mathfrak{H}$ is a uniformly positive subspace. $\mathfrak{L} \in \mathfrak{M}_+$. Next, we have $\mathfrak{H} = \mathfrak{L} + \operatorname{Ker} A$. In this case, $\operatorname{Ker} A \subset \operatorname{Ker} A^{n_0}$, which implies that $\operatorname{Ker} A$ is a uniformly nonnegative subspace [15]. In view of Theorem 1.1 [12], we obtain $\mathfrak{H} = \mathfrak{L} \dot{+} \operatorname{Ker} A$, where $\operatorname{Ker} A \in \mathfrak{M}_-$.

Let $\mathcal{J}(x_n) = 1$, and let $\mathcal{J}(Ax_n) \to 0$ as $n \to \infty$. We have $x_n = y_n + z_n$, where $y_n \in \mathfrak{L}$ and $z_n \in \operatorname{Ker} A$, which implies that $Ax_n = Ay_n$ and $\mathcal{J}(Ay_n) \to 0$ as $n \to \infty$. Since $\mathfrak{L}$ is uniformly positive, we see that $Ay_n \to 0$ as $n \to \infty$. Since $A\mathfrak{L} = \mathfrak{L}$, in view of the Banach theorem, $y_n \to 0$ as $n \to \infty$. Since $\|x_n\| \geq 1$, we obtain $\|z_n\| \geq \frac{1}{2}$ starting from some n. We set $t_n = \frac{x_n}{\|z_n\|}$, $u_n = \frac{y_n}{\|z_n\|}$, and $v_n = \frac{z_n}{\|z_n\|}$. We see that $t_n - v_n \to 0$ as $n \to \infty$, $\|v_n\| = 1$, which, in view of Proposition 1 [17], implies that $|\mathcal{J}(t_n) - \mathcal{J}(v_n)| \to 0$ as $n \to \infty$. On the other hand, we have $\mathcal{J}(t_n) > 0$ and $-\mathcal{J}(v_n) \geq c\|v_n\|^2 = c > 0$. This is a contradiction. □

Proposition 6. *Suppose that $\mathfrak{H}$ is a Krein space and A is a strict plus-operator acting in $\mathfrak{H}$. Then all the operators $(A^n)_{11}$, where $n \in \mathbb{N}$, are either bijective or not bijective.*

Proof. The proof can be obtained by using Theorem 5 [18]. □

We note that this statement also holds for Banach spaces of some classes studied in [18].

From Propositions 5 and 6, using Theorem 4.17 [15], we obtain the following assertion.

Theorem 2. *Under the conditions of Proposition 5, any operator A^n, where $n \in \mathbb{N}$, is bistrict.*

Now we consider a more general case of strict plus-operators in Krein spaces.

It is not difficult to construct a Krein space and a nonstrict plus-operator A in this space such that, for any $n \geq 2$, the plus-operator A^n is strict.

Example 1. Let $\{e_i^+\}_{i\in\mathbb{N}}$ be an orthonormal basis of the space $\mathfrak{H}_1$, and let $\mathfrak{H}_2 = \mathfrak{Lin}\{e^-\}$, where $[e^-, e^-] = -1$. It suffices to define a continuous operator A by the relations $Ae_1^+ = e_2^+ + e^-$, $Ae_i^+ = e_{i+1}^+$ for $i \geq 2$, and $Ae^- = 0$.

In Example 1, the plus-operator A is completely nonstrict, i.e., $A\mathfrak{p}_{++} \cap \mathfrak{p}_0 \neq \varnothing$. But modifying this example, we can easily make the operator A be semistrict, i.e., in this case, the operator takes any positive vector into a positive vector.

Example 2. Let $\mathfrak{H}^m$ be the Krein space from Example 1, $m \in \mathbb{N}$. We define a plus-operator A_m in the space $\mathfrak{H}^m$ by the following equalities: $A_m e^- = 0$, $A_m e_1^+ = e_2^+ + \left(1 - \frac{1}{m}\right) e^-$, and $A_m e_k^+ = e_{k+1}^+$, $k = 2, 3, \ldots$. We set $B = \bigoplus_{m\in\mathbb{N}} A_m$ and B^n is strict for each $n \geq 2$.

But in the case of a Krein space $\mathfrak{H}$ with a finite-dimensional $\mathfrak{H}_1$, we have the following assertion.

Theorem 3. *Let $\dim \mathfrak{H}_1 < \infty$, and let A be a plus-operator. Then all of the operators A^n, where $n \in \mathbb{N}$, are either nonstrict or bistrict. Next, in the first case, either all of these operators are semistrict or all of them are not semistrict.*

Proof. The first assertion in this theorem follows from the fact that, in the case $\dim \mathfrak{H}_1 < \infty$, any strict plus-operator is bistrict [19].

To prove the second assertion, it suffices to prove that, in the case of a completely nonstrict plus-operator A, any plus-operator A^n is also completely nonstrict. Without loss of generality, we assume that $\mathfrak{p}_{++} \cap \operatorname{Ker} A = \varnothing$; we also assume that $a \in \mathfrak{p}_{++}$ and $Aa \in \mathfrak{p}_0 \backslash \{0\}$. Let $\operatorname{Ker} A^n|_{\mathfrak{H}_1} = \{0\}$ (otherwise, the proof of the statement is complete). Then $\dim A^n\mathfrak{H}_1 = \dim \mathfrak{H}_1$, and because $A\mathfrak{H} \subset \mathfrak{p}_+$ [15, 16] and hence $\dim A\mathfrak{H} \leq \dim \mathfrak{H}_1$ [15, Chapter I, Corollary 4.3], we obtain $A^n\mathfrak{H}_1 = A\mathfrak{H}$. Thus, $Aa \in A^n\mathfrak{H}_1$, and the plus-operator A^n is completely nonstrict. The proof of the theorem is complete. □

Now we study the problem of whether the condition $\dim \mathfrak{H}_2 = \infty$ is essential in the construction of Example 2. We show that, for a rather wide class of plus-operators, this condition cannot be omitted in such constructions.

Proposition 7. *Suppose that $\mathfrak{H}$ is a Krein space,* $\dim \mathfrak{H}_2 < \infty$, *$A$ is a plus-operator,* $\operatorname{Ker} A = \{0\}$, *$A|_{\mathfrak{H}_1}$ is a homeomorphism, and $A\mathfrak{H}$ is a nondegenerate lineal. Then A is a strict plus-operator.*

Proof. We assume that the plus-operator A is nonstrict. Then the lineal $A\mathfrak{H}$ is nonnegative [15, 16]. Because of [15, Chapter I, Proposition 1.17°], the nonnegative nondegenerate lineal $A\mathfrak{H}$ is positive. Therefore, the form $[\cdot, \cdot]$ determines the norm on $A\mathfrak{H}$: $\|\cdot\|_1 = \sqrt{[\cdot, \cdot]}$.

Because the mapping $A|_{\mathfrak{H}_1}$ is homeomorphic, the lineal $A\mathfrak{H}_1$ is a (closed) positive subspace. Since $\dim \mathfrak{H}_2 < \infty$, this subspace is uniformly positive [15, Chapter I, Theorem 9.6], and hence the topology of the norm $\|\cdot\|_1$ coincides on $A\mathfrak{H}_1$ with the topology of the original norm $\|\cdot\|$. Because $A\mathfrak{H}_1$ is complete and $A\mathfrak{H}_2$ is finite dimensional, these topologies coincide on the entire $A\mathfrak{H}$.

Since $\dim \mathfrak{H}_2 < \infty$ and $\operatorname{Ker} A = \{0\}$, the mapping A, homeomorphic on $\mathfrak{H}_1$, is homeomorphic on the entire space $\mathfrak{H}$.

Since the plus-operator A is nonstrict, there exists a sequence of vectors x_n such that $\mathcal{J}(x_n) = 1$ for $n \in \mathbb{N}$ and $\mathcal{J}(Ax_n) \to 0$ as $n \to \infty$. We have $\|Ax_n\|_1 \to 0$ as $n \to \infty$ and $\|Ax_n\| \to 0$ as $n \to \infty$. Hence $\|x_n\| \to 0$ as $n \to \infty$ and $\mathcal{J}(x_n) \to 0$ as $n \to \infty$. This is a contradiction. □

Remark 4. It is easy to show that if any of the three conditions imposed on the plus-operator in Proposition 7 is removed, then its assertion does not hold any more.

Theorem 4. *Suppose that $\mathfrak{H}$ is a Krein space,* $\dim \mathfrak{H}_2 < \infty$, *$A$ is a plus-operator,* $\operatorname{Ker} A = \{0\}$, *and $A\mathfrak{H}$ is a nondegenerate lineal. Then either all of the plus-operators A^n, where $n \in \mathbb{N}$, are strict or all of them are nonstrict.*

Proof. Obviously, the fact that the plus-operator A is strict implies that all of the plus-operators A^n, where $n \in \mathbb{N}$, are strict. Let A^{n_0}, where $n_0 \in \mathbb{N}$, be a strict plus-operator. Then $A|_{\mathfrak{H}_1}$ is a homeomorphism, and the conditions of Proposition 7 are satisfied. □

Remark 5. Any plus-operator A in Theorem 4 is semistrict. Indeed, as was shown in the proof of Proposition 7, the lineal $A\mathfrak{H}$ is positive. Therefore, since $\operatorname{Ker} A = \{0\}$, we obtain $A\mathfrak{p}_{++} \subset A(\mathfrak{H}\backslash\{0\}) \subset \mathfrak{p}_{++}$.

We note that the results of the second part of our paper naturally continue and develop the results of [20].

In conclusion, we consider several applications to dynamical systems.

In [21]–[26], the exponential dichotomy conditions compatible with the signature for dynamical systems of the form

$$\dot{x}(t) = A(t)x(t), \qquad x(0) = x_0, \tag{4}$$

were studied in the Hilbert space $\mathfrak{H}$ under the condition that their trajectories $x(t)$ are described by the evolution operator $u(t,s)$, $s \le t$, $t,s \in \mathbb{R}^+$. In the general case, $u(t,s)$ satisfies the "transition conditions"

$$u(t,s) = u(t,\tau) \cdot u(\tau,s), \qquad 0 \le s \le \tau \le t.$$

In the special case of autonomous system (4) ($A(t) = A$, where A is a constant operator), the evolution operator u depends on a single parameter and determines the one-parameter semigroup

$$u(t+\tau) = u(t) \cdot u(\tau), \quad t,\tau \in \mathbb{R}^+, \qquad u(0) = I.$$

On the other hand, it follows from Theorem 1 that if all the elements of a continuous one-parameter semigroup are plus-operators, then they are bistrict plus-operators.

Therefore, the additional operator restrictions ensuring the nonemptiness of the operator sets under study, which were imposed in the preceding papers, are automatically satisfied in the case of autonomous systems.

The discrete-time dynamical systems of the form

$$y_{n+1} = A_n y_n, \qquad y_n \in \mathfrak{H}, \quad n = 0,1,2,\ldots,$$

where A_n are linear continuous operators in $\mathfrak{H}$ studied in [27]. Moreover, on the one hand, [27] presents an analog of the dichotomy conditions studied in [21–26]; on the other hand, [27] presents an example of an autonomous ($A_n = A$, where A is a constant operator) nondichotomous system. The last property dramatically illustrates that the relationship between the elements of a discrete one-parameter semigroup is significantly less than that between the elements of a continuous semigroup. This fact was studied in detail in the second part of the present paper.

References

[1] V. Khatskevich, S. Reich S., and D. Shoikhet, *Schroeder's functional equation and the Koenigs embedding property.* Nonlin. Anal.**47** (2001), 3977–3988.

[2] V. Khatskevich, S. Reich, and D. Shoikhet, *Abel–Schroeder equations for linear fractional mappings and the Koenigs embedding problem.* Acta Sci. Math. (Szeged) **69** (2003), 67–98.

[3] V. Khatskevich and V. Senderov, *The Koenigs problem for fractional-linear mappings.* Dokl. RAN **403** (2005), no. 5, 607–609 (in Russian).

[4] M. Elin and V. Khatskevich, *The Koenigs embedding problem for operator affine mappings.* Contemporary Math. **382** (2005), 113–120.

[5] M. Elin and V. Khatskevich, *Triangular plus-operators in Banach spaces: applications to the Koenigs embedding problem.* J. Nonlinear and Convex Analysis **6** (2005), no. 1, 173–185.

[6] F.F. Bonsall, *Indefinitely isometric linear operators in a reflexive Banach space.* Quart. J. Math. Oxford **2** (1955), no. 6, 175–187.

[7] M.L. Brodskii, *On the properties of the operator that maps the nonnegative part of a space with an indefinite metric into itself.* Usp. Matem. Nauk **14(I)** (1959),147–152 (in Russian).

[8] I.S. Iokhvidov, *Banach spaces with J-metrics: J-nonnegative operators.* Dokl. Akad. Nauk SSSR **169** (1966), no. 2, 259–261 (in Russian).

[9] I.S. Iokhvidov, *J-Nondilating operators in a Banach space.* Dokl. Akad. Nauk SSSR **169** (1966), no. 3, 501–504 (in Russian).

[10] I.S. Iokhvidov, *On Banach spaces with J-metrics and some classes of linear operators in these spaces.* Izv. Akad. Nauk Mold. SSR, Ser. Fiz.–Tekhn. Mat. Nauk (1968), no. 1, 60–80 (in Russian).

[11] V.A. Khatskevich and V.A. Senderov, *On properties of linear operators of certain classes in rigged spaces with indefinite metric.* Integral Equations and Operator Theory **15** (1992), 301–324.

[12] V.A. Senderov and V.A. Khatskevich, *On normed J-spaces and some classes of linear operators in these spaces.* Matem. Issledovaniya, Kishinev, Akad. Nauk Mold. SSR, **8** (1973), no. 3, 56–75 (in Russian).

[13] I.C. Gohberg and M.G. Krein, *The basic propositions on defect numbers, root numbers, and indices of linear operators.* Usp. Matem. Nauk, **12(2)** (1957), 43–118 (in Russian).

[14] V.A. Senderov and V.A. Khatskevich, *On strict plus-operators and their conjugate operators.* Funktsional'nyi Analiz, Ulyanovsk **35** (1994), 82–91 (in Russian).

[15] T.Ya. Azizov and I.S. Iokhvidov, *Linear Operators in Spaces with an Indefinite Metric.* Chichester, John Wiley and Sons, 1989.

[16] M.G.Krein and Yu.L. Shmul'yan, *Plus-operators in a space with indefinite metric.* Amer. Math. Soc. Transl. (2) **85** (1969), 93–113.

[17] T.Ya. Azizov and V.A. Senderov, *On the spectrum and invariant subspaces of unitary operators in Banach spaces with indefinite metric.* Dokl. Akad. Nauk Armyan. SSR **LIV** (1972), no. 1, 13–16 (in Russian).

[18] V.A. Khatskevich and V.A. Senderov, *Fundamentals of the theory of plus-index.* Far East J. Math. Sci. (FJMS) **3** (2001), no. 1, 1–13.

[19] Yu.P. Ginzburg, *On J-contractive operators in Hilbert space.* Nauch. Zap. Fiz.-Mat. Fak. Odessk. Gos. Ped. Inst. **22** (1958), no. 1 (in Russian).

[20] V.A. Khatskevich and V.A. Senderov, *Powers of plus-operators.* Integral Equations and Operator Theory **15** (1992), no. 5, 784–795.

[21] V.A. Khatskevich and V. Shulman, *Operator fractional-linear transformations: convexity and compactness of image; applications.* Studia Math. **116** (1995), 189–195.

[22] V.A. Khatskevich and L. Zelenko, *Indefinite metrics and dichotomy of solutions of linear differential equations in Hilbert spaces.* Chinese J. Math. **24** (1996), no. 2, 99–112.

[23] V.A. Khatskevich and L. Zelenko, *The fractional-linear transformations of the operator ball and dichotomy of solutions to evolution equations.* In: *Recent Developments in Optimization Theory and Nonlinear Analysis* (*Jerusalem, 1995*), Contemp. Math. **204** (1997), 149–154, Amer. Math. Soc., Providence, RI.

[24] V.A. Khatskevich and L. Zelenko, *Bistrict plus-operators in Krein spaces and dichotomous behavior of irreversible dynamical systems.* In: *Operator Theory and Related Topics, V. II* (*Odessa,* 1997), Oper. Theory Adv. Appl. **118** (2000), 191–203, Birkhäuser, Basel.

[25] V.A. Khatskevich, *Generalized fractional-linear transformations; convexity and compactness of the image and the pre-image; applications.* Studia Math. **137** (1999), 169–175.

[26] V.A. Khatskevich and V.A. Senderov, *Fractional-linear transformation of operator balls; applications to dynamical systems.* Acta Math. Sinica, English Series **22** (2006), no. 6, 1687–1694.

[27] V.A. Khatskevich and L. Zelenko, *Plus-operators in Krein spaces and dichotomous behavior of irreversible dynamical systems with a discrete time.* Studia Mathematica **177** (2006), 195–210.

V.A. Khatskevich
Braude College
College Campus P.O. Box 78
Karmiel 21982, Israel
e-mail: `victor_kh@hotmail.com`

V.A. Senderov
Pyatnitskoe highway, 23-2-156
Moscow, 125430, Russia
e-mail: `senderov@mccme.ru`

Operator Theory:
Advances and Applications, Vol. 188, 207–235

G-Self-adjoint Operators in Almost Pontryagin Spaces

Friedrich Philipp and Carsten Trunk

Abstract. An Almost Pontryagin space $(\mathcal{H}, [\cdot\,,\cdot])$ admits a decomposition

$$\mathcal{H} = \mathcal{H}_+[\dot{+}]\mathcal{H}_-[\dot{+}]\mathcal{H}^\circ,$$

where $(\mathcal{H}_+, [\cdot\,,\cdot])$ and $(\mathcal{H}_-, -[\cdot\,,\cdot])$ are Hilbert spaces and $\mathcal{H}_-$ as well as $\mathcal{H}^\circ$ are finite dimensional. Based on the theory of linear relations we introduce the notion of G-self-adjoint operators in Almost Pontryagin spaces and study their spectral properties. In particular, we construct a spectral function for G-self-adjoint operators in Almost Pontryagin spaces. Finally, we apply our results to the Klein-Gordon equation.

Mathematics Subject Classification (2000). Primary 46C20; Secondary 46C05.

Keywords. Pontryagin space, indefinite inner product, self-adjoint, Klein-Gordon equation.

1. Introduction

Pontryagin spaces are inner product spaces which can be written as a direct and orthogonal sum of a Hilbert space and a finite-dimensional anti-Hilbert space.

An Almost Pontryagin space is an inner product space $(\mathcal{H}, [\cdot\,,\cdot])$ which can be written as the direct and orthogonal sum of a Hilbert space, a finite-dimensional anti-Hilbert space and a finite-dimensional neutral space. Such a decomposition defines a Hilbert space topology $\mathcal{O}$ on $\mathcal{H}$ in a natural way such that the inner product $[\cdot\,,\cdot]$ is continuous with respect to $\mathcal{O}$.

Conversely, if $(\mathcal{H}, (\cdot\,,\cdot))$ is a Hilbert space and if G is a bounded self-adjoint operator in $\mathcal{H}$ such that $\sigma(G) \cap (-\infty, \varepsilon)$ consists of finitely many eigenvalues of G with finite multiplicities for some $\varepsilon > 0$, then $(\mathcal{H}, [\cdot\,,\cdot])$,

$$[\cdot\,,\cdot] := (G\,\cdot\,,\cdot),$$

equipped with the Hilbert space topology induced by $(\cdot\,,\cdot)$ is an Almost Pontryagin space. The main subject of this paper are Almost Pontryagin spaces and operators therein.

We call a densely defined linear operator A in an Almost Pontryagin space $\mathcal{H}$ G-symmetric if $[Ax, y] = [x, Ay]$ holds for all $x, y \in \operatorname{dom} A$, or, what is the same, if GA is a symmetric operator in the Hilbert space $(\mathcal{H}, (\cdot\,,\cdot))$. This is equivalent to $A \subset A^+$, where A^+ is the adjoint of A, i.e.,

$$A^+ = \left\{ \begin{pmatrix} u \\ v \end{pmatrix} : [Ax, u] = [x, v] \text{ for all } x \in \operatorname{dom} A \right\}. \tag{1.1}$$

The adjoint A^+ in (1.1) is not an operator, in general, but a closed linear relation in $\mathcal{H} \times \mathcal{H}$, cf. [DS]. Obviously, if the isotropic part $\mathcal{H}^\circ$ of $\mathcal{H}$ is not trivial, we have for all operators A in $\mathcal{H}$

$$A \neq A^+.$$

The question arises if there is a class of symmetric operators in Almost Pontryagin spaces which complies with that of self-adjoint operators in Pontryagin spaces. We call a densely defined operator A in $\mathcal{H}$ G-self-adjoint if $GA = (GA)^*$ and if A is closed. Obviously, if $(\mathcal{H}, [\cdot\,,\cdot])$ is a Pontryagin space (i.e., $\mathcal{H}^\circ = \{0\}$), an operator A in $\mathcal{H}$ is G-self-adjoint if and only if $A = A^+$. Such operators are necessarily closed. However, in an Almost Pontryagin space there are densely defined operators A with $GA = (GA)^*$ which are not closable (see Example 4.9 below).

We show that G-self-adjoint operators in Almost Pontryagin spaces have similar properties as self-adjoint operators in Pontryagin spaces. For instance, their spectrum is real, with possible exception of a finite set of non-real eigenvalues with finite multiplicities. Moreover, they possess a spectral function with finitely many singularities.

Almost Pontryagin spaces and operators therein were considered in various situations, we mention only [KW1, KW2, KW3, LMT, W1, W2]. Here, we consider as an application the Klein-Gordon equation with assumptions leading to a setting with an Almost Pontryagin space.

We proceed as follows. In Section 2 we recall some commonly known notions which are related to inner product spaces. Section 3 starts with the definition of Almost Pontryagin spaces. Some properties concerning the geometry are collected and we recall some results from [KWW]. We focus especially on the differences between Almost Pontryagin spaces and Pontryagin spaces. E.g., given a subspace $\mathcal{L}$ in a Pontryagin space $(\Pi, [\cdot\,,\cdot])$ with $\mathcal{L} = \mathcal{L}_+[\dot{+}]\mathcal{L}_-[\dot{+}]\mathcal{L}^\circ$ we have the well-known decomposition

$$\Pi = \mathcal{L}_+[\dot{+}]\mathcal{L}_-[\dot{+}](\mathcal{L}^\circ \dot{+} \mathcal{P})[\dot{+}]\mathcal{M},$$

where $\mathcal{L}^\circ \dot{+} \mathcal{P}$ is non-degenerate, cf. [B], and $\mathcal{L}^{[\perp]} = \mathcal{L}^\circ[\dot{+}]\mathcal{M}$. Here $[\dot{+}]$ denotes the direct and orthogonal (with respect to $[\cdot\,,\cdot]$) sum of two subspaces. We generalize this fact to a subspace $\mathcal{L}$ in an Almost Pontryagin space $(\mathcal{H}, [\cdot\,,\cdot])$ and obtain a decomposition

$$\mathcal{H} = \mathcal{L}_+[\dot{+}]\mathcal{L}_-[\dot{+}]\mathcal{L}_{00}[\dot{+}](\mathcal{L}_{01} \dot{+} \mathcal{P})[\dot{+}]\mathcal{M},$$

where $\mathcal{L}_{00} = \mathcal{L}^\circ \cap \mathcal{H}^\circ$, $\mathcal{L}^\circ = \mathcal{L}_{00}[\dot{+}]\mathcal{L}_{01}$, $\mathcal{L}_{01} \dot{+} \mathcal{P}$ is non-degenerate and $\mathcal{L}^{[\perp]} = \mathcal{L}^\circ[\dot{+}]\mathcal{M}$.

In Section 4 we define G-symmetric and G-self-adjoint operators in Almost Pontryagin spaces and give several necessary and sufficient conditions for G-self-adjointness. In particular, we show that a G-self-adjoint operator A admits a matrix representation with respect to any fundamental decomposition $\mathcal{H} = \mathcal{H}^\circ[\dot{+}]\Pi$,

$$A = \begin{pmatrix} A_0 & A_{12} \\ 0 & \widetilde{A} \end{pmatrix}, \quad \operatorname{dom} A = \mathcal{H}^\circ \dot{+} \operatorname{dom} \widetilde{A}, \tag{1.2}$$

where $\widetilde{A}$ is a self-adjoint operator in the Pontryagin space $(\Pi, [\cdot\,,\cdot])$. Moreover, we prove that this is equivalent to

$$\mathcal{H}^\circ \subset \operatorname{dom} A \quad \text{and} \quad A^+ = A^{++}.$$

Finally, we describe the spectrum of G-self-adjoint operators and show in Section 5 that G-self-adjoint operators in Almost Pontryagin spaces possess a spectral function with singularities.

In Section 6 we apply the results of Sections 3–5 to the Klein-Gordon equation which describes the motion of a relativistic spinless particle of mass m and charge e in an electrostatic field with potential q,

$$\left(\left(\frac{\partial}{\partial t} - ieq\right)^2 - \Delta + m^2\right)\psi(t, \vec{x}) = 0, \quad t \in \mathbb{R}, \ \vec{x} \in \mathbb{R}^n.$$

We rewrite this equation as in [LNT] and obtain a first-order differential equation for $\mathbf{x} = \binom{x}{y}$

$$\frac{\mathrm{d}\mathbf{x}}{\mathrm{d}t} = i\hat{A}\mathbf{x}, \quad \hat{A} = \begin{pmatrix} 0 & I \\ H_0 - V^2 & 2V \end{pmatrix}, \tag{1.3}$$

where H_0 is a strictly positive self-adjoint operator and V a symmetric operator in a Hilbert space. As $\hat{A}$ may not even be densely defined nor closed, suitable assumptions have to be imposed (see Section 6 below) on V such that we can associate a closed operator A with the block operator matrix $\hat{A}$. In general, $\hat{A}$ does not exhibit symmetry in any Hilbert space. But it is symmetric with respect to a so-called energy inner product which is in general an indefinite inner product.

For the Klein-Gordon equation the formal operator-matrix $\hat{A}$ in the energy inner product has been studied in a number of papers. We mention here only [J, Kk, N1, N2, N3, LNT] and the references given in [LNT].

In [LNT] the assumptions on V were chosen such that the operator A is self-adjoint in some Pontryagin space. Then (cf. [LNT]) the differential equation (1.3) is solvable for any given initial value. We weaken the assumptions from [LNT] in such a way that A is self-adjoint in an Almost Pontryagin space and we obtain with the help of the spectral function for self-adjoint operators in Almost Pontryagin spaces (Section 5) similar results as in [LNT].

As a final remark we mention that some of our results can also be obtained by considering the factor space $\mathcal{H}/\mathcal{H}^\circ$, which is a Pontryagin space, see, e.g., [IKL]. E.g., the operator A induces a self-adjoint operator in this Pontryagin space which corresponds to $\widetilde{A}$ in (1.2). However, it is the aim of this paper to show that one can obtain a spectral theory for G-self-adjoint operators in Almost Pontryagin spaces $\mathcal{H}$ without changing the underlying space $\mathcal{H}$.

2. Preliminaries

In this section we introduce basic notions related to inner product spaces and collect some properties. In particular we consider spaces with an inner product which is continuous with respect to some Hilbert space scalar product.

An inner product space $(\mathcal{H}, [\cdot\,,\cdot])$ is a complex vector space $\mathcal{H}$ with an inner product $[\cdot\,,\cdot] : \mathcal{H} \times \mathcal{H} \to \mathbb{C}$ such that for $x, y, z \in \mathcal{H}$ and $\alpha, \beta \in \mathbb{C}$ we have

$$[\alpha x + \beta y, z] = \alpha[x, z] + \beta[y, z] \quad \text{and} \quad \overline{[x, y]} = [y, x].$$

Let $(\mathcal{H}, [\cdot\,,\cdot])$ be an inner product space. If $\mathcal{L} \subset \mathcal{H}$ is a linear manifold, the *orthogonal companion of* $\mathcal{L}$ *(in* $\mathcal{H}$*)* is defined by

$$\mathcal{L}^{[\perp]} := \{x \in \mathcal{H} : [x, \ell] = 0 \ \text{ for all } \ \ell \in \mathcal{L}\}.$$

The *isotropic part of* $\mathcal{L}$ is the set of all vectors in $\mathcal{L}$ which are $[\cdot\,,\cdot]$-orthogonal to $\mathcal{L}$, i.e.,

$$\mathcal{L}^{\circ} := \mathcal{L} \cap \mathcal{L}^{[\perp]}.$$

We call the linear manifold $\mathcal{L}$ *non-degenerate* if $\mathcal{L}^{\circ} = \{0\}$. If $\mathcal{N} \subset \mathcal{H}$ is a linear manifold with $\mathcal{N} \subset \mathcal{L}^{[\perp]}$ we write $\mathcal{N}[\perp]\mathcal{L}$. If, in addition, $\mathcal{N} \cap \mathcal{L} = \{0\}$, then by $\mathcal{N}[\dot{+}]\mathcal{L}$ we denote the *direct orthogonal sum of* $\mathcal{N}$ *and* $\mathcal{L}$.

A vector $x \in \mathcal{H}$ is called positive (negative, neutral) if $[x, x] > 0$ (resp. $[x, x] < 0$, $[x, x] = 0$), and nonnegative (nonpositive) if x is not negative (resp. not positive). A linear manifold $\mathcal{L} \subset \mathcal{H}$ is called positive (negative, neutral, nonnegative, nonpositive) if all vectors in $\mathcal{L} \setminus \{0\}$ are positive (resp. negative, neutral, nonnegative, nonpositive). The linear manifold $\mathcal{L}$ is called maximal positive if it is positive and if there is no positive linear manifold $\mathcal{L}' \neq \mathcal{L}$ containing $\mathcal{L}$.

Lemma 2.1. *Let* $(\mathcal{H}, [\cdot\,,\cdot])$ *be a nonnegative inner product space. Then a linear manifold* $\mathcal{L} \subset \mathcal{H}$ *is maximal positive if and only if* $\mathcal{H} = \mathcal{L} \dot{+} \mathcal{H}^{\circ}$. *If there exists a maximal positive linear manifold* $\mathcal{H}_+ \subset \mathcal{H}$ *such that* $(\mathcal{H}_+, [\cdot\,,\cdot])$ *is a Hilbert space, then for all maximal positive linear manifolds* $\mathcal{H}'_+ \subset \mathcal{H}$ *the inner product space* $(\mathcal{H}'_+, [\cdot\,,\cdot])$ *is a Hilbert space.*

Proof. If $\mathcal{H} = \mathcal{L} \dot{+} \mathcal{H}^{\circ}$ then $\mathcal{L}$ is maximal positive by [AI, 1.25, Chapter 1]. Let $\mathcal{L}$ be maximal positive in $\mathcal{H}$. If $x_0 \notin \mathcal{L} \dot{+} \mathcal{H}^{\circ}$, then with [AI, 1.17, Chapter 1] we obtain that for $x \in \operatorname{span}\{x_0\} \setminus \{0\}$ and $\ell \in \mathcal{L}$ the vector $x + l$ is positive. Hence, $\mathcal{L}' := \mathcal{L} \dot{+} \operatorname{span}\{x_0\}$ is positive which contradicts the fact that $\mathcal{L}$ is maximal positive. This implies $\mathcal{H} = \mathcal{L} \dot{+} \mathcal{H}^{\circ}$.

Let $\mathcal{H}'_+$ be maximal positive and let (x'_n) be a Cauchy sequence in the pre-Hilbert space $(\mathcal{H}'_+, [\cdot\,,\cdot])$. With $x_n^+ \in \mathcal{H}_+$ and $x_n^{\circ} \in \mathcal{H}^{\circ}$ such that $x'_n = x_n^+ + x_n^{\circ}$ it follows that (x_n^+) is a Cauchy sequence in $(\mathcal{H}_+, [\cdot\,,\cdot])$. Thus, there exists $x^+ \in \mathcal{H}_+$

such that $[x_n^+ - x^+, x_n^+ - x^+] \to 0$ as $n \to \infty$. As $\mathcal{H}'_+$ is maximal positive there exist vectors $x'_+ \in \mathcal{H}'_+$ and $x'_0 \in \mathcal{H}^\circ$ such that $x^+ = x'_+ + x'_0$. Thus, we have

$$\begin{aligned}[x'_n - x'_+, x'_n - x'_+] &= [x_n^+ + x_n^\circ - x'_+, x_n^+ + x_n^\circ - x'_+] = [x_n^+ - x'_+, x_n^+ - x'_+] \\ &= [x_n^+ - x^+ + x'_0, x_n^+ - x^+ + x'_0] = [x_n^+ - x^+, x_n^+ - x^+],\end{aligned}$$

which converges to zero as $n \to \infty$. □

Now, suppose that $\mathcal{O}$ is a Hilbert space topology on $\mathcal{H}$ and that the inner product $[\cdot\,,\cdot]$ is $\mathcal{O}$-continuous, i.e., for any Hilbert space norm $\|\cdot\|$ on $\mathcal{H}$ which induces $\mathcal{O}$ there exists some $c > 0$ such that

$$|[x, y]| \le c\,\|x\|\|y\| \quad \text{for all} \quad x, y \in \mathcal{H}.$$

In the following all topological notions are related to the Hilbert space topology $\mathcal{O}$. By a subspace we always understand a closed linear manifold. Note that $\mathcal{L}^{[\perp]}$ is a subspace for every linear manifold $\mathcal{L} \subset \mathcal{H}$, and we have $\mathcal{L}^{[\perp]} = \overline{\mathcal{L}}^{[\perp]}$. Recall (cf. [B, Theorem IV.5.2]) that a subspace $\mathcal{L} \subset \mathcal{H}$ always admits a decomposition

$$\mathcal{L} = \mathcal{L}_+[\dot{+}]\mathcal{L}_-[\dot{+}]\mathcal{L}^\circ, \tag{2.1}$$

where $\mathcal{L}_+$ is a positive subspace, $\mathcal{L}_-$ is a negative subspace and the projections in $\mathcal{L}$ onto $\mathcal{L}_+$, $\mathcal{L}_-$ and $\mathcal{L}^\circ$, respectively, corresponding to the decomposition (2.1) are continuous. We shall call a decomposition (2.1) with the above properties a *fundamental decomposition of* $\mathcal{L}$. It is easily seen that the numbers

$$\kappa_+(\mathcal{L}) := \dim \mathcal{L}_+, \;\; \kappa_-(\mathcal{L}) := \dim \mathcal{L}_- \;\; \text{and} \;\; \kappa_0(\mathcal{L}) := \dim \mathcal{L}^\circ$$

do not depend on the fundamental decomposition. We call them the *rank of positivity, rank of negativity* and *rank of degeneracy of* $\mathcal{L}$, respectively. Furthermore, we call the sums

$$\kappa_-(\mathcal{L}) + \kappa_0(\mathcal{L}) \;\; \text{and} \;\; \kappa_+(\mathcal{L}) + \kappa_0(\mathcal{L})$$

the *rank of non-positivity* and the *rank of non-negativity of* $\mathcal{L}$, respectively.

A positive (negative) linear manifold $\mathcal{L} \subset \mathcal{H}$ is called ($\mathcal{O}$-)*uniformly positive* (resp. ($\mathcal{O}$-)*uniformly negative*) if the topology on $\mathcal{L}$ which is induced by the norm $[x, x]^{1/2}$ (resp. $(-[x, x])^{1/2}$), $x \in \mathcal{L}$, coincides with the relative topology induced by $\mathcal{O}$. This is equivalent to the fact that for any norm $\|\cdot\|$ on $\mathcal{H}$ which induces $\mathcal{O}$ there exists a $\delta > 0$ such that

$$[x, x] \ge \delta\|x\|^2 \quad (\text{resp.} \; -[x, x] \ge \delta\|x\|^2) \;\; \text{for all} \;\; x \in \mathcal{L}.$$

A *subspace* $\mathcal{L} \subset \mathcal{H}$ is uniformly positive (uniformly negative) if and only if $(\mathcal{L}, [\cdot\,,\cdot])$ (resp. $(\mathcal{L}, -[\cdot\,,\cdot])$) is a Hilbert space. We will call a linear manifold *uniformly definite* if it is either uniformly positive or uniformly negative.

The linear manifold $\mathcal{L} \subset \mathcal{H}$ is called *ortho-complemented (in* $\mathcal{H}$*)* if

$$\mathcal{H} = \mathcal{L} + \mathcal{L}^{[\perp]}.$$

Lemma 2.2. *Let $\mathcal{L}_1, \mathcal{L}_2 \subset \mathcal{H}$ be uniformly definite subspaces with $\mathcal{L}_1[\perp]\mathcal{L}_2$. Then their sum is direct, $\mathcal{L} := \mathcal{L}_1[\dot{+}]\mathcal{L}_2$ is a subspace, and we have:*

(i) *$\mathcal{L}_1$, $\mathcal{L}_2$ and $\mathcal{L}$ are ortho-complemented.*
(ii) *If $\mathcal{L}_1$ and $\mathcal{L}_2$ both are uniformly positive (uniformly negative), then $\mathcal{L}$ is uniformly positive (resp. uniformly negative).*

Proof. We evidently have $\mathcal{L}_1 \cap \mathcal{L}_2 = \{0\}$. It was shown in [LMM, Lemma 2.1] that $\mathcal{L}_1$ and $\mathcal{L}_2$ are ortho-complemented. If by $\mathcal{M}$ we denote the orthogonal companion of $\mathcal{L}_2$ in $\mathcal{L}_1^{[\perp]}$, we have $\mathcal{H} = \mathcal{L}_1[\dot{+}]\mathcal{L}_2[\dot{+}]\mathcal{M}$. From this it follows that $\mathcal{L}$ is a subspace. Hence, $\mathcal{M} = \mathcal{L}^{[\perp]}$, and $\mathcal{L}$ is ortho-complemented. In order to show (ii) let $\mathcal{L}_1$ and $\mathcal{L}_2$ be uniformly positive and let $\|\cdot\|$ be a norm inducing $\mathcal{O}$. Then there exists a $\delta > 0$ such that $[x_i, x_i] \geq 2\delta\|x_i\|^2$ for all $x_i \in \mathcal{L}_i$ $(i = 1, 2)$. Consequently, $[x_1 + x_2, x_1 + x_2] = [x_1, x_1] + [x_2, x_2] \geq \delta(2\|x_1\|^2 + 2\|x_2\|^2) \geq \delta(\|x_1\| + \|x_2\|)^2 \geq \delta\|x_1 + x_2\|^2$ for all $x_1 \in \mathcal{L}_1$ and all $x_2 \in \mathcal{L}_2$. □

3. Geometry of Almost Pontryagin spaces

We consider now a special case of the spaces from the previous section.

Definition 3.1. Let $(\mathcal{H}, [\cdot\,,\cdot])$ be an inner product space and let $\mathcal{O}$ be a Hilbert space topology on $\mathcal{H}$. The triplet $(\mathcal{H}, \mathcal{O}, [\cdot\,,\cdot])$ is called an *Almost Pontryagin space with finite rank of non-positivity* (*non-negativity*) if

(i) the inner product $[\cdot\,,\cdot]$ is $\mathcal{O}$-continuous and
(ii) there exists a $\mathcal{O}$-uniformly positive (resp. $\mathcal{O}$-uniformly negative) $\mathcal{O}$-closed linear manifold $\mathcal{L} \subset \mathcal{H}$ with $\operatorname{codim} \mathcal{L} < \infty$.

In the sequel, we will collect some properties of Almost Pontryagin spaces. The following lemma justifies the term "with finite rank of non-positivity (non-negativity)" in Definition 3.1.

Lemma 3.2. *Let $(\mathcal{H}, \mathcal{O}, [\cdot\,,\cdot])$ be an Almost Pontryagin space with finite rank of non-positivity (non-negativity). Then $\kappa_-(\mathcal{H}) + \kappa_0(\mathcal{H}) < \infty$ (resp. $\kappa_+(\mathcal{H}) + \kappa_0(\mathcal{H}) < \infty$). Moreover, there exists a fundamental decomposition*

$$\mathcal{H} = \mathcal{H}_+[\dot{+}]\mathcal{H}_-[\dot{+}]\mathcal{H}^\circ \tag{3.1}$$

of $\mathcal{H}$ in which $\mathcal{H}_+$ (resp. $\mathcal{H}_-$) is uniformly positive (resp. uniformly negative). Obviously the fundamental decomposition (3.1) can be written in the following way

$$\mathcal{H} = \Pi[\dot{+}]\mathcal{H}^\circ, \tag{3.2}$$

where $(\Pi, [\cdot\,,\cdot])$, $\Pi = \mathcal{H}_+[\dot{+}]\mathcal{H}_-$, is a Pontryagin space.

Proof. Let $(\mathcal{H}, \mathcal{O}, [\cdot\,,\cdot])$ be an Almost Pontryagin space with finite rank of non-positivity. A similar reasoning applies to Almost Pontryagin spaces with finite rank of non-negativity. Let $\mathcal{L} \subset \mathcal{H}$ be a uniformly positive subspace with $\operatorname{codim} \mathcal{L} < \infty$ and let $\mathcal{N} := \mathcal{L}^{[\perp]}$. If $\mathcal{N} = \mathcal{N}_+[\dot{+}]\mathcal{N}_-[\dot{+}]\mathcal{N}^\circ$ is a fundamental decomposition of $\mathcal{N}$, then, since $\mathcal{L}$ is ortho-complemented (cf. Lemma 2.2), $\mathcal{N}$ is finite dimensional and with $\mathcal{H}_+ := \mathcal{L}[\dot{+}]\mathcal{N}_+$ and $\mathcal{H}_- := \mathcal{N}_-$ we obtain the desired fundamental decomposition of $\mathcal{H}$, cf. Lemma 2.2. □

Proposition 3.3. *Let $(\mathcal{H}, \mathcal{O}, [\cdot\,,\cdot])$ be an Almost Pontryagin space with finite rank of non-positivity (non-negativity) and let $\widetilde{\mathcal{O}}$ be another Hilbert space topology on $\mathcal{H}$, such that $[\cdot\,,\cdot]$ is $\widetilde{\mathcal{O}}$-continuous. Then also $(\mathcal{H}, \widetilde{\mathcal{O}}, [\cdot\,,\cdot])$ is an Almost Pontryagin space with the same rank of non-positivity (non-negativity).*

Proof. By Lemma 3.2 we retrieve a decomposition $\mathcal{H} = \mathcal{H}_+[\dot{+}]\mathcal{H}_-[\dot{+}]\mathcal{H}^\circ$ of $\mathcal{H}$ in which $(\mathcal{H}_+, [\cdot\,,\cdot])$ is a Hilbert space and finite-codimensional in $\mathcal{H}$ and $\mathcal{H}_-$ is uniformly negative. Let $\langle\cdot\,,\cdot\rangle$ be a Hilbert space inner product (on $\mathcal{H}$) inducing $\widetilde{\mathcal{O}}$ and set $\widehat{\mathcal{H}} := \mathcal{H}_+[\dot{+}]\mathcal{H}^\circ$. This linear manifold is $\widetilde{\mathcal{O}}$-closed since $\widehat{\mathcal{H}} = \mathcal{H}_-^{[\perp]}$ and $[\cdot\,,\cdot]$ is $\widetilde{\mathcal{O}}$-continuous. We now define the linear manifold

$$\mathcal{L} := (\mathcal{H}^\circ)^{\langle\perp\rangle} \cap \widehat{\mathcal{H}}.$$

Here, "$\langle\perp\rangle$" denotes the orthogonal complement (in $\mathcal{H}$) with respect to the scalar product $\langle\cdot\,,\cdot\rangle$. The linear manifold $\mathcal{L}$ is $\widetilde{\mathcal{O}}$-closed and finite-codimensional in $\mathcal{H}$. Furthermore, as $\widehat{\mathcal{H}} = \mathcal{L} \dot{+} \mathcal{H}^\circ$, $\mathcal{L}$ is maximal positive (in $\widehat{\mathcal{H}}$). Thus, by Lemma 2.1 $(\mathcal{L}, [\cdot\,,\cdot])$ is a Hilbert space. This implies that $\mathcal{L}$ is $\widetilde{\mathcal{O}}$- uniformly positive. □

If the Hilbert space inner product $(\cdot\,,\cdot)$ on $\mathcal{H}$ induces the topology $\mathcal{O}$, the bounded and self-adjoint operator G which is uniquely defined by

$$[x, y] = (Gx, y) \quad \text{for all} \quad x, y \in \mathcal{H}$$

is called the *Gram operator of* $[\cdot\,,\cdot]$ *with respect to* $(\cdot\,,\cdot)$. If $\mathcal{L} \subset \mathcal{H}$ is a linear manifold, then evidently

$$\mathcal{L}^{[\perp]} = (G\mathcal{L})^\perp = G^{-1}\left(\mathcal{L}^\perp\right) := \{x \in \mathcal{H} : Gx \in \mathcal{L}^\perp\}, \tag{3.3}$$

where "$\perp$" denotes the orthogonal complement in $\mathcal{H}$ with respect to the Hilbert space inner product $(\cdot\,,\cdot)$.

The following two results were shown in [KWW] for Almost Pontryagin spaces with finite rank of non-positivity. Similar statements hold for Almost Pontryagin spaces with finite rank of non-negativity.

Proposition 3.4. *Let $(\mathcal{H}, [\cdot\,,\cdot])$ be an inner product space which admits a decomposition*

$$\mathcal{H} = \mathcal{H}_+[\dot{+}]\mathcal{H}_-[\dot{+}]\mathcal{H}^\circ, \tag{3.4}$$

where $(\mathcal{H}_+, [\cdot\,,\cdot])$ is a Hilbert space, $\mathcal{H}_-$ is negative and $\mathcal{H}_-$ as well as $\mathcal{H}^\circ$ are finite dimensional. If P_+, P_- and P_0 denote the projections onto $\mathcal{H}_+$, $\mathcal{H}_-$ and $\mathcal{H}^\circ$, respectively, corresponding to the decomposition (3.4)*, then the inner product $(\cdot\,,\cdot)$, defined by*

$$(x, y) := [P_+x, P_+y] - [P_-x, P_-y] + (P_0x, P_0y)_0, \quad x, y \in \mathcal{H},$$

where $(\cdot\,,\cdot)_0$ is any positive definite inner product on $\mathcal{H}^\circ$, is a Hilbert space inner product on $\mathcal{H}$. If $\mathcal{O}$ denotes the Hilbert space topology induced by $(\cdot\,,\cdot)$, then

$(\mathcal{H}, \mathcal{O}, [\cdot\,,\cdot])$ is an Almost Pontryagin space, and the Gram operator of $[\cdot\,,\cdot]$ with respect to $(\cdot\,,\cdot)$ is given by $G = P_+ - P_-$.

For every Hilbert space topology $\widetilde{\mathcal{O}}$ on $\mathcal{H}$ such that $(\mathcal{H}, \widetilde{\mathcal{O}}, [\cdot\,,\cdot])$ is an Almost Pontryagin space and $\mathcal{H}_+$ in (3.4) is $\widetilde{\mathcal{O}}$-closed, we have $\widetilde{\mathcal{O}} = \mathcal{O}$.

Proposition 3.5. *Let $(\mathcal{H}, \mathcal{O}, [\cdot\,,\cdot])$ be an Almost Pontryagin space with finite rank of non-positivity, let $(\cdot\,,\cdot)$ be a Hilbert space inner product inducing $\mathcal{O}$ and denote by G the Gram operator of $[\cdot\,,\cdot]$ with respect to $(\cdot\,,\cdot)$. Then there exists a number $\varepsilon > 0$ such that the set $\sigma(G) \cap (-\infty, \varepsilon)$ consists of finitely many eigenvalues of G with finite multiplicities.*

Conversely, if $(\mathcal{H}, (\cdot\,,\cdot))$ is a Hilbert space with topology $\mathcal{O}$ and G is a bounded self-adjoint operator in $\mathcal{H}$ with the spectral properties described above, then $(\mathcal{H}, \mathcal{O}, (G\cdot,\cdot))$ is an Almost Pontryagin space with finite rank of non-positivity.

For the rest of this section, let $(\mathcal{H}, \mathcal{O}, [\cdot\,,\cdot])$ be an Almost Pontryagin space with finite rank of non-positivity $\kappa = \kappa_-(\mathcal{H}) + \kappa_0(\mathcal{H})$. For Almost Pontryagin spaces with finite rank of non-negativity similar statements hold. Moreover, we fix a Hilbert space inner product $(\cdot\,,\cdot)$ on $\mathcal{H}$ which defines the topology $\mathcal{O}$ and the norm $\|\cdot\|$.

Remark 3.6. In [KWW] it was shown that if $\dim \mathcal{H} = \infty$ and $\kappa_0(\mathcal{H}) > 0$, there exists another Hilbert space topology $\widetilde{\mathcal{O}} \neq \mathcal{O}$ on $\mathcal{H}$ such that $(\mathcal{H}, \widetilde{\mathcal{O}}, [\cdot\,,\cdot])$ is an Almost Pontryagin space with finite rank of non-positivity κ. That is the reason why we have to fix the topology $\mathcal{O}$ in Definition 3.1.

The topology $\widetilde{\mathcal{O}}$ can be constructed as follows: Choose a fundamental decomposition (3.1) as in Lemma 3.2 and set $\widehat{\mathcal{H}} := \mathcal{H}_+ \dotplus \mathcal{H}^\circ$. Choose a $\mathcal{O}$-*non*-closed linear manifold $\mathcal{H}_+' \subset \widehat{\mathcal{H}}$ such that $\widehat{\mathcal{H}} = \mathcal{H}_+' \dotplus \mathcal{H}^\circ$ (to this end let $x_0 \in \mathcal{H}^\circ$, $\|x_0\| = 1$, set $\mathcal{L} := \left(\{x_0\}^\perp \cap \mathcal{H}^\circ\right)^\perp \cap \widehat{\mathcal{H}}$ and choose a non-continuous linear functional ϕ on $\mathcal{L}$ with $\phi(x_0) \neq 1$. Then $\mathcal{H}_+' := \ker(\phi - (\cdot, x_0))$ is a linear manifold as desired). As a consequence of Lemma 2.1, $(\mathcal{H}_+', [\cdot\,,\cdot])$ is a Hilbert space. By applying Proposition 3.4 to the decomposition $\mathcal{H} = \mathcal{H}_+' \dotplus \mathcal{H}_- \dotplus \mathcal{H}^\circ$ we obtain a Hilbert space topology $\widetilde{\mathcal{O}}$ on $\mathcal{H}$ which does not coincide with $\mathcal{O}$ since $\mathcal{H}_+'$ is closed in $\widetilde{\mathcal{O}}$ but not closed in $\mathcal{O}$.

However, if $\kappa_0(\mathcal{H}) = 0$ it is well known that the topology $\mathcal{O}$ is the unique Banach space topology on $\mathcal{H}$, in which the inner product $[\cdot\,,\cdot]$ is continuous (cf. [L2, Proposition 1.2]).

Proposition 3.7. *If $\mathcal{L} \subset \mathcal{H}$ is a linear manifold, then we have*

$$\mathcal{L}^{[\perp][\perp]} = \overline{\mathcal{L}} + \mathcal{H}^\circ. \tag{3.5}$$

Proof. Since $\overline{\mathcal{L}}^{[\perp]} = \mathcal{L}^{[\perp]}$ we may assume that $\mathcal{L}$ is a subspace. Let $(\cdot\,,\cdot)$ be a Hilbert space inner product inducing $\mathcal{O}$ and let G denote the Gram operator of $[\cdot\,,\cdot]$ with respect to $(\cdot\,,\cdot)$. By Proposition 3.5 G has a closed range and an (at most) finite-dimensional kernel. Hence, $G\mathcal{L}$ is closed (see [Ka, Lemma IV.5.29]), and by (3.3) we have

$$\mathcal{L}^{[\perp][\perp]} = \left((G\mathcal{L})^{\perp}\right)^{[\perp]} = G^{-1}\left((G\mathcal{L})^{\perp}\right)^{\perp} = G^{-1}\left(G\mathcal{L}\right) = \mathcal{L} + \ker(G) = \mathcal{L} + \mathcal{H}^{\circ},$$

and the proposition is proved. □

Lemma 3.8. *For a linear manifold $\mathcal{L} \subset \mathcal{H}$ the following statements hold:*

(i) *If $\mathcal{L}$ is closed and positive, then $\mathcal{L}$ is uniformly positive;*
(ii) *If $\mathcal{L}$ is non-positive, then* $\dim \mathcal{L} \leq \kappa$;
(iii) *If $\mathcal{L}$ is negative, then* $\dim \mathcal{L} \leq \kappa_-(\mathcal{H})$;
(iv) *If $\mathcal{L}$ is maximal negative, then* $\dim \mathcal{L} = \kappa_-(\mathcal{H})$.

Proof. Fix a fundamental decomposition (3.1) for $\mathcal{H}$ and denote by P_+, P_- and P_0 the fundamental projections onto $\mathcal{H}_+$, $\mathcal{H}_-$ and $\mathcal{H}^{\circ}$, respectively. Let $\|\cdot\|$ be a norm on $\mathcal{H}$ which induces $\mathcal{O}$. In order to prove (i), we assume the contrary. Then there exists a sequence $(x_n) \subset \mathcal{L}$ with $\|x_n\| = 1$ and $\lim_{n\to\infty}[x_n, x_n] = 0$ which converges weakly to some $x_0 \in \mathcal{L}$. Since

$$|[x_0, x_0]| \leq |[x_0 - x_n, x_0]| + [x_n, x_n]^{1/2}[x_0, x_0]^{1/2}$$

$x_0 = 0$ follows. We set $u_n := P_+x_n$ and $v_n := (I - P_+)x_n$. Since $\mathcal{H}_- \dotplus \mathcal{H}^{\circ}$ is finite dimensional, $v_n \to 0$ as $n \to \infty$. Thus, $\lim_{n\to\infty} \|u_n\| = 1$. But $\lim_{n\to\infty}[u_n, u_n] = \lim_{n\to\infty}[x_n, x_n] = 0$ which implies $u_n \to 0$ as $n \to \infty$. A contradiction.

Let us now consider a non-positive subspace $\mathcal{L}$ with $(\kappa + 1)$ linearly independent vectors $e_1, \ldots, e_{\kappa+1} \in \mathcal{L}$. For $j = 1, \ldots, \kappa+1$ set $x_j^{\pm} := P_{\pm}e_j$ and $x_j^{\circ} := P_0 e_j$. Then there exist $\lambda_1, \ldots, \lambda_{\kappa+1} \in \mathbb{C}$ with at least one λ_i different from zero such that $\sum_{j=1}^{\kappa+1} \lambda_j(x_j^- + x_j^{\circ}) = 0$. But then $\sum_{j=1}^{\kappa+1} \lambda_j e_j \in \mathcal{H}_+$ and $\lambda_1 = \cdots = \lambda_{\kappa+1} = 0$ follows which is a contradiction.

If $\mathcal{L}$ is negative, then from (ii) we conclude that $\mathcal{L}$ is finite dimensional and thus uniformly negative and therefore ortho-complemented (cf. Lemma 2.2). Set $\mathcal{N} := \mathcal{L}^{[\perp]}$ and let $\mathcal{N} = \mathcal{N}_+[\dotplus]\mathcal{N}_-[\dotplus]\mathcal{N}^{\circ}$ be a fundamental decomposition of $\mathcal{N}$. Then we see that $\dim \mathcal{L} \leq \dim \mathcal{L} + \dim \mathcal{N}_- = \kappa_-(\mathcal{H})$. If $\mathcal{L}$ is even maximal negative, then assertion (iv) follows from $\mathcal{N}_- = \{0\}$. □

For a set $\Lambda \subset \mathcal{H}$ we denote by $\mathcal{O}_{\Lambda}$ the relative topology on Λ which is induced by $\mathcal{O}$. The following result is an immediate consequence of Lemma 3.8 (see also [KWW, Proposition 3.1]).

Corollary 3.9. *If $\mathcal{L} \subset \mathcal{H}$ is a subspace, then $(\mathcal{L}, \mathcal{O}_{\mathcal{L}}, [\cdot\,,\cdot])$ is an Almost Pontryagin space with finite rank of non-positivity $\kappa' \leq \kappa$.*

Recall, that if $(\Pi, [\cdot\,,\cdot])$ is a Pontryagin space and $\mathcal{L} \subset \Pi$ is a subspace with a fundamental decomposition $\mathcal{L} = \mathcal{L}_+[\dotplus]\mathcal{L}_-[\dotplus]\mathcal{L}^{\circ}$, we have

$$\Pi = \mathcal{L}_+[\dotplus]\mathcal{L}_-[\dotplus](\mathcal{L}^{\circ} \dotplus \mathcal{P})[\dotplus]\mathcal{M},$$

where $\mathcal{L}^{\circ} \dotplus \mathcal{P}$ is non-degenerate (cf. [B, Theorem IX.2.5]). The following lemma can be seen as a generalization of this fact.

Lemma 3.10. *Let $\mathcal{D} \subset \mathcal{H}$ be a dense linear manifold in $\mathcal{H}$. Furthermore, let $\mathcal{L}$ be a subspace, $\mathcal{L} \subset \mathcal{D}$. If $\mathcal{L} = \mathcal{L}_+[\dot{+}]\mathcal{L}_-[\dot{+}]\mathcal{L}^\circ$ is a fundamental decomposition for $\mathcal{L}$, then there exist subspaces $\mathcal{L}_{00}, \mathcal{L}_{01}, \mathcal{P} \subset \mathcal{D}$ and $\mathcal{M} \subset \mathcal{H}$ such that $\mathcal{H}$ admits a decomposition*

$$\mathcal{H} = \mathcal{L}_+[\dot{+}]\mathcal{L}_-[\dot{+}]\mathcal{L}_{00}[\dot{+}](\mathcal{L}_{01} \dot{+} \mathcal{P})[\dot{+}]\mathcal{M} \tag{3.6}$$

with the following properties:

(i) $\mathcal{L}_{00} = \mathcal{L}^\circ \cap \mathcal{H}^\circ$ *and* $\mathcal{L}^\circ = \mathcal{L}_{00}[\dot{+}]\mathcal{L}_{01}$;
(ii) $\mathcal{P}$ *is neutral and* $\mathcal{G} := \mathcal{L}_{01} \dot{+} \mathcal{P}$ *is non-degenerate;*
(iii) $\mathcal{P} \cap \mathcal{L}_{01} = \mathcal{P} \cap \mathcal{L}_{01}^{[\perp]} = \mathcal{P}^{[\perp]} \cap \mathcal{L}_{01} = \{0\}$;
(iv) $\kappa_+(\mathcal{G}) = \kappa_-(\mathcal{G}) = \dim \mathcal{P} = \dim \mathcal{L}_{01} < \infty$;
(v) $\mathcal{L}^{[\perp]} = \mathcal{L}^\circ[\dot{+}]\mathcal{M}$.

Proof. By Lemma 2.2 the subspace $\mathcal{L}_+[\dot{+}]\mathcal{L}_-$ is ortho-complemented. With $\mathcal{K} := (\mathcal{L}_+[\dot{+}]\mathcal{L}_-)^{[\perp]}$ we have $\mathcal{H} = \mathcal{L}_+[\dot{+}]\mathcal{L}_-[\dot{+}]\mathcal{K}$, and $\mathcal{D} \cap \mathcal{K}$ is dense in $\mathcal{K}$. Let now $\mathcal{K}_1$ and $\mathcal{L}_{01}$ be subspaces such that $\mathcal{K} = \mathcal{L}^\circ \dot{+} \mathcal{K}_1$ and $\mathcal{L}^\circ = \mathcal{L}_{00} \dot{+} \mathcal{L}_{01}$, where $\mathcal{L}_{00} := \mathcal{L}^\circ \cap \mathcal{H}^\circ$. Then we have $\mathcal{K} = \mathcal{L}_{00} \dot{+} \mathcal{L}_{01} \dot{+} \mathcal{K}_1$. Since $\dim \mathcal{L}_{01} < \infty$, the space $\mathcal{K}_2 := \mathcal{L}_{01} \dot{+} \mathcal{K}_1$ is a subspace, and $\mathcal{D}_2 := \mathcal{D} \cap \mathcal{K}_2$ is dense in $\mathcal{K}_2$. We observe that by construction we have

$$\mathcal{H} = \mathcal{L}_+[\dot{+}]\mathcal{L}_-[\dot{+}]\mathcal{L}_{00}[\dot{+}]\mathcal{K}_2, \tag{3.7}$$

which in particular implies $\mathcal{K}_2^\circ \subset \mathcal{H}^\circ$, and from $\mathcal{L}_{01} = \mathcal{L}_{01} \cap \mathcal{K}_2$ we conclude

$$\mathcal{L}_{01} \cap \mathcal{D}_2^{[\perp]} = \mathcal{L}_{01} \cap \mathcal{K}_2^{[\perp]} = \mathcal{L}_{01} \cap \mathcal{K}_2^\circ \subset \mathcal{L}_{01} \cap \mathcal{H}^\circ = \{0\}.$$

According to [B, Lemma I.10.4] there exist a basis $\{e_1, \dots, e_n\}$ of $\mathcal{L}_{01}$ and vectors $g_1, \dots, g_n$ in $\mathcal{D}_2$ such that

$$[e_i, g_j] = \delta_{ij} \text{ holds for all } i, j = 1, \dots, n.$$

It is easy to see that the subspace

$$\mathcal{G} := \operatorname{span}\{e_1, \dots, e_n, g_1, \dots, g_n\}$$

is non-degenerate. With a fundamental symmetry J in $\mathcal{G}$ (see, e.g., [L2]) define the neutral subspace $\mathcal{P} := J\mathcal{L}_{01}$. Finally, with $\mathcal{M} := \mathcal{G}^{[\perp]} \cap \mathcal{K}_2$, we have $\mathcal{K}_2 = (\mathcal{L}_{01} \dot{+} \mathcal{P})[\dot{+}]\mathcal{M}$, and with (3.7) the decomposition (3.6) follows. Now, it is not difficult to see that the statements (i)–(v) hold. □

The following propositions are consequences of Lemma 3.10.

Proposition 3.11. *A non-positive subspace $\mathcal{L}$ is maximal non-positive if and only if it has dimension κ.*

Proof. If $\dim \mathcal{L} = \kappa$, the statement follows from Lemma 3.8. Let $\mathcal{L}$ be maximal non-positive. We use the same notations as in Lemma 3.10. Since $\mathcal{H} = \mathcal{L}_-[\dot{+}]\mathcal{L}_{00}[\dot{+}](\mathcal{L}_{01} \dot{+} \mathcal{P})[\dot{+}]\mathcal{M}$, the subspace $\mathcal{M}$ must be positive and thus $\kappa = \dim \mathcal{L}_- + \dim \mathcal{L}_{00} + \kappa_-(\mathcal{G}) = \dim \mathcal{L}_- + \dim \mathcal{L}_{00} + \dim \mathcal{L}_{01} = \dim \mathcal{L}$. □

In a Pontryagin space with finite rank of negativity any dense linear manifold contains a maximal negative subspace (cf. [B, Theorem IX.1.4]). This also holds in an Almost Pontryagin space with finite rank of non-positivity. Indeed, if $\mathcal{H} = \Pi[\dot{+}]\mathcal{H}^\circ$ with a Pontryagin space Π and $\mathcal{D}$ is dense in $\mathcal{H}$, then $\mathcal{D} \cap \Pi$ is dense in Π. This implies the assertion. The analogue statement for maximal non-positive subspaces in Almost Pontryagin spaces is not true, in general.

Proposition 3.12. *Let $\mathcal{D} \subset \mathcal{H}$ be a dense linear manifold in $\mathcal{H}$. Then for every non-positive subspace $\mathcal{L} \subset \mathcal{D}$ we have*

$$\dim \mathcal{L} \le \kappa_-(\mathcal{H}) + \dim \mathcal{D}^\circ \le \kappa,$$

and there exists a non-positive subspace $\mathcal{L}' \subset \mathcal{D}$ with $\dim \mathcal{L}' = \kappa_-(\mathcal{H}) + \dim \mathcal{D}^\circ$. In particular, $\mathcal{D}$ contains a maximal non-positive subspace if and only if $\mathcal{H}^\circ \subset \mathcal{D}$.

Proof. Let $\mathcal{L} \subset \mathcal{D}$ be a non-positive subspace. If $\mathcal{L} = \mathcal{L}_-[\dot{+}]\mathcal{L}^\circ$ is a fundamental decomposition of $\mathcal{L}$, we obtain a decomposition

$$\mathcal{H} = \mathcal{L}_-[\dot{+}]\mathcal{L}_{00}[\dot{+}](\mathcal{L}_{01} \dot{+} \mathcal{P})[\dot{+}]\mathcal{M}$$

with the properties stated in Lemma 3.10. Then we have

(i) $\kappa_-(\mathcal{H}) = \dim \mathcal{L}_- + \dim \mathcal{L}_{01} + \kappa_-(\mathcal{M})$ and
(ii) $\mathcal{D} = \mathcal{L}_-[\dot{+}]\mathcal{L}_{00}[\dot{+}](\mathcal{L}_{01} \dot{+} \mathcal{P})[\dot{+}](\mathcal{D} \cap \mathcal{M})$.

From (ii) it easily follows that $\mathcal{D}^\circ = \mathcal{L}_{00} \dot{+} (\mathcal{D} \cap \mathcal{M}^\circ)$. This and (i) imply

$$\begin{aligned}\dim \mathcal{L} &= \dim \mathcal{L}_- + \dim \mathcal{L}_{00} + \dim \mathcal{L}_{01} \\ &= \kappa_-(\mathcal{H}) - \kappa_-(\mathcal{M}) + \dim \mathcal{L}_{00} \\ &\le \kappa_-(\mathcal{H}) + \dim \mathcal{L}_{00} \\ &\le \kappa_-(\mathcal{H}) + \dim \mathcal{D}^\circ.\end{aligned}$$

By $\mathcal{D}^\circ \subset \mathcal{H}^\circ$ this number does not exceed κ. In order to show the existence of a subspace $\mathcal{L}' \subset \mathcal{D}$ as above, choose a fundamental decomposition $\mathcal{H} = \Pi[\dot{+}]\mathcal{H}^\circ$ of $\mathcal{H}$ with a Pontryagin space $(\Pi, [\cdot\,,\cdot])$ with finite rank of negativity $\kappa_-(\mathcal{H})$ (see Lemma 3.2). The linear manifold $\mathcal{D}' := \mathcal{D} \cap \Pi$ is dense in Π. Thus, there exists a negative subspace $\mathcal{L}_- \subset \mathcal{D}'$ with $\dim \mathcal{L}_- = \kappa_-(\mathcal{H})$ and the subspace $\mathcal{L}' := \mathcal{L}_-[\dot{+}]\mathcal{D}^\circ$ is non-positive and has dimension $\kappa_-(\mathcal{H}) + \dim \mathcal{D}^\circ$. Hence, $\mathcal{D}$ contains a maximal non-positive (and thus κ-dimensional) subspace if and only if $\dim \mathcal{D}^\circ = \dim \mathcal{H}^\circ$, which, as $\mathcal{D}^\circ \subset \mathcal{H}^\circ$, is equivalent to $\mathcal{D}^\circ = \mathcal{H}^\circ$. But this is again equivalent to $\mathcal{H}^\circ \subset \mathcal{D}$. □

In a Pontryagin space a subspace $\mathcal{L}$ is ortho-complemented if and only if it is non-degenerate. In an Almost Pontryagin space we have the following

Proposition 3.13. *A subspace $\mathcal{L} \subset \mathcal{H}$ is ortho-complemented in $\mathcal{H}$ if and only if $\mathcal{L}^\circ \subset \mathcal{H}^\circ$.*

Proof. By applying Lemma 3.10 to $\mathcal{L}$ and using the same notations we obtain $\mathcal{L} + \mathcal{L}^{[\perp]} = \mathcal{L} \dot{+} \mathcal{M}$. Thus, $\mathcal{L}$ is ortho-complemented if and only if $\mathcal{P} = \{0\}$. But this is equivalent to $\mathcal{L}^\circ \subset \mathcal{H}^\circ$. □

4. G-symmetric and G-self-adjoint operators in Almost Pontryagin spaces

In this section let $(\mathcal{H}, \mathcal{O}, [\cdot\,,\cdot])$ be an Almost Pontryagin space with finite rank of non-positivity $\kappa = \kappa_-(\mathcal{H}) + \kappa_0(\mathcal{H})$ and let $(\cdot\,,\cdot)$ be a Hilbert space inner product inducing $\mathcal{O}$. By G we denote the Gram operator of $[\cdot\,,\cdot]$ with respect to $(\cdot\,,\cdot)$. The results of this section also hold for Almost Pontryagin spaces with finite rank of non-negativity.

We recall some notions related to linear relations in $\mathcal{H}$. A *linear relation in* $\mathcal{H}$ is a linear manifold $S \subset \mathcal{H} \times \mathcal{H}$. For the basic properties of linear relations we refer to [C, DS, H]. We only mention that for linear relations $S, T \subset \mathcal{H} \times \mathcal{H}$ one defines

$$\operatorname{dom} S := \left\{ x : \begin{pmatrix} x \\ y \end{pmatrix} \in S \right\}, \qquad \textit{the domain of } S,$$

$$\operatorname{ran} S := \left\{ y : \begin{pmatrix} x \\ y \end{pmatrix} \in S \right\}, \qquad \textit{the range of } S,$$

$$\operatorname{mul} S := \left\{ y : \begin{pmatrix} x \\ y \end{pmatrix} \in S \right\}, \qquad \textit{the multi-valued part of } S,$$

$$S^{-1} := \left\{ \begin{pmatrix} y \\ x \end{pmatrix} : \begin{pmatrix} x \\ y \end{pmatrix} \in S \right\}, \qquad \textit{the inverse of } S,$$

$$S + T := \left\{ \begin{pmatrix} x \\ y+z \end{pmatrix} : \begin{pmatrix} x \\ y \end{pmatrix} \in S, \begin{pmatrix} x \\ z \end{pmatrix} \in T \right\}, \qquad \textit{the sum of } S \textit{ and } T$$

and the *product of S and T*,

$$ST := \left\{ \begin{pmatrix} x \\ z \end{pmatrix} : \text{there exists } y \in \mathcal{H} \text{ with } \begin{pmatrix} y \\ z \end{pmatrix} \in S, \begin{pmatrix} x \\ y \end{pmatrix} \in T \right\}.$$

The elements in a linear relation S will usually be written as column vectors $\binom{x}{y}$, where $x \in \operatorname{dom} S$ and $y \in \operatorname{ran} S$. Linear operators are identified as linear relations via their graphs.

Further on, we define

$$S^+ := \left\{ \begin{pmatrix} u \\ v \end{pmatrix} \in \mathcal{H} \times \mathcal{H} : [y, u] = [x, v] \text{ for all } \begin{pmatrix} x \\ y \end{pmatrix} \in S \right\}$$

and call this linear relation the $[\cdot\,,\cdot]$-*adjoint of* S. If S is (the graph of) an operator, we have

$$S^+ := \left\{ \begin{pmatrix} u \\ v \end{pmatrix} \in \mathcal{H} \times \mathcal{H} : [Sx, u] = [x, v] \text{ for all } x \in \operatorname{dom} S \right\}.$$

Lemma 4.1. *If* $S \subset \mathcal{H} \times \mathcal{H}$ *is a linear relation, then* S^+ *is a closed linear relation with* $\mathcal{H}^\circ \times \mathcal{H}^\circ \subset S^+$ *and*

$$S^{++} = \overline{S} + (\mathcal{H}^\circ \times \mathcal{H}^\circ). \tag{4.1}$$

If $\operatorname{dom} S$ *is dense in* $\mathcal{H}$ *we have*

$$\operatorname{mul} S^+ = \mathcal{H}^\circ.$$

The relation S^+ *is a linear operator if and only if* S *is densely defined and* $\mathcal{H}^\circ = \{0\}$.

Proof. It follows from the definition of S^+ that S^+ is closed and $\mathcal{H}^\circ \times \mathcal{H}^\circ \subset S^+$. As

$$(\operatorname{dom} S)^{[\perp]} = \operatorname{mul} S^+,$$

it remains to show (4.1). By $\mathcal{O} \times \mathcal{O}$ we denote the product topology on $\mathcal{H} \times \mathcal{H}$. We introduce an inner product $[\cdot\,,\cdot]$ on $\mathcal{H} \times \mathcal{H}$,

$$\left[\begin{pmatrix} x \\ y \end{pmatrix}, \begin{pmatrix} u \\ v \end{pmatrix} \right] := [x, u] + [y, v], \quad x, y, u, v \in \mathcal{H}. \tag{4.2}$$

By [KWW, Proposition 3.1], $(\mathcal{H} \times \mathcal{H}, \mathcal{O} \times \mathcal{O}, [\cdot\,,\cdot])$ is an Almost Pontryagin space with finite rank of non-positivity and isotropic part $\mathcal{H}^\circ \times \mathcal{H}^\circ$. It is easily seen that $(S^{-1})^{[\perp]} = (S^{[\perp]})^{-1}$, where the orthogonal companion is now with respect to the inner product defined in (4.2). We have

$$S^+ = \left((-S)^{[\perp]} \right)^{-1}.$$

Applying this twice we obtain

$$S^{++} = \left((-S^+)^{[\perp]} \right)^{-1} = \left(((S^{[\perp]})^{-1})^{[\perp]} \right)^{-1} = S^{[\perp][\perp]} = \overline{S} + (\mathcal{H}^\circ \times \mathcal{H}^\circ)$$

by Proposition 3.7. □

We now introduce the notion of G-symmetric operators where the symmetry is understood with respect to the inner product $[\cdot\,,\cdot]$.

Definition 4.2. A densely defined linear operator A in $\mathcal{H}$ is called G-*symmetric* if the operator GA is symmetric in the Hilbert space $(\mathcal{H}, (\cdot\,,\cdot))$.

Evidently, for a densely defined linear operator A in $\mathcal{H}$ the following statements are equivalent.

(i) A is G-symmetric;
(ii) $A \subset A^+$;
(iii) $[Ax, y] = [x, Ay]$ for all $x, y \in \operatorname{dom} A$.

We say that a closed and densely defined operator T in $\mathcal{H}$ is a Φ_+-operator if $\dim\ker(T) < \infty$ and if $\operatorname{ran}(T)$ is closed.

Lemma 4.3. *Let A be a closed and densely defined, G-symmetric operator in $\mathcal{H}$. Then for all $\lambda \in \mathbb{C}\setminus\mathbb{R}$ the operator $A-\lambda$ is a Φ_+-operator. Moreover,* $\ker(A-\lambda) \neq \{0\}$ *only holds for at most finitely many $\lambda \in \mathbb{C}\setminus\mathbb{R}$.*

Proof. Let $\mathcal{H} = \mathcal{H}_+[\dot{+}]\mathcal{H}_-[\dot{+}]\mathcal{H}^\circ$ be a fundamental decomposition of $\mathcal{H}$. Then, since $\mathcal{H}_+$ is uniformly positive, there exists a $\delta > 0$ such that $[x,x] \geq \delta\|x\|^2$ for $x \in \mathcal{H}_+$. Thus, for all $\lambda \in \mathbb{C}\setminus\mathbb{R}$ and all $x \in \mathcal{H}_+ \cap \operatorname{dom} A$ we have

$$\begin{aligned}\delta|\mathrm{Im}\lambda|\|x\|^2 &\leq |\mathrm{Im}\lambda|[x,x] = |\mathrm{Im}[\lambda x,x]| = |\mathrm{Im}[(A-\lambda)x,x]| \\ &\leq |[(A-\lambda)x,x]| \leq \|G\|\|(A-\lambda)x\|\|x\|,\end{aligned}$$

which implies

$$\|(A-\lambda)x\| \geq \delta\|G\|^{-1}|\mathrm{Im}\lambda|\|x\|.$$

This shows that $A-\lambda$ is a Φ_+-operator for all $\lambda \in \mathbb{C}\setminus\mathbb{R}$. And as the subspace $\overline{\operatorname{span}}\{\ker(A-\lambda) : \lambda \in \mathbb{C}^+\}$ is neutral, the additional assertion follows from Lemma 3.8(ii). □

In [LMM] an everywhere defined and bounded linear operator A was called G-self-adjoint if $[Ax,y] = [x,Ay]$ holds for all $x,y \in \mathcal{H}$. This is obviously equivalent to the condition $(GA)^* = GA$. We extend this notion to unbounded operators.

Definition 4.4. A linear densely defined operator A in $\mathcal{H}$ is called G-self-adjoint if $GA = (GA)^*$ and if A is closed.

Obviously, if $(\mathcal{H}, [\cdot\,,\cdot])$ is a Pontryagin space (i.e., $\mathcal{H}^\circ = \{0\}$), an operator A in $\mathcal{H}$ is G-self-adjoint if and only if $A = A^+$. Such operators are necessarily closed. However, in an Almost Pontryagin space there are densely defined operators A with $GA = (GA)^*$ which are not closable (see Example 4.9 below).

If $\langle\cdot\,,\cdot\rangle$ is a second Hilbert space inner product inducing $\mathcal{O}$ and $\widetilde{G}$ is the Gram operator of $[\cdot\,,\cdot]$ with respect to $\langle\cdot\,,\cdot\rangle$, it can be easily seen that an operator A is G-symmetric (G-self-adjoint) if and only if it is $\widetilde{G}$-symmetric (resp. $\widetilde{G}$-self-adjoint). That is, Definition 4.4 does not depend on the choice of the particular Hilbert space inner product (resp. on G).

Theorem 4.5. *For a densely defined operator A in $\mathcal{H}$ the following statements are equivalent:*

(i) $GA = (GA)^*$;
(ii) $A \subset A^+$ *and* $\operatorname{dom} A = \operatorname{dom} A^+$;
(iii) $A^+ = A \dot{+} (\{0\} \times \mathcal{H}^\circ)$;
(iv) *with respect to any fundamental decomposition $\mathcal{H} = \mathcal{H}^\circ[\dot{+}]\Pi$ (cf. (3.2)) the operator A admits a matrix representation*

$$A = \begin{pmatrix} A_0 & A_{12} \\ 0 & \widetilde{A} \end{pmatrix}, \qquad \operatorname{dom} A = \mathcal{H}^\circ \dot{+} \operatorname{dom}\widetilde{A}, \tag{4.3}$$

where $\widetilde{A}$ is a self-adjoint operator in the Pontryagin space $(\Pi, [\cdot\,,\cdot])$.

If A is additionally assumed to be closed, then (ii)–(iv) *are equivalent to the G-self-adjointness of the operator A. Moreover, in this case,* (i)–(iv) *are equivalent to one of the following statements.*

(v) $\mathcal{H}^\circ \subset \operatorname{dom} A$ *and* $A^+ = A^{++}$*;*

(vi) *The operator A is G-symmetric and the sets $\rho(A) \cap \mathbb{C}^+$ and $\rho(A) \cap \mathbb{C}^-$ are not empty.*

Proof. The implication (i)⇒(ii) is evident. Suppose that (ii) holds. Then, as $A \subset A^+$ and $\{0\} \times \mathcal{H}^\circ \subset A^+$ (cf. Lemma 4.1), we have $A + (\{0\} \times \mathcal{H}^\circ) \subset A^+$. Let now $\binom{u}{v} \in A^+$. Then $u \in \operatorname{dom} A^+ = \operatorname{dom} A$ and

$$\begin{pmatrix} u \\ v \end{pmatrix} = \begin{pmatrix} u \\ Au \end{pmatrix} + \begin{pmatrix} 0 \\ v - Au \end{pmatrix}.$$

Since for all $x \in \operatorname{dom} A$ we have $[x, v - Au] = [x, v] - [Ax, u] = 0$, it follows that $v - Au \in \mathcal{H}^\circ$ and (iii) follows.

Assume now that (iii) holds and let $\mathcal{H} = \mathcal{H}^\circ[\dot{+}]\Pi$ be a fundamental decomposition of $\mathcal{H}$ as in (3.2). By Lemma 4.1 $\mathcal{H}^\circ \subset \operatorname{dom} A^+ = \operatorname{dom} A$, hence

$$\operatorname{dom} A = \mathcal{H}^\circ \dot{+} (\operatorname{dom} A \cap \Pi) \quad \text{and} \quad A\mathcal{H}^\circ \subset \mathcal{H}^\circ.$$

It is now clear that A admits a representation (4.3). By P we denote the projection in $\mathcal{H}$ onto Π with $\ker P = \mathcal{H}^\circ$. Then we have $\widetilde{A} = P(A \upharpoonright \operatorname{dom} \widetilde{A})$, where $\operatorname{dom} \widetilde{A} = \operatorname{dom} A \cap \Pi$. It is easy to see that $\widetilde{A}$ is symmetric in $(\Pi, [\cdot\,,\cdot])$. If $u, z \in \Pi$ such that $[\widetilde{A}x, u] = [x, z]$ for all $x \in \operatorname{dom} \widetilde{A}$, then one easily verifies $[Ax, u] = [x, z]$ for all $x \in \operatorname{dom} A$ which implies $\binom{u}{z} \in A^+ = A \dot{+} (\{0\} \times \mathcal{H}^\circ)$ and thus $u \in \operatorname{dom} A \cap \Pi = \operatorname{dom} \widetilde{A}$. This shows that $\widetilde{A}$ is self-adjoint in the Pontryagin space $(\Pi, [\cdot\,,\cdot])$.

Let us show the implication (iv)⇒(i). To this end we set $\Pi := \operatorname{ran} G$. Then $\mathcal{H} = \mathcal{H}^\circ[\dot{+}]\Pi$ is a fundamental decomposition of $\mathcal{H}$, cf. Proposition 3.5, and A admits a representation (4.3) with respect to this decomposition. With the projection P in $\mathcal{H}$ onto Π with $\ker P = \mathcal{H}^\circ$ and $\widetilde{G} := P(G \upharpoonright \Pi)$ we have

$$G = \begin{pmatrix} 0 & 0 \\ 0 & \widetilde{G} \end{pmatrix} \quad \text{and} \quad GA = \begin{pmatrix} 0 & 0 \\ 0 & \widetilde{G}\widetilde{A} \end{pmatrix}.$$

Thus, it remains to show that $\widetilde{G}\widetilde{A}$ is self-adjoint in the Hilbert space $(\Pi, (\cdot\,,\cdot))$. But this is an easy consequence of the fact that $\widetilde{A}$ is self-adjoint in the Pontryagin space $(\Pi, [\cdot\,,\cdot])$ and $\widetilde{G}$ is the Gram operator of $[\cdot\,,\cdot]$ with respect to $(\cdot\,,\cdot)$ in Π.

Let now A be a closed operator. Then (4.1) implies

$$A^{++} = A + (\mathcal{H}^\circ \times \mathcal{H}^\circ)$$

and the equivalence of (iii) and (v) follows with Lemma 4.1.

Let A be a G-self-adjoint operator. The matrix representation (4.3) of A implies

$$\sigma(A) = \sigma(A_0) \cup \sigma(\widetilde{A}), \tag{4.4}$$

hence (iv) implies (vi). Now, suppose that both the upper and the lower half-plane contain points from $\rho(A)$. Then by Lemma 4.3 $\mathbb{C} \setminus \mathbb{R}$ belongs with at most finitely many exceptional points to $\rho(A)$. Thus, there exist numbers $\lambda_0, \overline{\lambda_0} \in (\mathbb{C}\backslash\mathbb{R})\cap\rho(A)$. Let $\{h_1, \dots, h_n\}$ be a basis of $\mathcal{H}^\circ$. Then for all $j = 1, \dots, n$ and all $y \in \mathcal{H}$ we have

$$[(A - \lambda_0)^{-1}h_j, y] = [h_j, (A - \overline{\lambda_0})^{-1}y] = 0.$$

Hence $(A-\lambda_0)^{-1}h_j \in \mathcal{H}^\circ$. But these vectors are linearly independent, and therefore $\mathcal{H}^\circ = \operatorname{span}\{(A - \lambda_0)^{-1}h_j : j = 1, \dots, n\} \subset \operatorname{dom} A$. If $u \in \operatorname{dom} A^+$, there exists a $v \in \mathcal{H}$ such that $[Ax, u] = [x, v]$ for all $x \in \operatorname{dom} A$. Thus, for all $x \in \operatorname{dom} A$ we have

$$\begin{aligned}&[u - (A - \lambda_0)^{-1}(v - \lambda_0 u), (A - \overline{\lambda_0})x]\\ &= [u, Ax] - [u, \overline{\lambda_0}x] - [v - \lambda_0 u, x] = 0.\end{aligned}$$

But this implies $u - (A - \lambda_0)^{-1}(v - \lambda_0 u) \in \mathcal{H}^\circ$, which yields $u \in \operatorname{dom} A$. Hence

$$\operatorname{dom} A = \operatorname{dom} A^+$$

and (ii) holds. □

Corollary 4.6. *Let A be a densely defined operator in $\mathcal{H}$ satisfying one of the conditions* (i)–(iv) *of Proposition* 4.5. *Then the following statements are equivalent:*

(i) *A is closed;*
(ii) *A is closable;*
(iii) *A_{12} is $\widetilde{A}$-bounded.*

Proof. Let A be closable. Let $(x_n) \subset \operatorname{dom} A$ be a sequence with $x_n \to x$ and $Ax_n \to y$ as $n \to \infty$. Then for $u \in \operatorname{dom} A$ we have

$$[Au, x] = \lim_{n\to\infty} [Au, x_n] = \lim_{n\to\infty} [u, Ax_n] = [u, y],$$

which implies $\binom{x}{y} \in A^+$. Therefore, by Theorem 4.5, $x \in \operatorname{dom} A^+ = \operatorname{dom} A$. As A is closable we have $Ax = y$, i.e., A is closed and (i) is equivalent to (ii).

Let λ be a complex number in $\rho(\widetilde{A}) \cap \rho(A_0)$. Then it is easy to see that $A-\lambda$ is bijective with

$$(A - \lambda)^{-1} = \begin{pmatrix} (A_0 - \lambda)^{-1} & -(A_0 - \lambda)^{-1}A_{12}(\widetilde{A} - \lambda)^{-1} \\ 0 & (\widetilde{A} - \lambda)^{-1} \end{pmatrix}$$

and that A_{12} is $\widetilde{A}$-bounded if and only if $A_{12}(\widetilde{A} - \lambda)^{-1}$ is bounded. If A is closed then $\lambda \in \rho(A)$ which implies that $(A_0 - \lambda)^{-1}A_{12}(\widetilde{A} - \lambda)^{-1}$ is bounded. Hence, A_{12} is $\widetilde{A}$-bounded. If $A_{12}(\widetilde{A} - \lambda)^{-1}$ is bounded, then obviously $(A - \lambda)^{-1}$ is bounded and thus $\lambda \in \rho(A)$ which shows that A is closed. □

Note that in general the operator A_{12} in representation (4.3) need not be closed or closable.

Corollary 4.7. *The spectrum of a G-self-adjoint operator A in $\mathcal{H}$ is the union of the spectrum of a self-adjoint operator in a Pontryagin space and a finite set in $\mathbb{C}$. Moreover, there exists a maximal non-positive (and hence κ-dimensional) subspace $\mathcal{L} \subset \operatorname{dom} A$ which is invariant under A.*

Proof. The first statement of the corollary follows from representation (4.3) and the closedness of A (see (4.4)). By a theorem of Pontryagin there exists a maximal non-positive (in Π) subspace $\widetilde{\mathcal{L}} \subset \operatorname{dom} \widetilde{A}$ which is invariant under $\widetilde{A}$ and it is easily seen that the subspace $\mathcal{L} := \widetilde{\mathcal{L}} \dot{+} \mathcal{H}^\circ \subset \operatorname{dom} A$ is maximal non-positive (in $\mathcal{H}$) and invariant under A. □

Remark 4.8. Let A be a densely defined operator in $\mathcal{H}$ with $GA = (GA)^*$ and let $\Pi, \Pi' \subset \mathcal{H}$ be two subspaces such that $\mathcal{H} = \mathcal{H}^\circ[\dot{+}]\Pi = \mathcal{H}^\circ[\dot{+}]\Pi'$. Then, by Theorem 4.5, A has the representations

$$A = \begin{pmatrix} A_0 & A_{12} \\ 0 & \widetilde{A} \end{pmatrix} \quad \text{and} \quad A = \begin{pmatrix} A_0 & A'_{12} \\ 0 & \widetilde{A}' \end{pmatrix}$$

with respect to the particular decompositions. Let P and P' be the projections in $\mathcal{H}$ with $\operatorname{ran} P = \Pi$, $\operatorname{ran} P' = \Pi'$ and $\ker P = \ker P' = \mathcal{H}^\circ$ and set $U := P' \restriction \Pi$. Then U is a bounded operator. It is easily seen that $U : \Pi \to \Pi'$ is bijective and $U^{-1} = U^+ = P \restriction \Pi'$. We have

$$\widetilde{A}' = U\widetilde{A}U^+ \quad \text{and} \quad A'_{12} = (AP' - P'A)U^+.$$

Example 4.9. In contrast to the case that $(\mathcal{H}, [\cdot\,,\cdot])$ is a Pontryagin space, a densely defined operator A in an Almost Pontryagin space with $GA = (GA)^*$ may not even be closable. As an example, let $(\Pi, [\cdot\,,\cdot]_\Pi)$ be a Pontryagin space with fundamental symmetry J, scalar product $(\cdot\,,\cdot)_\Pi = [J\cdot,\cdot]_\Pi$ and associated norm $\|\cdot\|_\Pi$. Suppose that $\widetilde{A}$ is a bounded self-adjoint operator in $(\Pi, [\cdot\,,\cdot]_\Pi)$ and set $\mathcal{H} := \mathbb{C} \oplus \Pi$. Furthermore, define the inner products $[\cdot\,,\cdot]$ and $(\cdot\,,\cdot)$ on $\mathcal{H}$ by

$$\left[\begin{pmatrix} x \\ z \end{pmatrix}, \begin{pmatrix} y \\ w \end{pmatrix}\right] := [x,y]_\Pi \quad \text{and} \quad \left(\begin{pmatrix} x \\ z \end{pmatrix}, \begin{pmatrix} y \\ w \end{pmatrix}\right) := (x,y)_\Pi + z\overline{w}$$

for $x, y \in \Pi$ and $z, w \in \mathbb{C}$. Then $(\mathcal{H}, \mathcal{O}, [\cdot\,,\cdot])$ is an Almost Pontryagin space (where $\mathcal{O}$ is the topology induced by $(\cdot\,,\cdot)$), and the operator

$$G := \begin{pmatrix} 0 & 0 \\ 0 & J \end{pmatrix}$$

is the Gram operator of $[\cdot\,,\cdot]$ with respect to $(\cdot\,,\cdot)$ in $\mathcal{H}$. If we choose a linear functional $\Phi : \Pi \to \mathbb{C}$ which is not bounded, it is easily seen that the operator A, defined by

$$A := \begin{pmatrix} 0 & \Phi \\ 0 & \widetilde{A} \end{pmatrix},$$

satisfies $GA = (GA)^*$.

Now, let $(\widetilde{x}_n) \subset \Pi$ be a sequence with the properties $\lim_{n\to\infty} \|\widetilde{x}_n\|_\Pi = 0$ and $\liminf_{n\to\infty} |\Phi\widetilde{x}_n| > 0$. Without loss of generality we may assume $\Phi\widetilde{x}_n \to z \in \mathbb{C} \setminus \{0\}$ as $n \to \infty$. The sequence $\binom{0}{\widetilde{x}_n}$ then tends to zero in $\mathcal{H} = \Pi \oplus \mathbb{C}$ and $A\binom{0}{\widetilde{x}_n}$ to $\binom{z}{0}$ in $\mathcal{H}$ as $n \to \infty$, which implies that A is not closable.

5. The spectral function of a G-self-adjoint operator

As in the previous section, let $(\mathcal{H}, \mathcal{O}, [\cdot\,,\cdot])$ be an Almost Pontryagin space with finite rank of non-positivity. Furthermore, let $(\cdot\,,\cdot)$ be a Hilbert space inner product inducing $\mathcal{O}$ and let G be the Gram operator of $[\cdot\,,\cdot]$ with respect to $(\cdot\,,\cdot)$. The results of this section also hold for Almost Pontryagin spaces with finite rank of non-negativity.

With the help of the spectral function of a self-adjoint operator in a Pontryagin space we will construct the spectral function for a G-self-adjoint operator A in an Almost Pontryagin space. By $L(\mathcal{H})$ we denote the set of all bounded linear operators in $\mathcal{H}$.

Theorem 5.1. *Let A be a G-self-adjoint operator in the Almost Pontryagin space $(\mathcal{H}, \mathcal{O}, [\cdot\,,\cdot])$. Then A possesses a spectral function with finitely many singularities, i.e., there exists a finite set $c(A) \subset \mathbb{R}$ and a mapping $E : \mathcal{R}(A) \to L(\mathcal{H})$ (here, $\mathcal{R}(A)$ is the collection of all bounded Borel-sets $M \subset \mathbb{R}$ with $\partial M \cap c(A) = \varnothing$ and their complements in $\overline{\mathbb{R}}$) with the following properties ($\Delta, \Delta_1, \Delta_2, \ldots \in \mathcal{R}(A)$):*

(i) *$E(\Delta)$ is a G-self-adjoint projection,*
(ii) *$E(\varnothing) = 0$,*
(iii) *$E(\Delta_1 \cap \Delta_2) = E(\Delta_1)E(\Delta_2)$,*
(iv) *if $\Delta_1, \Delta_2, \ldots$ are pairwise disjoint and if $\Delta = \bigcup_{i=1}^{\infty} \Delta_i$ is in $\mathcal{R}(A)$, then*

$$E(\Delta) = \sum_{i=1}^{\infty} E(\Delta_i),$$

where the series converges in the strong operator topology,
(v) *$E(\Delta)$ is in the double commutant of the resolvent of A,*
(vi) *if Δ is bounded then $A \upharpoonright E(\Delta)\mathcal{H}$ is bounded,*
(vii) *$\sigma(A \upharpoonright E(\Delta)\mathcal{H}) \subset \overline{\Delta}$.*
(viii) *If $\Delta \cap c(A) = \varnothing$, then $(E(\Delta)\mathcal{H}, [\cdot\,,\cdot])$ is a Hilbert space.*

Proof. We fix a fundamental decomposition $\mathcal{H} = \mathcal{H}^\circ [\dot{+}] \Pi$ of $\mathcal{H}$ with a Pontryagin space Π as in (3.2). By Proposition 3.5 we may assume without loss of generality that

$$\Pi = (\mathcal{H}^\circ)^\perp, \tag{5.1}$$

where the orthogonal complement is understood with respect to the Hilbert space inner product $(\cdot\,,\cdot)$ inducing $\mathcal{O}$. Then with Theorem 4.5 we have a block operator representation

$$A = \begin{pmatrix} A_0 & A_{12} \\ 0 & \widetilde{A} \end{pmatrix}, \quad \operatorname{dom} A = \mathcal{H}^\circ \dotplus (\operatorname{dom} A \cap \Pi)$$

of A where $\widetilde{A}$ is self-adjoint in $(\Pi, [\cdot\,,\cdot])$. For $\lambda \in \rho(A)$ we have

$$(A-\lambda)^{-1} = \begin{pmatrix} (A_0-\lambda)^{-1} & -(A_0-\lambda)^{-1}A_{12}(\widetilde{A}-\lambda)^{-1} \\ 0 & (\widetilde{A}-\lambda)^{-1} \end{pmatrix}.$$

The Riesz-Dunford projection P_0 corresponding to the non-real spectrum of A satisfies for $x = x_1 + x_2$, $y = y_1 + y_2$ with $x_1, y_1 \in \mathcal{H}^\circ$ and $x_2, y_2 \in \Pi$

$$\begin{aligned}
[P_0x, y] &= -\frac{1}{2\pi i}\int_{\mathcal{C}} \left[\begin{pmatrix} (A_0-\lambda)^{-1}x_1 - (A_0-\lambda)^{-1}A_{12}(\widetilde{A}-\lambda)^{-1}x_2 \\ (\widetilde{A}-\lambda)^{-1}x_2 \end{pmatrix}, \begin{pmatrix} y_1 \\ y_2 \end{pmatrix} \right] d\lambda \\
&= -\frac{1}{2\pi i}\int_{\mathcal{C}} \left[(\widetilde{A}-\lambda)^{-1}x_2, y_2\right] d\lambda = \left[\widetilde{E}_0 x_2, y_2\right],
\end{aligned}$$

where $\mathcal{C}$ is a closed, smooth curve enclosing the non-real spectrum of A and $\widetilde{E}_0$ is the Riesz-Dunford projection corresponding to the non-real spectrum of $\widetilde{A}$. Hence, P_0 is a bounded, G-symmetric operator, that is, P_0 is G-self-adjoint in $\mathcal{H}$. The subspace $(I - P_0)\mathcal{H}$ is an Almost Pontryagin space (Corollary 3.9) and the restriction of A to $(I - P_0)\mathcal{H}$ is also G-self-adjoint, see Theorem 4.5 (vi). Therefore we will assume that $\sigma(A)$ is real.

Let $\lambda_1, \ldots, \lambda_n$ be the (real) eigenvalues of A_0 ($\lambda_i \neq \lambda_j$ for $i \neq j$), $m := \dim \mathcal{H}^\circ$, and let $\widetilde{E}$ be the spectral function of $\widetilde{A}$ in Π with the set of critical points $c(\widetilde{A})$ (see [L1, L2]). We set

$$c(A) := \{\lambda_1, \ldots, \lambda_n\} \cup c(\widetilde{A}).$$

It follows from Remark 4.8 that $c(A)$ does not depend on the chosen decomposition (3.2) of the Almost Pontryagin space $\mathcal{H}$. By $E_0^{(j)}$ denote the Riesz-Dunford projection of A_0 in $\mathcal{H}^\circ$ corresponding to the eigenvalue λ_j ($j = 1, \ldots, n$) of A_0. Let $\Delta \in \mathcal{R}(A)$. For $j \in \{1, \ldots, n\}$ we define $\widetilde{A}_\Delta := \widetilde{A} \restriction \widetilde{E}(\Delta)\Pi$ and, if $\lambda_j \notin \Delta$,

$$E_{12}^{(j)}(\Delta) := E_0^{(j)} \left(\sum_{k=1}^{m} (A_0 - \lambda_j)^{k-1} A_{12} (\widetilde{A}_\Delta - \lambda_j)^{-k} \right) \widetilde{E}(\Delta).$$

Otherwise ($\lambda_j \in \Delta$) we set

$$E_{12}^{(j)}(\Delta) := -E_{12}^{(j)}(\Delta^c), \quad \text{where } \Delta^c := \overline{\mathbb{R}} \setminus \Delta.$$

Note that $E_{12}^{(j)}(\Delta)$ is well defined since $\lambda_j \notin \partial\Delta$ and that it is a bounded operator from Π to $\mathcal{H}^\circ$.

Now we define

$$E_{12}(\Delta) := \sum_{j=1}^{n} E_{12}^{(j)}(\Delta), \quad E_0(\Delta) := \sum_{\lambda_j \in \Delta} E_0^{(j)} \quad \text{and}$$

$$E(\Delta) := \begin{pmatrix} E_0(\Delta) & E_{12}(\Delta) \\ 0 & \widetilde{E}(\Delta) \end{pmatrix}.$$

Obviously, $E(\Delta)$ is a bounded and, by Theorem 4.5, a G-self-adjoint operator in $\mathcal{H}$ for every $\Delta \in \mathcal{R}(A)$.

In the following, we will show (iii), which also implies that $E(\Delta)$ is a projection. To this end it suffices to show that for $\Delta_1, \Delta_2 \in \mathcal{R}(A)$ and $j \in \{1, \dots, n\}$ we have

$$E_{12}^{(j)}(\Delta_1 \cap \Delta_2) = E_0(\Delta_1)E_{12}^{(j)}(\Delta_2) + E_{12}^{(j)}(\Delta_1)\widetilde{E}(\Delta_2). \tag{5.2}$$

For the sake of simplicity we assume $\lambda_j = 0$, which is no restriction. If $0 \notin \Delta_1$, then

$$\begin{aligned} E_{12}^{(j)}(\Delta_1 \cap \Delta_2) &= \sum_{k=1}^{m} E_0^{(j)} A_0^{k-1} A_{12} \widetilde{A}_{\Delta_1\cap\Delta_2}^{-k} \widetilde{E}(\Delta_1 \cap \Delta_2) \\ &= \sum_{k=1}^{m} E_0^{(j)} A_0^{k-1} A_{12} \widetilde{A}_{\Delta_1}^{-k} \widetilde{E}(\Delta_1)\widetilde{E}(\Delta_2) \\ &= E_{12}^{(j)}(\Delta_1)\widetilde{E}(\Delta_2). \end{aligned}$$

We have $E_0(\Delta_1)E_0^{(j)} = 0$ and, therefore, $E_0(\Delta_1)E_{12}^{(j)}(\Delta_2) = 0$, hence, (5.2) holds.

In the case $0 \in \Delta_1 \setminus \Delta_2$ we have

$$\begin{aligned} E_{12}^{(j)}(\Delta_1 \cap \Delta_2) &= \sum_{k=1}^{m} E_0^{(j)} A_0^{k-1} A_{12} \widetilde{A}_{\Delta_1\cap\Delta_2}^{-k} \widetilde{E}(\Delta_1 \cap \Delta_2) \\ &= E_{12}^{(j)}(\Delta_2)\widetilde{E}(\Delta_1) = E_{12}^{(j)}(\Delta_2) - E_{12}^{(j)}(\Delta_2)\widetilde{E}(\Delta_1^c) \\ &= E_{12}^{(j)}(\Delta_2) - \sum_{k=1}^{m} E_0^{(j)} A_0^{k-1} A_{12} \widetilde{A}_{\Delta_2}^{-k} \widetilde{E}(\Delta_2)\widetilde{E}(\Delta_1^c) \\ &= E_{12}^{(j)}(\Delta_2) - \sum_{k=1}^{m} E_0^{(j)} A_0^{k-1} A_{12} \widetilde{A}_{\Delta_1^c}^{-k} \widetilde{E}(\Delta_1^c)\widetilde{E}(\Delta_2) \\ &= E_{12}^{(j)}(\Delta_2) + E_{12}^{(j)}(\Delta_1)\widetilde{E}(\Delta_2) \\ &= E_0(\Delta_1)E_{12}^{(j)}(\Delta_2) + E_{12}^{(j)}(\Delta_1)\widetilde{E}(\Delta_2), \end{aligned}$$

as $0 \in \Delta_1$. Now, let $0 \in \Delta_1 \cap \Delta_2$. First of all we observe that

$$\begin{aligned} \widetilde{E}((\Delta_1 \cap \Delta_2)^c) &= I - \widetilde{E}(\Delta_1 \cap \Delta_2) = \widetilde{E}(\Delta_2) + \widetilde{E}(\Delta_2^c) - \widetilde{E}(\Delta_1)\widetilde{E}(\Delta_2) \\ &= (I - \widetilde{E}(\Delta_1))\widetilde{E}(\Delta_2) + \widetilde{E}(\Delta_2^c) = \widetilde{E}(\Delta_1^c)\widetilde{E}(\Delta_2) + \widetilde{E}(\Delta_2^c). \end{aligned}$$

This implies

$$\begin{aligned} E_{12}^{(j)}(\Delta_1 \cap \Delta_2) &= -\sum_{k=1}^{m} E_0^{(j)} A_0^{k-1} A_{12} \widetilde{A}_{\Delta_1^c\cup\Delta_2^c}^{-k} \big(\widetilde{E}(\Delta_2^c) + \widetilde{E}(\Delta_1^c)\widetilde{E}(\Delta_2)\big) \\ &= -\sum_{k=1}^{m} E_0^{(j)} A_0^{k-1} A_{12} \big(\widetilde{A}_{\Delta_2^c}^{-k} \widetilde{E}(\Delta_2^c) + \widetilde{A}_{\Delta_1^c}^{-k} \widetilde{E}(\Delta_1^c)\widetilde{E}(\Delta_2)\big) \\ &= E_{12}^{(j)}(\Delta_2) + E_{12}^{(j)}(\Delta_1)\widetilde{E}(\Delta_2) \\ &= E_0(\Delta_1)E_{12}^{(j)}(\Delta_2) + E_{12}^{(j)}(\Delta_1)\widetilde{E}(\Delta_2), \end{aligned}$$

as $0 \in \Delta_1$ and the proof of (5.2) is complete.

Let $\Delta, \Delta_1, \Delta_2, \dots \in \mathcal{R}(A)$ be as in (iv). In order to prove (iv) it suffices to show that $E_{12}^{(j)}(\Delta)x = \sum_{i=1}^{\infty} E_{12}^{(j)}(\Delta_i)x$ holds for all $x \in \Pi$ and all $j \in \{1, \dots, n\}$. If $\lambda_j = 0$ and $0 \notin \Delta$, then, in fact, we have

$$\begin{aligned}\sum_{i=1}^{N} E_{12}^{(j)}(\Delta_i) &= \sum_{k=1}^{m} E_0^{(j)} A_0^{k-1} A_{12} \sum_{i=1}^{N} \widetilde{A}_{\Delta_i}^{-k} \widetilde{E}(\Delta_i) \\ &= \sum_{k=1}^{m} E_0^{(j)} A_0^{k-1} A_{12} \widetilde{A}_{\Delta}^{-k} \sum_{i=1}^{N} \widetilde{E}(\Delta_i).\end{aligned}$$

As $\sum_{i=1}^{N} \widetilde{E}(\Delta_i)$ converges strongly to $\widetilde{E}(\Delta)$, see [L2], the above sum converges strongly to $E_{12}^{(j)}(\Delta)$ as $N \to \infty$.

Let $0 \in \Delta$. Without loss of generality we may assume $0 \in \Delta_1$. Then $0 \notin \widetilde{\Delta} := \bigcup_{i=2}^{\infty} \Delta_i$ and by

$$\widetilde{E}(\Delta_1^c) = \widetilde{E}(\widetilde{\Delta}) + \widetilde{E}(\Delta^c)$$

we see that

$$E_{12}^{(j)}(\Delta_1) + E_{12}^{(j)}(\widetilde{\Delta}) = \sum_{k=1}^{m} E_0^{(j)} A_0^{k-1} A_{12} \widetilde{A}_{\Delta_1^c \cup \widetilde{\Delta}}^{-k} (\widetilde{E}(\widetilde{\Delta}) - \widetilde{E}(\Delta_1^c)) = E_{12}^{(j)}(\Delta)$$

holds. Thus, (iv) follows from the proof of the case $0 \notin \Delta$.

It suffices to show (v) only for bounded and closed intervals $\Delta \in \mathcal{R}(A)$. Let $\Delta = [a, b]$ be such an interval, and let $\eta > 0$ be a fixed number. For $\varepsilon > 0$ let $\mathcal{C}_{\Delta}^{\varepsilon}$ be the positively oriented and piecewise linear, closed curve which parameterizes the boundary of the set

$$\{z \in \mathbb{C} : a - \varepsilon \leq \mathrm{Re}z \leq b + \varepsilon,\ |\mathrm{Im}z| \leq \eta\}.$$

For δ with $0 < \delta < \eta$ we denote by $\mathcal{C}_{\Delta}^{\varepsilon,\delta}$ the curve $\mathcal{C}_{\Delta}^{\varepsilon}$ without the line segment connecting the points $a - \varepsilon + i\delta$ and $a - \varepsilon - i\delta$ and without the line segment connecting the points $b + \varepsilon - i\delta$ and $b + \varepsilon + i\delta$. We will show

$$E(\Delta) = \lim_{\varepsilon \downarrow 0} -\frac{1}{2\pi i} \int_{\mathcal{C}_{\Delta}^{\varepsilon}}' (A - \lambda)^{-1}\, d\lambda := -\frac{1}{2\pi i} \cdot \lim_{\varepsilon \downarrow 0} \lim_{\delta \downarrow 0} \int_{\mathcal{C}_{\Delta}^{\varepsilon,\delta}} (A - \lambda)^{-1}\, d\lambda,$$

where the limits exist in the strong operator topology. It follows from the properties of the spectral functions E_0 and $\widetilde{E}$ that we only have to prove

$$E_{12}^{(j)}(\Delta) = \lim_{\varepsilon \downarrow 0} \frac{1}{2\pi i} \int_{\mathcal{C}_{\Delta}^{\varepsilon}}' E_0^{(j)} (A_0 - \lambda)^{-1} A_{12} (\widetilde{A} - \lambda)^{-1}\, d\lambda \tag{5.3}$$

for $j = 1, \dots, n$. Let $j \in \{1, \dots, n\}$ be given. Considering the Jordan structure of A_0, we obtain for $\lambda \in \rho(A)$

$$E_0^{(j)} (A_0 - \lambda)^{-1} = -E_0^{(j)} \sum_{k=1}^{m} (\lambda - \lambda_j)^{-k} (A_0 - \lambda_j)^{k-1}.$$

Hence, as A_{12} is $\widetilde{A}$-bounded (see Corollary 4.6), (5.3) reduces to

$$E_{12}^{(j)}(\Delta) = E_0^{(j)} \sum_{k=1}^{m} (A_0 - \lambda_j)^{k-1} A_{12} \left(\lim_{\varepsilon \downarrow 0} -\frac{1}{2\pi i} \int_{C_\Delta^\varepsilon}' (\lambda - \lambda_j)^{-k} (\widetilde{A} - \lambda)^{-1} d\lambda \right).$$

But this holds since by [L2, p. 33] we have

$$\lim_{\varepsilon \downarrow 0} -\frac{1}{2\pi i} \int_{C_\Delta^\varepsilon}' (\lambda - \lambda_j)^{-k} (\widetilde{A} - \lambda)^{-1}\, d\lambda = \begin{cases} (\widetilde{A}_\Delta - \lambda_j)^{-k} \widetilde{E}(\Delta), & \text{if } \lambda_j \notin \Delta \\ -(\widetilde{A}_{\Delta^c} - \lambda_j)^{-k} \widetilde{E}(\Delta^c), & \text{if } \lambda_j \in \Delta. \end{cases}$$

Now, (v) follows immediately. We note that an operator $B \in L(\mathcal{H})$ commutes with the resolvent of A if and only if B commutes with A in the following sense: if $x \in \operatorname{dom} A$ then $Bx \in \operatorname{dom} A$ and $ABx = BAx$. Therefore, (v) implies in particular that all the operators $E(\Delta)$, $\Delta \in \mathcal{R}(A)$, commute with A. This leads to the identity

$$A_0 E_{12}(\Delta)x + A_{12}\widetilde{E}(\Delta)x = E_0(\Delta)A_{12}x + E_{12}(\Delta)\widetilde{A}x, \; x \in \operatorname{dom} A, \tag{5.4}$$

which we will need below. If $\Delta \in \mathcal{R}(A)$ is bounded, then $\widetilde{E}(\Delta)\Pi \subset \operatorname{dom} \widetilde{A} = \operatorname{dom} A_{12}$. Hence $E(\Delta)\mathcal{H} \subset \operatorname{dom} A$ which implies (vi) as A is closed.

We show (vii). Let $\Delta \in \mathcal{R}(A)$ and $\lambda \notin \overline{\Delta}$. We have

$$E(\Delta)\mathcal{H} = \left\{ \begin{pmatrix} u \\ v \end{pmatrix} \in \mathcal{H} : \; u = E_0(\Delta)u + E_{12}(\Delta)v, \;\; v = \widetilde{E}(\Delta)v \right\}. \tag{5.5}$$

Let $x = \binom{u}{v} \in (E(\Delta)\mathcal{H}) \cap \operatorname{dom} A$ with $(A - \lambda)x = 0$. Then $(\widetilde{A}_\Delta - \lambda)v = 0$ implies $v = 0$. Thus, $u = E_0(\Delta)u$ and $(A_0 - \lambda)u = 0$. Then $\lambda \notin \overline{\Delta}$ implies $u = 0$. Thus, $x = 0$, and $(A \upharpoonright E(\Delta)\mathcal{H}) - \lambda$ is injective. In order to show that this operator is also surjective, choose some $y = \binom{w}{z} \in E(\Delta)\mathcal{H}$ and set $x := \binom{u}{v}$, where

$$\begin{aligned} v &:= (\widetilde{A}_\Delta - \lambda)^{-1} z \text{ and} \\ u &:= E_{12}(\Delta)v + \big((A_0 \upharpoonright E_0(\Delta)\mathcal{H}^\circ) - \lambda\big)^{-1} E_0(\Delta)(w - A_{12}v). \end{aligned}$$

Obviously, $x \in \operatorname{dom} A$. The relations (5.2) and $E_{12}(\Delta) = -E_{12}(\Delta^c)$ imply

$$(I - E_0(\Delta))u = E_0(\Delta^c)E_{12}(\Delta)v = E_{12}(\Delta)\widetilde{E}(\Delta)v = E_{12}(\Delta)v.$$

Hence $x \in E(\Delta)\mathcal{H}$. Finally, by (5.4) we have

$$\begin{aligned} (A - \lambda)x &= \begin{pmatrix} (A_0 - \lambda)E_{12}(\Delta)v + E_0(\Delta)(w - A_{12}v) + A_{12}v \\ z \end{pmatrix} \\ &= \begin{pmatrix} E_{12}(\Delta)(\widetilde{A} - \lambda)v - A_{12}\widetilde{E}(\Delta)v + E_0(\Delta)w + A_{12}v \\ z \end{pmatrix} \\ &= \begin{pmatrix} E_{12}(\Delta)z + E_0(\Delta)w \\ z \end{pmatrix} = \begin{pmatrix} w \\ z \end{pmatrix} = y, \end{aligned}$$

which shows (vii).

Let $\Delta \in \mathcal{R}(A)$ with $\Delta \cap c(A) = \varnothing$. In order to show (viii) we prove the existence of $\delta > 0$ with

$$[x, x] \geq \delta\|x\|^2 \quad \text{for all } x \in E(\Delta)\mathcal{H}. \tag{5.6}$$

By [L2], $(\widetilde{E}(\Delta)\Pi, [\cdot\,,\cdot])$ is a Hilbert space. Hence, there exists some $\delta_1 > 0$ such that $[x_1, x_1] \geq \delta_1\|x_1\|^2$ for all $x_1 \in \widetilde{E}(\Delta)\Pi$. Set $\delta := \delta_1(1 + \|E_{12}(\Delta)\|^2)^{-1}$. By $E_0(\Delta) = 0$ and (5.5) we have for $x = \binom{u}{v} \in E(\Delta)\mathcal{H}$, $u \in \mathcal{H}^\circ$, $v \in \Pi$:

$$u = E_{12}(\Delta)v \quad \text{and} \quad v = \widetilde{E}(\Delta)v.$$

Therefore,

$$[x, x] = [v, v] \geq \delta_1\|v\|^2 = \delta(1 + \|E_{12}(\Delta)\|^2)\|v\|^2 \geq \delta\|v\|^2 + \delta\|E_{12}(\Delta)v\|^2 = \delta\|x\|^2,$$

where the last equality follows from (5.1), and thus (5.6) holds. □

6. Application to the Klein-Gordon equation

The motion of a relativistic spinless particle of mass m and charge e in an electrostatic field with potential q is described by the Klein-Gordon equation

$$\left(\left(\frac{\partial}{\partial t} - ieq\right)^2 - \Delta + m^2\right)\psi = 0, \tag{6.1}$$

where ψ is a function of $t \in \mathbb{R}$ and $\vec{x} \in \mathbb{R}^n$. As in [LNT] we obtain an abstract model of Equation (6.1) if we replace the expression $-\Delta + m^2$ by a strictly positive self-adjoint operator H_0 in a Hilbert space $(\mathcal{H}, (\cdot\,,\cdot))$ such that $(H_0 x, x) \geq m^2\|x\|^2$ for all $x \in \operatorname{dom} H_0$ and if we replace the operator of multiplication by the function eq by a symmetric operator V in $\mathcal{H}$:

$$\left(\left(\frac{\mathrm{d}}{\mathrm{d}t} - iV\right)^2 + H_0\right)u = 0,$$

where u is a function of t with values in $\mathcal{H}$. By substituting

$$x = u \quad \text{and} \quad y = -i\dot{u}$$

we obtain a first-order differential equation for $\mathbf{x} = \binom{x}{y}$

$$\frac{\mathrm{d}\mathbf{x}}{\mathrm{d}t} = i\hat{A}\mathbf{x}, \quad \hat{A} = \begin{pmatrix} 0 & I \\ H_0 - V^2 & 2V \end{pmatrix}. \tag{6.2}$$

In order to associate a well-defined operator to $\hat{A}$, we make the following assumptions (cf. [LNT]):

(i) $\operatorname{dom} H_0^{1/2} \subset \operatorname{dom} V$ and
(ii) there exist $S_0, S_1 \in \mathcal{L}(\mathcal{H})$ with $\|S_0\| < 1$ and S_1 compact such that

$$V H_0^{-1/2} = S_0 + S_1.$$

We emphasize that the condition $1 \in \rho(S^*S)$ from [LNT] is omitted here. Assumption (i) assures that the operator

$$S := VH_0^{-1/2}$$

is bounded. Assumption (ii) implies $I - S^*S = (I - S_0^*S_0) + K$ with a compact operator K in $\mathcal{H}$. Moreover, $I - S_0^*S_0$ is strictly positive. Thus, there is an $\varepsilon > 0$ such that $\sigma(I - S^*S) \cap (-\infty, \varepsilon)$ consists of at most finitely many eigenvalues with finite multiplicities. In particular, $\operatorname{ran}(I - S^*S)$ is closed and $\dim\ker(I - S^*S) < \infty$.

By $\mathcal{H}_{1/2}$ we denote the Hilbert space $\left(\operatorname{dom} H_0^{1/2}, (\cdot\,,\cdot)_{1/2}\right)$, where

$$(x, y)_{1/2} := \left(H_0^{1/2}x, H_0^{1/2}y\right) \quad \text{for } x, y \in \operatorname{dom} H_0^{1/2}.$$

Now, we define the operator $H : \mathcal{H}_{1/2} \to \mathcal{H}$ by

$$H := H_0^{1/2}(I - S^*S)H_0^{1/2}, \quad \operatorname{dom} H := \left\{x \in \mathcal{H}_{1/2} : (I - S^*S)H_0^{1/2}x \in \operatorname{dom} H_0^{1/2}\right\}.$$

Lemma 6.1. *The operator H, $H : \mathcal{H}_{1/2} \to \mathcal{H}$, is densely defined and closed.*

Proof. The operator $(I - S^*S)H_0^{1/2} : \mathcal{H}_{1/2} \to \mathcal{H}$ is bounded and $H_0^{1/2}$ is a closed operator in $\mathcal{H}$. This shows the closedness of H. Since $H_0^{1/2}$ (from $\mathcal{H}_{1/2}$ to $\mathcal{H}$) is bounded and boundedly invertible, H is densely defined if and only if the linear manifold

$$M := \left\{y \in \mathcal{H} : (I - S^*S)y \in \operatorname{dom} H_0^{1/2}\right\}$$

is dense in $\mathcal{H}$. We have

$$\mathcal{H} = \ker(I - S^*S) \oplus \operatorname{ran}(I - S^*S).$$

Choose $y = y_0 + \widetilde{y} \in \mathcal{H}$ with $y_0 \in \ker(I - S^*S)$ and $\widetilde{y} \in \operatorname{ran}(I - S^*S)$. As $\operatorname{codim}\operatorname{ran}(I - S^*S) < \infty$, the linear manifold $\operatorname{dom} H_0^{1/2} \cap \operatorname{ran}(I - S^*S)$ is dense in $\operatorname{ran}(I - S^*S)$. Thus, there exists a sequence $(x_n) \subset \operatorname{dom} H_0^{1/2} \cap \operatorname{ran}(I - S^*S)$ with $x_n \to (I - S^*S)\widetilde{y}$ as $n \to \infty$. Observe that the restriction of $I - S^*S$ to $\operatorname{ran}(I - S^*S)$ is a bounded and boundedly invertible operator in $\operatorname{ran}(I - S^*S)$. Therefore

$$\left((I - S^*S) \upharpoonright \operatorname{ran}(I - S^*S)\right)^{-1} x_n \to \widetilde{y} \quad \text{as } n \to \infty.$$

That is, $\widetilde{y}$ belongs to the closure of M and, as $y_0 \in M$, the lemma is proved. □

Now, we define the state space

$$\mathcal{G} := \mathcal{H}_{1/2} \times \mathcal{H}.$$

Let $\mathcal{O}$ be the natural Hilbert space topology on $\mathcal{G}$ induced by the Hilbert space inner products $(\cdot\,,\cdot)$ and $(\cdot\,,\cdot)_{1/2}$. With the bounded and self-adjoint operator

$$G := \begin{pmatrix} H_0^{-1/2}(I - S^*S)H_0^{1/2} & 0 \\ 0 & I \end{pmatrix}$$

in $\mathcal{G}$ the so-called energy inner product is defined as follows: For $x, x' \in \mathcal{H}_{1/2}$ and $y, y' \in \mathcal{H}$ we set

$$\begin{aligned}\left[\begin{pmatrix} x \\ y \end{pmatrix}, \begin{pmatrix} x' \\ y' \end{pmatrix}\right] &:= \left(G\begin{pmatrix} x \\ y \end{pmatrix}, \begin{pmatrix} x' \\ y' \end{pmatrix}\right)_{\mathcal{G}} \\ &= ((I - S^*S)H_0^{1/2}x, H_0^{1/2}x') + (y, y') \\ &= (x, x')_{1/2} + (y, y') - (Vx, Vx').\end{aligned}$$

With the formal matrix in (6.2) we associate (see [LNT]) the operator A in $\mathcal{G}$ defined by

$$A := \begin{pmatrix} 0 & I \\ H & 2V \end{pmatrix}, \qquad \operatorname{dom} A := \operatorname{dom} H \times \operatorname{dom} H_0^{1/2} \subset \mathcal{G}.$$

As V, considered as an operator from $\mathcal{H}_{1/2}$ into $\mathcal{H}$, is bounded and as H is closed (Lemma 6.1), it follows that A is a closed and densely defined operator in $\mathcal{G}$.

Recall that a closed and densely defined linear operator B in a Banach space is called *Fredholm* if the dimension of the kernel of B and the codimension of the range of B are finite. In particular, a Fredholm operator has a closed range. The set

$$\sigma_{ess}(B) := \{\lambda \in \mathbb{C} \mid B - \lambda I \text{ is not Fredholm}\}$$

is called the *essential spectrum* of B. The main result of this section is the following

Theorem 6.2. *Under the assumptions* (i) *and* (ii) *the following assertions hold:*

(a) *The triplet $\mathcal{K} := (\mathcal{G}, \mathcal{O}, [\cdot, \cdot])$ is an Almost Pontryagin space with finite rank of non-positivity κ, where κ is the dimension of the spectral subspace corresponding to the non-positive eigenvalues of the operator $I - S^*S$;*

(b) *A is G-self-adjoint in $\mathcal{K}$;*

(c) *A has a spectral function with at most finitely many critical points.*

(d) *The non-real spectrum of A consists of at most finitely many eigenvalues with finite multiplicities.*

(e) *The essential spectrum $\sigma_{\mathrm{ess}}(A)$ is real and*

$$\sigma_{\mathrm{ess}}(A) \cap (-\alpha, \alpha) = \varnothing,$$

where $\alpha := (1 - \|S_0\|)m$.

(f) *The operator iA is the generator of a strongly continuous group $\left(e^{itA}\right)_{t\in\mathbb{R}}$ in $\mathcal{G}$. Hence, the Cauchy problem*

$$\frac{d\mathbf{x}}{dt} = iA\mathbf{x}, \quad \mathbf{x}(0) = \mathbf{x}_0,$$

has a unique solution $\mathbf{x}(t) = e^{itA}\mathbf{x}_0$, $t \in \mathbb{R}$, for all initial values $\mathbf{x}_0 \in \mathcal{G}$.

Proof of Theorem 6.2. We have $\sigma(G) = \sigma(I - S^*S) \cup \{1\}$. Thus, (a) follows from Proposition 3.5. In the following we show that the operator

$$GA = \begin{pmatrix} 0 & H_0^{-1/2}(I - S^*S)H_0^{1/2} \\ H & 2V \end{pmatrix}, \qquad \operatorname{dom} GA = \operatorname{dom} A,$$

is self-adjoint in $(\mathcal{G}, (\cdot\,,\cdot)_{\mathcal{G}})$. Then (b) follows. For $\binom{x}{y} \in \operatorname{dom} A$ we have

$$\begin{aligned}\left(GA\begin{pmatrix} x \\ y \end{pmatrix}, \begin{pmatrix} x \\ y \end{pmatrix}\right) &= \left[\begin{pmatrix} H_0^{-1/2}(I - S^*S)H_0^{1/2}y \\ Hx + 2Vy \end{pmatrix}, \begin{pmatrix} x \\ y \end{pmatrix}\right] \\ &= \left((I - S^*S)H_0^{1/2}y, H_0^{1/2}x\right) + (Hx + 2Vy, y) \\ &= (y, Hx) + (Hx, y) + 2(Vy, y),\end{aligned}$$

and this is a real value. Thus, $GA \subset (GA)^*$. Now, let $\binom{u}{v} \in \operatorname{dom}((GA)^*)$. Then the linear functional

$$\begin{pmatrix} x \\ y \end{pmatrix} \mapsto \left(GA\begin{pmatrix} x \\ y \end{pmatrix}, \begin{pmatrix} u \\ v \end{pmatrix}\right) = (H_0^{1/2}y, (I - S^*S)H_0^{1/2}u) + 2(Vy, v) + (Hx, v)$$

is continuous on $\operatorname{dom} A = \operatorname{dom} H \times \operatorname{dom} H_0^{1/2}$. In particular, the linear functionals

$$\begin{aligned} &f : x \mapsto (Hx, v), \quad x \in \operatorname{dom} H \subset \mathcal{H}_{1/2}, \quad \text{and} \\ &g : y \mapsto \left(H_0^{1/2}y, (I - S^*S)H_0^{1/2}u\right) + 2(Vy, v), \quad y \in \operatorname{dom} H_0^{1/2} \subset \mathcal{H} \end{aligned}$$

are continuous, where f is continuous with respect to the norm in $\mathcal{H}_{1/2}$ and g is continuous with respect to the norm in $\mathcal{H}$. We show that the continuity of f implies $v \in \operatorname{dom} H_0^{1/2}$. There exists a finite-dimensional subspace $X \subset \operatorname{dom} H_0^{1/2}$ such that

$$\mathcal{H} = X \dotplus \operatorname{ran}(I - S^*S) \quad \text{and} \quad \operatorname{dom} H_0^{1/2} = X \dotplus \left(\operatorname{dom} H_0^{1/2} \cap \operatorname{ran}(I - S^*S)\right).$$

Let P_1 be the projection in $\mathcal{H}$ onto X along $\operatorname{ran}(I - S^*S)$ and set $P_2 := I - P_1$. Then $H_0^{1/2}P_1$ is a bounded operator in $\mathcal{H}$. By T denote the bounded and boundedly invertible restriction of $I - S^*S$ to $\operatorname{ran}(I - S^*S) = P_2\mathcal{H}$. For $z \in \operatorname{dom} H_0^{1/2}$ we have

$$\begin{aligned}\left(H_0^{1/2}z, v\right) &= \left(H_0^{1/2}P_1z, v\right) + \left(H_0^{1/2}(I - S^*S)H_0^{1/2}H_0^{-1/2}T^{-1}P_2z, v\right) \\ &= \left(H_0^{1/2}P_1z, v\right) + f\left(H_0^{-1/2}T^{-1}P_2z\right),\end{aligned}$$

which implies $v \in \operatorname{dom}\left(\left(H_0^{1/2}\right)^*\right)$ as $H_0^{-1/2}T^{-1}P_2$ is a bounded operator from $\mathcal{H}$ to $\mathcal{H}_{1/2}$. But $H_0^{1/2}$ is self-adjoint in $\mathcal{H}$, and therefore $v \in \operatorname{dom} H_0^{1/2}$. By (i), this yields $v \in \operatorname{dom} V$ and thus, as g is continuous on $\operatorname{dom} H_0^{1/2}$, it follows that $(I - S^*S)H_0^{1/2}u \in \operatorname{dom} H_0^{1/2}$, hence $\binom{u}{v} \in \operatorname{dom} A$ and (b) is proved.

Statements (c) and (d) hold due to the results of Sections 4 and 5 while (f) is also an easy consequence of Section 5, Theorem 5.1, since with the help of the spectral function the operator A can be written as a direct sum $A_1[\dotplus]A_2$ of a bounded operator in an Almost Pontryagin space and a self-adjoint operator in a

Hilbert space. Obviously, $e^{itA} := e^{itA_1} \, [\dot{+}] \, e^{itA_2}$ is a strongly continuous group of bounded operators in $\mathcal{G}$, and it is easily seen that iA is its generator.

Let us prove (e). To this end we define the quadratic pencils

$$L(\lambda) := I - S^*S + 2\lambda H_0^{-1/2}S - \lambda^2 H_0^{-1} \quad \text{and}$$
$$\widetilde{L}(\lambda) := I - S_0^*S_0 + 2\lambda H_0^{-1/2}S_0 - \lambda^2 H_0^{-1}$$

of bounded operators in $\mathcal{H}$, $\lambda \in \mathbb{C}$. If $\lambda \in \mathbb{R}$, $|\lambda| < (1 - \|S_0\|)m$, then

$$\begin{aligned} &\|S_0^*S_0 - 2\lambda H_0^{-1/2}S_0 + \lambda^2 H_0^{-1}\| \\ &\quad < \|S_0\|^2 + 2(1 - \|S_0\|)m\frac{1}{m}\|S_0\| + (1 - \|S_0\|)^2 m^2 \frac{1}{m^2} = 1. \end{aligned}$$

This implies that $\widetilde{L}(\lambda)$ is boundedly invertible for any $\lambda \in (-\alpha, \alpha)$. And as $L(\lambda)$ is a compact perturbation of $\widetilde{L}(\lambda)$ it follows that $L(\lambda)$ is a (bounded) Fredholm operator for all $\lambda \in (-\alpha, \alpha)$ (see [Ka, Theorem IV.5.26]). We will now show that $A - \lambda$ is a Fredholm operator for $\lambda \in (-\alpha, \alpha)$ which then completes the proof of (e). Let $\lambda \in (-\alpha, \alpha)$. For $\binom{x}{y} \in \operatorname{dom} A$ we have

$$(A - \lambda)\binom{x}{y} = \begin{pmatrix} y - \lambda x \\ H_0^{1/2}L(\lambda)H_0^{1/2}x + (2SH_0^{1/2} - \lambda)(y - \lambda x) \end{pmatrix}. \tag{6.3}$$

From this it is immediately seen that $\ker(A - \lambda)$ is finite-dimensional since $\ker L(\lambda)$ is finite-dimensional. We show that $\operatorname{ran}(A - \lambda)$ is closed. In order to see this let $(\binom{x_n}{y_n})$ be a sequence in $\operatorname{dom} A$ such that $(A - \lambda)\binom{x_n}{y_n}$ converges in $\mathcal{G}$ to some $\binom{w}{z} \in \mathcal{G}$. Then from (6.3) it follows that $H_0^{1/2}L(\lambda)H_0^{1/2}x_n$ converges in $\mathcal{H}$ to $z - (2SH_0^{1/2} - \lambda)w$. As $L(\lambda)$ is a Fredholm operator, we find some $x \in \operatorname{dom} H$ such that

$$L(\lambda)H_0^{1/2}x = H_0^{-1/2}\left(z - (2SH_0^{1/2} - \lambda)w\right).$$

With $y := w + \lambda x$ we have $\binom{x}{y} \in \operatorname{dom} A$ and (see (6.3))

$$(A - \lambda)\binom{x}{y} = \begin{pmatrix} w \\ H_0^{1/2}H_0^{-1/2}\left(z - (2SH_0^{1/2} - \lambda)w\right) + (2SH_0^{1/2} - \lambda)w \end{pmatrix} = \begin{pmatrix} w \\ z \end{pmatrix},$$

and hence $\binom{w}{z} \in \operatorname{ran}(A - \lambda)$.

We have just shown that for all $\lambda \in (-\alpha, \alpha)$ the operator $A - \lambda$ is a Φ_+-operator. By Corollary 4.7 and [Ka, IV §5.6] the operator $A - \lambda$ is even a Fredholm operator of index 0 for all $\lambda \in (-\alpha, \alpha)$. This completes the proof of the theorem. □

Remark 6.3. In [LNT, Remark 5.4] it was already mentioned that if (i) and (ii) are fulfilled, the operator A induces a self-adjoint operator $\widetilde{A}$ in the factor space $\widetilde{\mathcal{G}} := \mathcal{G}/\mathcal{G}^\circ$, which is a Pontryagin space. This operator corresponds to the $\widetilde{A}$ in Proposition 4.5.

References

[AI] T.Ya. Azizov and I.S. Iokhvidov, *Linear Operators in Spaces with an Indefinite Metric*, John Wiley & Sons, 1989.

[B] J. Bognar, *Indefinite Inner Product Spaces*, Springer, 1974.

[C] Cross, R., *Multivalued Linear Operators*, Monographs and Textbooks in Pure and Applied Mathematics **213**, Marcel Dekker, Inc., New York, 1998.

[DS] A. Dijksma and H.S.V. de Snoo, *Symmetric and selfadjoint relations in Krein spaces I*, Oper. Theory Adv. Appl. **24** (1987), 145–166.

[H] M. Haase, *The Functional Calculus for Sectorial Operators*, Oper. Theory Adv. Appl. **169**, Birkhäuser, 2006.

[IKL] I.S. Iohvidov, M.G. Krein and H. Langer, *Introduction to the Spectral Theory of Operators in Spaces with an Indefinite Metric*, Akademie-Verlag, 1982.

[J] P. Jonas, *On local wave operators for definitizable operators in Krein space and on a paper by T. Kako*, Preprint P-46/79 Zentralinstitut für Mathematik und Mechanik der AdW DDR, Berlin, 1979.

[Kk] T. Kako, *Spectral and scattering theory for the J-selfadjoint operators associated with the perturbed Klein-Gordon type equations*, J. Fac. Sci. Univ. Tokyo Sect. IA Math. **23** (1976), 199–221.

[Ka] T. Kato, *Perturbation Theory for Linear Operators*, Second Edition, Springer, 1976.

[KWW] M. Kaltenbäck, H. Winkler and H. Woracek, *Almost Pontryagin Spaces*, Oper. Theory Adv. Appl. **160** (2005), 253–271.

[KW1] M. Kaltenbäck and H. Woracek, *Selfadjoint extensions of symmetric operators in degenerated inner product spaces*, Integral Equations Operator Theory **28** (1997), 289–320.

[KW2] M. Kaltenbäck and H. Woracek, *On extensions of Hermitian functions with a finite number of negative squares*, J. Operator Theory **40** (1998), 147–183.

[KW3] M. Kaltenbäck and H. Woracek, *The Krein formula for generalized resolvents in degenerated inner product spaces*, Monatsh. Math. **127** (1999), 119–140.

[L1] H. Langer, *Spektraltheorie linearer Operatoren in J-Räumen und einige Anwendungen auf die Schar $L(\lambda) = \lambda^2 + \lambda B + C$*, Habilitationsschrift, Technische Universität Dresden, 1965.

[L2] H. Langer, *Spectral functions of definitizable operators in Krein spaces*, Lect. Notes Math. **948** (1982), 1–46.

[LMM] H. Langer, A. Markus and V. Matsaev, *Locally definite operators in indefinite inner product spaces*, Math. Ann. **308** (1997), 405–424.

[LMT] H. Langer, R. Mennicken, and C. Tretter, *A self-adjoint linear pencil $Q - \lambda P$ of ordinary differential operators*, Methods Funct. Anal. Topology **2** (1996), 38–54.

[LNT] H. Langer, B. Najman and C. Tretter, *Spectral theory of the Klein-Gordon equation in Pontryagin spaces*, Commun. Math. Phys. **267**, No. 1 (2006), 159–180.

[N1] B. Najman, *Spectral properties of the operators of Klein-Gordon type*, Glas. Mat. Ser. III, **15**(35) (1980), 97–112.

[N2] B. Najman, *Localization of the critical points of Klein-Gordon type operators*, Math. Nachr. **99** (1980), 33–42.

[N3] B. Najman, *Eigenvalues of the Klein-Gordon equation* Proc. Edinburgh Math. Soc. **26** (1983), 181–190.

[W1] H. Woracek, *An operator theoretic approach to degenerated Nevanlinna-Pick interpolation*, Math. Nachr. **176** (1995), 335–350.

[W2] H. Woracek, *Resolvent matrices in degenerated inner product spaces*, Math. Nachr. **213** (2000), 155–175.

Friedrich Philipp
Technische Universität Berlin
Sekretariat MA 6-4
Straße des 17. Juni 136
D-10623 Berlin, Germany
e-mail: philipp@math.tu-berlin.de

Carsten Trunk
Technische Universität Ilmenau
Institut für Mathematik
Postfach 100565
D-98684 Ilmenau, Germany
e-mail: carsten.trunk@tu-ilmenau.de

Operator Theory:
Advances and Applications, Vol. 188, 237–243

On a Kreĭn Criterion

Adrian Sandovici

Abstract. The set of all nonnegative self-adjoint extensions of a nonnegative linear relation (multi-valued linear operator) is described by using a partial order defined on the set of the corresponding quadratic forms. This characterization leads to a generalization of a result due to M.G. Kreĭn.

Mathematics Subject Classification (2000). Primary 47A06; Secondary 47A20, 47B25, 47A63.

Keywords. Hilbert space, nonnegative linear relation, nonnegative self-adjoint extension.

1. Introduction

This paper deals with the problem of the extension theory of nonnegative linear relations (nonnegative multi-valued operators) in Hilbert spaces (see [1] for the case of closed nonnegative operators and [16] for the case of bounded nonnegative operators). The main result of this note – namely Theorem 3.1 – extends the applicability of a characterization of all nonnegative self-adjoint operator extensions of a nonnegative operator (proved in [15]) to the case of nonnegative linear relations (see also [2, 3, 4, 7, 9, 10, 11, 12, 13, 14] for related papers). This characterization is also translated into the language of quadratic forms.

The paper is organized as follows. Section 2 contains a short introduction to linear relations in Hilbert spaces. Closed nonnegative forms are discussed, and nonnegative self-adjoint extensions of nonnegative linear relations are presented. A partial order on the class of all nonnegative linear relations is also introduced. Certain results concerning the class of all nonnegative self-adjoint extensions of a nonnegative linear relation are obtained in Section 3. In particular, a Kreĭn's criteria is extended to the case of nonnegative linear relations.

2. Linear relations and forms in Hilbert spaces

2.1. Linear relations in Hilbert spaces

Let $\mathfrak{H}$ be a complex Hilbert space. A linear subspace A in the Cartesian product $\mathfrak{H} \times \mathfrak{H}$ is called a linear relation in $\mathfrak{H}$. Its domain, range, kernel, and multi-valued part are denoted by $\operatorname{dom} A$, $\operatorname{ran} A$, $\ker A$, and $\operatorname{mul} A$:

$$\begin{aligned}
\operatorname{dom} A &= \{\, f \in \mathfrak{H} : \{f, f'\} \in A \text{ for some } f' \in \mathfrak{H} \,\}, \\
\operatorname{ran} A &= \{\, f' \in \mathfrak{H} : \{f, f'\} \in A \text{ for some } f \in \mathfrak{H} \,\}, \\
\ker A &= \{\, f \in \mathfrak{H} : \{f, 0\} \in A \,\}, \\
\operatorname{mul} A &= \{\, f' \in \mathfrak{H} : \{0, f'\} \in A \,\}.
\end{aligned}$$

The closures of $\operatorname{dom} A$ and $\operatorname{ran} A$ are denoted by $\overline{\operatorname{dom}}\, A$ and $\overline{\operatorname{ran}}\, A$, respectively. When the relation A is closed, then $\ker A$ and $\operatorname{mul} A$ are automatically closed. A linear operator will be identified with its graph. A relation A has a formal inverse $A^{-1} = \{\, \{f', f\} : \{f, f'\} \in A \,\}$. Let A and B be linear relations in $\mathfrak{H}$. Then the product BA is the linear relation defined by

$$BA = \{\, \{f, g\} \in \mathfrak{H} \times \mathfrak{H} : \{f, \varphi\} \in A,\ \{\varphi, g\} \in B \text{ for some } \varphi \in \mathfrak{H} \,\}.$$

This definition agrees with the usual one for operators. Let P be the orthogonal projection from $\mathfrak{H}$ onto $(\operatorname{mul} A)^{\perp}$. Then each $\{f, f'\} \in A$ can be uniquely decomposed as

$$\{f, f'\} = \{f, Pf'\} + \{0, (I - P)f'\}.$$

The linear relation

$$A_s = \{\, \{f, f'\} : \{f, f'\} \in A,\ f' = Pf' \,\} = \{\, \{f, Pf'\} : \{f, f'\} \in A \,\}$$

is called the (orthogonal) operator part of A: it is the graph of an operator from $\mathfrak{H}$ to $P\mathfrak{H} \subset \mathfrak{H}$. In the sense of multiplication of relations, A_s and A are related by $A_s = PA$.

The adjoint A^* of a linear relation A in $\mathfrak{H}$ is the linear relation in $\mathfrak{H}$, defined by

$$A^* = \{\, \{f', f\} \in \mathfrak{H} \times \mathfrak{H} : \langle \{f', f\}, \{h, h'\} \rangle = 0,\ \{h, h'\} \in A \,\},$$

where

$$\langle \{f', f\}, \{h, h'\} \rangle = (f, h) - (f', h'), \quad \{f, f'\}, \{h, h'\} \in \mathfrak{H} \times \mathfrak{H}.$$

The adjoint A^* is automatically closed and linear.

A linear relation A in $\mathfrak{H}$ is said to be symmetric if $(f', f) \in \mathbb{R}$ for all $\{f, f'\} \in A$, or, equivalently, if $A \subset A^*$. The relation A is said to be self-adjoint if $A = A^*$. If the relation A is self-adjoint, then $\overline{\operatorname{dom}}\, A = (\operatorname{mul} A)^{\perp}$ and A_s is a (densely defined) self-adjoint operator in $\overline{\operatorname{dom}}\, A$.

A linear relation A in a Hilbert space $\mathfrak{H}$ is said to be nonnegative, for short $A \geq 0$, if

$$(f', f) \geq 0, \quad \{f, f'\} \in A.$$

Clearly, every nonnegative relation is symmetric.

Let A be a nonnegative self-adjoint relation in a Hilbert space $\mathfrak{H}$. The square root $A^{\frac{1}{2}}$ of A is the unique nonnegative self-adjoint relation B in $\mathfrak{H}$ such that $B^2 = A$, as follows from the definition of the product, cf. [7].

2.2. Closed nonnegative forms

Let $\mathfrak{t} = \mathfrak{t}[\cdot,\cdot]$ be a nonnegative form in the Hilbert space $\mathfrak{H}$ with domain $\operatorname{dom}\mathfrak{t}$, cf. [8, Chapter VI]. The notation $\mathfrak{t}[h]$ will be used to denote $\mathfrak{t}[h,h]$, $h \in \operatorname{dom}\mathfrak{t}$. For the following definitions see [8, Chapter VI]. The nonnegative form $\mathfrak{t}$ is closed if

$$h_n \to h, \quad \mathfrak{t}[h_n - h_m] \to 0, \quad h_n \in \operatorname{dom}\mathfrak{t}, \quad h \in \mathfrak{H}, \quad m,n \to \infty,$$

imply that $h \in \operatorname{dom}\mathfrak{t}$ and $\mathfrak{t}[h_n - h] \to 0$. The nonnegative form $\mathfrak{t}$ is closable if it has a closed extension; in this case the closure of $\mathfrak{t}$ is the smallest closed extension of $\mathfrak{t}$. The inequality $\mathfrak{t}_1 \geq \mathfrak{t}_2$ for nonnegative forms $\mathfrak{t}_1$ and $\mathfrak{t}_2$ is defined by

$$\operatorname{dom}\mathfrak{t}_1 \subset \operatorname{dom}\mathfrak{t}_2, \quad \mathfrak{t}_1[h] \geq \mathfrak{t}_2[h], \quad h \in \operatorname{dom}\mathfrak{t}_1.$$

If the forms $\mathfrak{t}_1$ and $\mathfrak{t}_2$ are closable, the inequality $\mathfrak{t}_1 \geq \mathfrak{t}_2$ is preserved by their closures. It was proved in [7] that there is a one-to-one correspondence between all closed nonnegative forms $\mathfrak{t}$ in $\mathfrak{H}$ and all nonnegative self-adjoint relations A in $\mathfrak{H}$ via

$$\operatorname{dom} A \subset \operatorname{dom}\mathfrak{t},$$

and

$$\mathfrak{t}[f,g] = (A_s f, g), \quad f \in \operatorname{dom} A, \quad g \in \operatorname{dom}\mathfrak{t}.$$

Furthermore, let the nonnegative form $\mathfrak{t}$ and the nonnegative self-adjoint relation A be connected as above. If $\mathfrak{t} \geq 0$ or, equivalently, $A \geq 0$, then

$$\operatorname{dom}\mathfrak{t} = \operatorname{dom} A_s^{1/2},$$

and

$$\mathfrak{t}[f,g] = (A_s^{1/2} f, A_s^{1/2} g), \quad f,g \in \operatorname{dom}\mathfrak{t}.$$

2.3. Nonnegative self-adjoint extensions of nonnegative relations

Let S be a not necessarily closed nonnegative linear relation in a Hilbert space $\mathfrak{H}$, so that also the closure $\operatorname{clos} S$ of S is nonnegative. Since the defect numbers of S and thus of $\operatorname{clos} S$, are equal, there exist self-adjoint extensions of S in $\mathfrak{H}$. In particular, one nonnegative self-adjoint extension can be constructed as follows. Let $\{f,f'\},\{h,h'\} \in S$ and define $\mathfrak{s}[f,h] = (f',h)$, so that $\mathfrak{s}$ is a nonnegative form on $\operatorname{dom}\mathfrak{s} = \operatorname{dom} S$. The form $\mathfrak{s}$ is form-closable, cf. [8, VI Theorem 1.27]. The closure $\mathfrak{t}$ of the form $\mathfrak{s}$ is nonnegative (and is equal to the form obtained by starting with the closure of S) and gives rise to a nonnegative self-adjoint relation which is called the Friedrichs extension S_F of S. Furthermore, the so-called Kreĭn-von Neumann extension S_N of S is defined by

$$S_N = ((S^{-1})_F)^{-1},$$

cf. [1], [6] for the case that S is not densely defined.

The Kreĭn-von Neumann and the Friedrichs extensions are extreme nonnegative self-adjoint extensions of S: if H is any nonnegative self-adjoint extension of S, then

$$S_N \leq H \leq S_F,$$

where the inequalities are in the sense of the corresponding resolvent operators:

$$(S_F + a)^{-1} \leq (H + a)^{-1} \leq (S_N + a)^{-1}, \quad a > 0,$$

or, equivalently, in the sense of the corresponding closed nonnegative sesquilinear forms:

$$\mathfrak{t}_{S_N} \leq \mathfrak{t}_H \leq \mathfrak{t}_{S_F},$$

cf. [7] (see also [2, 6, 13]).

2.4. A partial order on the class of all nonnegative self-adjoint relations

Assume that A_1 and A_2 are two nonnegative self-adjoint linear relations in the Hilbert space $\mathfrak{H}$. It is said that $A_1 \prec A_2$ if the following two conditions are satisfied:

$$\operatorname{dom} A_2^{\frac{1}{2}} \subset \operatorname{dom} A_1^{\frac{1}{2}},$$

and

$$\|A_{1s}^{\frac{1}{2}} h\| \leq \|A_{2s}^{\frac{1}{2}} h\|, \quad \text{for all} \quad h \in \operatorname{dom} A_2^{\frac{1}{2}}.$$

The relation $\prec$ is a partial order on the class of all nonnegative self-adjoint linear relations in $\mathfrak{H}$. In particular, $A_1 = A_2$ if and only if $A_1 \prec A_2$ and $A_2 \prec A_1$. It can be easily seen that $A_1 \prec A_2$ if and only if $\operatorname{dom} A_2 \subset \operatorname{dom} A_1^{\frac{1}{2}}$ and $\|A_{1s}^{\frac{1}{2}} h\| \leq (h', h)$ for all $\{h, h'\} \in A_2$.

3. Certain characterizations of nonnegative self-adjoint extensions

Given a nonnegative linear relation S in a Hilbert space $\mathfrak{H}$ define $EXT(S)$ as the class of all nonnegative self-adjoint linear relations A in $\mathfrak{H}$ such that $S \subset A$. Note that if S itself is a nonnegative self-adjoint linear relation in $\mathfrak{H}$ then $EXT(S) = \{S\}$.

3.1. A general criteria

Theorem 3.1. *Assume that S is a nonnegative linear relation in a Hilbert space $\mathfrak{H}$ and $A_1 \in EXT(S)$. If A_2 is a nonnegative self-adjoint linear relation in $\mathfrak{H}$ satisfying the following three conditions:*

(i) $\operatorname{dom} S \subset \operatorname{dom} A_2^{\frac{1}{2}}$,
(ii) $\|A_{2s}^{\frac{1}{2}} h\|^2 \leq (h', h)$ *for all* $\{h, h'\} \in S$,
(iii) $A_1 \prec A_2$,

then $A_2 \in EXT(S)$.

Proof. Define the map $\beta_2 : \operatorname{dom} A_2^{\frac{1}{2}} \times \operatorname{dom} A_2^{\frac{1}{2}} \to \mathbb{C}$ by

$$\beta_2(h_1, h_2) = \left(A_{2s}^{\frac{1}{2}} h_1, A_{2s}^{\frac{1}{2}} h_2\right) - \left(A_{1s}^{\frac{1}{2}} h_1, A_{1s}^{\frac{1}{2}} h_2\right), \quad h_1, h_2 \in \operatorname{dom} A_2^{\frac{1}{2}}.$$

It follows from (iii) that β_2 is a semi-inner product. By (i), (ii), (iii) and $A_1 \in EXT(S)$ one has

$$\|A_{2s}^{\frac{1}{2}} h\|^2 \le (h', h) = \|A_{1s}^{\frac{1}{2}} h\|^2 = \|A_{2s}^{\frac{1}{2}} h\|^2,$$

for all $\{h, h'\} \in S$. Thus $\beta_2(h, h) = 0$ for all $h \in \operatorname{dom} S$. This and the Schwarz inequality applied for the semi-inner product β_2 leads to

$$|\beta_2(h_1, h_2)| \le \beta_2(h_1, h_1)\beta_2(h_2, h_2) = 0$$

for all $h_1 \in \operatorname{dom} S$ and $h_2 \in \operatorname{dom} A_2^{\frac{1}{2}}$. Let now $\{f, f'\} \in A_2^{\frac{1}{2}}$ and $\{g, g'\} \in S$. Then

$$\left(f', A_{2s}^{\frac{1}{2}} g\right) = \left(A_{2s}^{\frac{1}{2}} f, A_{2s}^{\frac{1}{2}} g\right) = \left(A_{1s}^{\frac{1}{2}} f, A_{1s}^{\frac{1}{2}} g\right) = (f, A_{1s} g) = (f, g'),$$

which leads to

$$\langle \{f, f'\}, \{A_{2s}^{\frac{1}{2}} g, g'\} \rangle = 0, \quad \text{for all} \quad \{f, f'\} \in A_2^{\frac{1}{2}}.$$

This shows that

$$\{A_{2s}^{\frac{1}{2}} g, g'\} \in \left(A_2^{\frac{1}{2}}\right)^\star = A_2^{\frac{1}{2}}. \tag{3.1}$$

On the other hand, $\{g, A_{2s}^{\frac{1}{2}} g\} \in A_{2s}^{\frac{1}{2}} \subset A_2^{\frac{1}{2}}$. This and (3.1) leads to $\{g, g'\} \in A_2$, which shows that $S \subset A_2$. Thus, A_2 is a nonnegative self-adjoint extension of S. □

Corollary 3.2. *Assume that S is a nonnegative linear relation in a Hilbert space $\mathfrak{H}$. A nonnegative self-adjoint linear relation A extends S if and only if the following three conditions hold true:*

(i) $\operatorname{dom} S \subset \operatorname{dom} A^{\frac{1}{2}}$,
(ii) $\|A_s^{\frac{1}{2}} f\|^2 \le (f', f)$ *for all* $\{f, f'\} \in S$,
(iii) $S_N \prec A$.

Proof. With $A_1 = S_N$, this is a direct consequence of Theorem 3.1 □

The result in Corollary 3.2 may be translated into the language of quadratic forms as follows.

Corollary 3.3. *Assume that S is a nonnegative linear relation in a Hilbert space $\mathfrak{H}$, and $\mathfrak{t}$ is a closed nonnegative quadratic form in $\mathfrak{H}$. The following two items are equivalent:*

(i) *there exists* $A \in EXT(S)$ *such that* $\mathfrak{t}_A = \mathfrak{t}$;
(ii) $\operatorname{dom} S \subset \operatorname{dom} \mathfrak{t}$, $\mathfrak{t}[f] \le (f', f)$ *for all* $\{f, f'\} \in S$, *and* $\mathfrak{t}_N \le \mathfrak{t}$.

3.2. Kreĭn's criteria

M.G. Kreĭn gave a complete description of the class $EXT(S)$ in the case S is a densely defined nonnegative operator (cf. [9, 10]). The next result extends this characterization to the case of nonnegative linear relations.

Theorem 3.4. *Assume that S is a nonnegative linear relation in a Hilbert space $\mathfrak{H}$. If A_1, $A_2 \in EXT(S)$, and A is a nonnegative self-adjoint linear relation in $\mathfrak{H}$ such that $A_1 \prec A \prec A_2$ then $A \in EXT(S)$.*

Proof. Since $\operatorname{dom} S \subset \operatorname{dom} A_2 \subset \operatorname{dom} A_1^{\frac{1}{2}}$, and

$$\|A_s^{\frac{1}{2}}h\|^2 \leq \|A_{2s}^{\frac{1}{2}}h\|^2 = (A_{2s}h, h) = (h', h)$$

for all $\{h, h'\} \in S$, one can apply Theorem 3.1 to complete the proof. □

Corollary 3.5. *Let S be a nonnegative linear relation in a Hilbert space $\mathfrak{H}$ and A be a nonnegative self-adjoint linear relation in $\mathfrak{H}$. Then A belongs to $EXT(S)$ if and only if there exists $B \in EXT(S)$ such that $S_N \prec A \prec B$.*

Proof. With $A_1 = S_N$, this is a direct consequence of Theorem 3.4 □

Conclude this note with a result which is due to Kreĭn [9]. It is an immediate consequence of Corollary 3.5.

Corollary 3.6. *Let S be a densely defined nonnegative linear operator in a Hilbert space $\mathfrak{H}$. Then $EXT(S)$ consists of all nonnegative self-adjoint operators A in $\mathfrak{H}$ such that $S_N \prec A \prec S_F$.*

References

[1] T. Ando and K. Nishio, Positive selfadjoint extensions of positive symmetric operators, Tôhoku Math. J., 22 (1970), 65–75.

[2] Yu.M. Arlinskiĭ, Extremal extensions of sectorial linear relations, Mat. Stud., 7 (1997), No.1, 81–96.

[3] Yu.M. Arlinskiĭ, S. Hassi, Z. Sebestyén, and H.S.V. de Snoo, On the class of extremal extensions of a nonnegative operator, Oper. Theory: Adv. Appl. (B. Sz.-Nagy memorial volume), 127 (2001), 41–81.

[4] Yu.M. Arlinskiĭ and E.R. Tsekanovskiĭ, On von Neumann's problem in extension theory of nonnegative operators, Proc. Amer. Math. Soc., 131 (2003), 3143–3154.

[5] E.A. Coddington, Extension theory of formally normal and symmetric subspaces, Mem. Amer. Math. Soc., 134, 1973.

[6] E.A. Coddington and H.S.V. de Snoo, Positive selfadjoint extensions of positive symmetric subspaces, Math. Z., 159 (1978), 203–214.

[7] S. Hassi, A. Sandovici, H.S.V. de Snoo, and H. Winkler, Form sums of nonnegative selfadjoint operators, Acta Math. Hungar., 111 (2006), 81–105.

[8] T. Kato, Perturbation theory for linear operators, Springer-Verlag, Berlin, 1980.

[9] M.G. Kreĭn, Theory of selfadjoint extensions of semibounded operators and its applications I, Mat. Sb., 20 (1947), 431–495.

[10] M.G. Kreĭn, Theory of selfadjoint extensions of semibounded operators and its applications II, Mat. Sb., 21 (1947), 365–404.

[11] V. Prokaj and Z. Sebestyén, On Friedrichs extensions of operators, Acta Sci. Math. (Szeged), 62 (1996), 243–246.

[12] V. Prokaj and Z. Sebestyén, On extremal positive operator extensions, Acta Sci. Math. (Szeged), 62 (1996), 458–491.

[13] F.S. Rofe–Beketov, The numerical range of a linear relation and maximum relations, (Russian), Teor. Funktsiĭ Funktsional. Anal. i Prihozhen., No. 44 (1985), 103–112.

[14] Z. Sebestyén and J. Stochel, Restrictions of positive self-adjoint operators, Acta Sci. Math. (Szeged), 55 (1991), 149–154.

[15] Z. Sebestyén and J. Stochel, Characterizations of positive selfadjoint extensions, Proc. Amer. Math. Soc., 135 (2007), 1389–1397.

[16] Z. Sebestyén, Restrictions of positive operators, Acta Sci. Math. (Szeged), 46 (1983), 299–301.

Adrian Sandovici
Department of Mathematics
University of Bacău
Str. Spiru Haret, nr. 8
600114 Bacău, Romania
e-mail: adrian.sandovici@gmail.com

Operator Theory:
Advances and Applications, Vol. 188, 245–250

Two-sided Weighted Shifts Are 'Almost Krein' Normal

Franciszek Hugon Szafraniec

To the memory of Peter Jonas (1941–2007), *in supporting the conjecture that friendship is immortal*

Abstract. In this essay we try to explain what may happen if one wants a two-sided weighted shift in a Hilbert space to be Krein normal. As in principle this is not the case it turns out to be so provided the definition of a Krein space is extended in a way which is both natural and provocative; the extended notion may be looked at as a sort of 'complexification' of the classical one. The aforesaid desire has come out of an attempt at classifying the odd solution of the commutation relation of the q-oscillator, which appears in the most innocent case of $0 < q < 1$. A more prosaic motivation is as follows: the two-sided shift is unitary, hence normal; what kind of normality may be attributed to a two-sided weighed shift?

Mathematics Subject Classification (2000). Primary 47B37; Secondary 47B50.

Keywords. Weighted shift, Krein space, self-adjoint operator, normal operator, q-deformed quantum harmonic oscillator.

1. Among ways of defining (or rather dealing with) Krein space is this which starts with a Hilbert space in which a so-called 'fundamental symmetry' is given. The latter is an operator which is both unitary and self-adjoint, and which leads to the indefinite inner product with all the consequences which usually follow. If one takes one of the typical fundamental symmetries and modifies it a bit by replacing its bottom right entry I by $\mathrm{i}I$, one gets instead

$$\begin{pmatrix} 0 & I & 0 \\ I & 0 & 0 \\ 0 & 0 & \mathrm{i}I \end{pmatrix}. \tag{1}$$

The author was supported by the MNiSzW grant N201 026 32/1350. He also would like to acknowledge an assistance of the EU Sixth Framework Programme for the Transfer of Knowledge "Operator theory methods for differential equations" (TODEQ) # MTKD-CT-2005-030042.

The indefinite inner product the matrix (1) originates is no longer Hermitean symmetric (this is due to the fact the operator (1) ceases to be symmetric) though the geometry of the space does not change. The example of Section 2 shows the geometrical situation may become more involved. Nevertheless, the idea of dropping symmetry may be tempting enough [1], at least for those who do not hesitate to leave well-trodden paths. Let us open a chance.

A complex linear space $\mathcal{K}$ with an inner product $[\cdot,-]$, that is a mapping $\mathcal{K}\times\mathcal{K}\ni(f,g)\mapsto[f,g]\in\mathbb{C}$ which is linear in the first variable and conjugate linear in the other, is said to be a *S-space* [2] if there is a Hilbert space structure in $\mathcal{K}$ with the positive definite inner product $\langle\cdot,-\rangle$ and a unitary operator U in the Hilbert space $(\mathcal{K},\langle\cdot,-\rangle)$ such that

$$[f,g]=\langle Uf,g\rangle,\quad f,g\in\mathcal{K};\tag{2}$$

at that point we refer to $[\cdot,-]$ as to the *S-inner product* of $\mathcal{K}$. It is convenient to call the triple $(\mathcal{K},\langle\cdot,-\rangle,U)$ a *Hilbert space realization* of the S-space $(\mathcal{K},[\cdot,-])$. Thus, in fact and a bit roughly [3], an S-space $(\mathcal{K},[\cdot,-])$ is an equivalence class of Hilbert space structures in $\mathcal{K}$ subject to the formula (2). The very first thing which can test this point of view is topology of the S-space $(\mathcal{K},[\cdot,-])$: this is the topology in $\mathcal{K}$ which is inherited from some (and, fortunately, from any) its Hilbert space realizations. In accordance to this we can think of continuous, densely defined, closed and closable operators; all these notions being independent of a particular choice of a Hilbert space realization of a given S-space. We can define in a unique way an S-space adjoint $A^\natural$ of a densely defined operator A with two key features of the Hilbert space theory to be maintained: the characterization of the domain of $A^\natural$ and that of dense definiteness of $A^\natural$. Passing to a Hilbert space realization $(\mathcal{K},\langle\cdot,-\rangle,U)$, which one eventually has to do, one gets [4]

(a) $\mathcal{D}(A^\natural)=U\mathcal{D}(A^*)$ and $U^*A^\natural\subset A^*U^*$.

This allows to define S-normality, like in a Hilbert space, in any of the two equivalent ways [5]

(b) N is closed and $N^\natural N=NN^\natural$,

(c) $\mathcal{D}(N)=\mathcal{D}(N^\natural)$ and $[Nf,Ng]=[N^\natural f,N^\natural g]$ for $f,g\in\mathcal{D}(N)$.

---

[1] This is a follow-up of [10]. We hope to be able to present a more mature form of [10] in a not too distant future.

[2] The author was suggested to change his former designation 'K-space' to the present 'S-space'. Therefore it is not his responsibility for any gossiping this brave decision may cause hereafter.

[3] This fact, as well as some others, is not required for the arguments exploited in this note to work. However all those are going to find their detailed and methodological argumentation in a paper writing of which is in progress, cf. footnote 1.

[4] This may be considered as an *ad hoc* definition of $A^\natural$ which, due to (2), is independent of a particular Hilbert space realization. Nevertheless, the content of footnote 3 applies here as well.

[5] Again, as mentioned in footnote 4, a definition of S-normality may be adapted via a Hilbert space realization.

Among very few references we can offer for Krein normality to be in there are [1], [2], [4], [6] and [7]. This makes our invitation to an ‘off-road excursion’ even more exiting.

2. Let $\mathcal{H}$ be a separable Hilbert space and let $(e_n)_{n\in\mathbb{Z}}$ be its orthonormal basis. A linear operator S is said to be a *two-sided*[6] (or, bilateral) *weighted shift* if $\mathcal{D}(S) \stackrel{\text{def}}{=} \operatorname{lin}\{e_n\colon\ n \in \mathbb{Z}\}$ and $Se_n = \sigma_n e_{n+1}$ for $n \in \mathbb{Z}$, where σ_n, $n \in \mathbb{Z}$, are positive numbers usually referred to as the weight of S. If a basis is arranged[7] as $(e_n)_{n\in\mathbb{N}}$ an operator S acting as $Se_n = \sigma_n e_n$ for $n \in \mathbb{N}$ is called a *one-sided* (or, unilateral) *weighted shift*. Weighted shifts are bounded if and only if the set of all its weights is bounded. If this happens we use the same character to denote its closure assuming tacitly its domain is considered to be the whole Hilbert space $\mathcal{H}$.

The linear space $\operatorname{lin}\{e_n\colon\ n \in \mathbb{N}\}$ is obviously invariant for a two-sided forward weighted shift, which results in a trivial observation that a bounded one-sided weighted shift may be a restriction of an unbounded two-sided one to its invariant subspace.

The two-sided backward shift U acting apparently as $Ue_n = e_{n-1}$, $n \in \mathbb{Z}$, is unitary and gives rise to define in $\mathcal{H}$ a suitable S-space structure. The diagonal operator $D \stackrel{\text{def}}{=} \operatorname{diag}(\sigma_n)_{n\in\mathbb{Z}}$ with $\mathcal{D}(D) = \mathcal{D}(S)$, which is symmetric, helps us to write the two sided-weighted shift S as

$$S = U^* D. \tag{3}$$

By this we have

$$S^* = DU \text{ and } S^\natural = UD. \tag{4}$$

Consequently,

$$[S^\natural f, S^\natural g] = \langle UUDf, UDg\rangle = \langle UDf, Dg\rangle,$$
$$[Sf, Sg] = \langle UU^*Df, U^*D\rangle = \langle Df, U^*Dg\rangle = \langle UDf, Dg\rangle.$$

This results in[8]

Conclusion. *The closure of a two-sided weighted shift is S-normal. Therefore, the closure of a one-sided weighted shift is a restriction of an S-normal operator to its invariant subspace*[9].

Notice that (4) indicates that (the closure of) S cannot be S-self-adjoint unless $D = I$.

---

[6] Sometimes we add the word *forward*, especially when we want to distinguish them from their adjoints which shift the basic vectors *backwardly*.

[7] We prefer $\mathbb{N}$ to start with (and include) 0.

[8] While the domain of $\overline{S}$ may differ from that of S^* (see [9]) fortunately and much to our surprise $\mathcal{D}(\overline{S}) = \mathcal{D}(S^\natural)$, cf. footnote 6.

[9] This touches the notion of S-subnormality which seems to be a delicate question even in a Krein space as the property of being such may not be inherited by closed subspaces.

3. We have come to the above conclusion trying to categorize the solutions of the commutation relation

$$S^*S - qSS^* = I \tag{5}$$

of the q-deformed version of the quantum harmonic oscillator, which appears in the most conclusive case of $0 < q < 1$; see [3], [5], [8] and for further comments also [11]. The basic, well-tailored solution of (5) is a one-sided weighted shift with the weights [10] $\sqrt{[n+1]_q}$, $n = 0, 1, \ldots$. It behaves like the creation operator, in fact it is considered as a q-creation. It is also a <u>bounded</u> subnormal operator in the underlying Hilbert space and this property distinguishes it among all the solutions of (5); cf. [11]. Surprisingly besides that there is [11] a solution which is an <u>unbounded</u> two-sided weighted shift with the weights $\sqrt{\alpha q^{n+N} + [n+N]_q}$, $n \in \mathbb{Z}$, and with the parameters $\alpha > (1-q)^{-1}$ and $N \in \mathbb{Z}$, and which is <u>not</u> subnormal. Thus our conclusion uncovers another feature of this somehow peculiar solution.

4. Instead of building an S-space by means of the unitary operator U which is, just to remaind, the two-sided backward shift one may use alternatively its adjoint U^*, the-two sided forward shift. Using the same kind of argument as before it turns out to be totally clear that now in the resulting S-space the backward weighted shift S^* becomes S-normal. This is a complementary (or rather dual) conclusion to that of Section 2.

However, much wants more [12]. Consider the Hilbert space $\mathcal{K} \stackrel{\text{def}}{=} \mathcal{H} \oplus \mathcal{H}$ and the operator

$$J \stackrel{\text{def}}{=} \begin{pmatrix} 0 & U^* \\ U & 0 \end{pmatrix}$$

in it. Thus J is a self-adjoint unitary operator in $\mathcal{H}$ and, consequently, it may serve as a fundamental symmetry for a Krein space [13]; denote this space shortly by $(\mathcal{K}, J)$.

Suppose, besides the two-sided forward weighted shift S, we are given a two-sided backward weighted shift T; both with respect to the same basis. What we have said so far supports the following: the operator

$$A \stackrel{\text{def}}{=} \begin{pmatrix} 0 & S \\ T & 0 \end{pmatrix}$$

with $\mathcal{D}(A) \stackrel{\text{def}}{=} \mathcal{D}(S) \oplus \mathcal{D}(T) = \operatorname{lin}\{e_n\colon\ n \in \mathbb{Z}\} \oplus \operatorname{lin}\{e_n\colon\ n \in \mathbb{Z}\}$ is very likely to be Krein essentially normal. However, it turns out, somehow amazingly, we can claim even more.

Theorem. *The closure of the operator A is self-adjoint in the Krein space $(\mathcal{H}, J)$.*

[10] Here $[0]_q \stackrel{\text{def}}{=} 0$, $[1]_q \stackrel{\text{def}}{=} 1$ and $[k]_q \stackrel{\text{def}}{=} (1-q^k)(1-q)^{-1}$, $k = 2, 3, \ldots$ are the *basic* (or q-) numbers.

[11] An interesting thing is that when $q \geqslant 1$ this kind of solution disappears and the only one remaining is the q-creation which becomes unbounded whatsoever.

[12] Let me confess the paper is organized in a the way the findings have evolved in time.

[13] At a glance it looks much different than those customary considered. The very end Remark makes this statement even stronger.

Proof. Write, in addition to (3), T as

$$T = EU \tag{6}$$

where E is a diagonal operator with respect to the same basis as D is. Using these two one can check in a completely direct way that for $f, g \in \mathcal{D}(A)$

$$A^{\natural} f = Af. \tag{7}$$

Because S and T are weighted shifts, $\text{lin}\{e_n\colon\ n \in \mathbb{Z}\}$ is a core[14] of both S^* and T^*. Hence $\mathcal{D}(A)$ is a core of A^* and consequently that of $A^{\natural}$. Applying the same sort of argument which is behind proving what is stated in footnote 6, one can check that $\mathcal{D}(\overline{A}) = \mathcal{D}(A^{\natural})$. Now using the fact that $\mathcal{D}(A)$ is a core of $A^{\natural}$ and making proper approximation in (7) one can check that (7) holds for $f, g \in \mathcal{D}(\overline{A}) = \mathcal{D}(A^{\natural})$ too, with $\overline{A}$ replacing A, of course. □

The theorem shows that a couple of two-sided weighted shifts twisted properly composes a self-adjoint operator in a ‘true’ Krein space. This finding is not only thought-provoking but also may be considered as a bridge leading to the circumstances under which the ‘almost Krein’ sense works well. Or, in a different manner, the theorem splits into two equivalent copies of conclusion.

Remark. Both conclusion and theorem remain true if S and T factorize as in (3) and (6), resp. with U being unitary and D, E self-adjoint; all this provided their domains behave in a suitable way. The proof can be done with ease though in course of it the domains indeed have to be taken care of except when both D and E are bounded. Because our persistence here is in spreading ideas rather than fabricating far going generalizations we leave this issue untouched unless any need really appears.

References

[1] T.Ya. Azizov and P. Jonas, On compact perturbations of normal operators in a Krein space, *Ukrainskiĭ Matem. Žurnal* **42** (1990), 1299–1306.

[2] T.Ya. Azizov and V.A. Strauss, Spectral decompositions for special classes of self-adjoint and normal operators on Krein spaces, *Theta Series in Advanced Mathematics, Spectral Analysis and its Applications*, 45–67, The Theta Foundation, Bucharest, **2004**.

[3] M. Chaichian, H. Grosse, and P. Presnajder, Unitary representations of the q-oscillator algebra, *J. Phys. A: Math. Gen.*, **27**(1994), 2045–2051.

[4] I. Gohberg and B. Reichstein, On classification of normal matrices in an indefinite scalar product, *Integral Equations Operator Theory*, **13** (1990), 364–394.

[5] K.-D. Kürsten and E. Wagner, Invariant integration theory on non-compact quantum spaces: Quantum $(n, 1)$-matrix ball, arXiv:QA/0305380v1.

[6] H. Langer and F.H. Szafraniec, Bounded normal operators in a Pontryagin space, *Operator Theory: Advances and Applications*, 162 (2005), 231–251.

[14] Notice that the notion of ‘core’ is topological.

[7] Ch. Mehl, A. Ran and L. Rodman: Semidefinite invariant subspaces: degenerate inner products, *Operator Theory: Advances and Applications*, **149** (2004), 467–486.

[8] K. Schmüdgen and E. Wagner, Hilbert space representations of cross product algebras II, *Algebr. Represent. Theor*, **9** (2006), 431–464.

[9] J. Stochel and F.H. Szafraniec, A few assorted questions about unbounded subnormal operators, *Univ. Iagel. Acta Math.*, **28** (1991), 163–170.

[10] F.H. Szafraniec, A look at Krein space: new thoughts and old truths, talk given at '5th Workshop Operator Theory in Krein Spaces and Differential Equations', Technische Universität, Berlin, December 16–18, **2005**.

[11] ———, Operators of the q-oscillator, in: *Noncommutative Harmonic Analysis with Applications to Probability*, Banach Center Publ. 78, Inst. Math. Polish Acad. Sci., Warszawa, **2007**, 293–307.

Franciszek Hugon Szafraniec
Instytut Matematyki
Uniwersytet Jagielloński
ul. Reymonta 4
PL-30059 Kraków, Poland
e-mail: `umszafra@cyf-kr.edu.pl`

Advances in Partial Differential Equations (APDE)

BIRKHÄUSER

Edited by
Bert-Wolfgang Schulze (Potsdam), **Sergio Albeverio** (Bonn), **Michael Demuth** (Clausthal), **Jerome Goldstein** (Memphis), **Nobuyuki Tose** (Yokohama)

The subseries *APDE* is intended to report on recent developments in all areas of partial differential equations, in particular, on those belonging to the tradition defined by the series *Operator Theory: Advances and Applications* and new areas. The *Advances* volumes will consist primarily of expository research articles, presenting both an overview of the state of a field and new results, in such areas as microlocal analysis and its applications to asymptotic phenomena, mathematical physics, spectral theory, as well as symplectic geometry, the analysis of singularities of solutions, and geometric analysis. Selected articles of this kind may also appear as a separate monographic volume. A certain number of short communications will also be published.

Vol. 184: Qin, Y., Nonlinear Parabolic-Hyperbolic Coupled Systems and Their Attractors (2008).
ISBN 978-3-7643-8813-3

Vol. 183: Nazaikinskii, V.E.. / Savin, A.Yu. / Sternin, B.Yu., Elliptic Theory and Noncommutative Geometry (2008).
This comprehensive yet concise book deals with nonlocal elliptic differential operators, whose coefficients involve shifts generated by diffeomorophisms of the manifold on which the operators are defined. The main goal of the study is to relate analytical invariants (in particular, the index) of such elliptic operators to topological invariants of the manifold itself. This problem can be solved by modern methods of noncommutative geometry.
This is the first and so far the only book featuring a consistent application of methods of noncommutative geometry to the index problem in the theory of nonlocal elliptic operators. Although the book provides important results, which are in a sense definitive, on the above-mentioned topic, it contains all the necessary preliminary material, such as $C*$-algebras and their K-theory or cyclic homology. Thus the material is accessible for undergraduate students of mathematics (third year and beyond). It is also undoubtedly of interest for post-graduate students and scientists specializing in geometry, the theory of differential equations, functional analysis, etc.
The book can serve as a good introduction to noncommutative geometry, which is one of the most powerful modern tools for studying a wide range of problems in mathematics and theoretical physics.
ISBN 978-3-7643-8774-7

Vol. 166: De Gosson, M., Symplectic Geometry and Quantum Mechanics (2006).
ISBN 978-3-7643-7574-4

Vol. 159: Reissig, M. / Schulze, B.-W. (Eds.), New Trends in the Theory of Hyperbolic Functions (2005).
ISBN 978-3-7643-7283-5

Vol. 151: Gil, J.B. / Krainer, T. / Witt, I. (Eds.), Aspects of Boundary Problems in Analysis and Geometry (2004).
ISBN 978-3-7643-7069-5

Vol. 145: Albeverio, S. / Demuth, M. / Schrohe, E. / Schulze, B.-W. (Eds.), Nonlinear Hyperbolic Equations, Spectral Theory, and Wavelet Transformations (2003).
ISBN 978-3-7643-2168-0

Vol. 138: Albeverio, S. / Demuth, M. / Schrohe, E. / Schulze, B.-W. (Eds.), Parabolicity, Volterra Calculus, and Conical Singularities (2002).
ISBN 978-3-7643-6906-4

Vol. 125: Gil, J.B. / Grieser, D. / Lesch, M. (Eds.), Approaches to Singular Analysis (2001).
ISBN 978-3-7643-6518-9

Operator Theory: Advances and Applications (OT)

Edited by
Israel Gohberg, Tel Aviv University, Israel

BIRKHÄUSER

This series is devoted to the publication of current research in operator theory, with particular emphasis on applications to classical analysis and the theory of integral equations, as well as to numerical analysis, mathematical physics and mathematical methods in electrical engineering.

OT 188: Behrndt, J. / Förster, K.-H. / Langer, H. / Trunk, C. (eds.), Spectral Theory in Inner Product Spaces and Applications (2008). ISBN 978-3-7643-8910-9

OT 187: Ando, T. / Curto, R.E. / Jung, I.B. / Lee, W.Y. (eds.), Recent Advances in Operator Theory and Applications (2008). ISBN 978-3-7643-8892-8

OT 186: Janas, J. / Kurasov, P. / Laptev, A. / Naboko, S. / Stolz, G. (eds.), Methods of Spectral Analysis in Mathematical Physics (2008). ISBN 978-3-7643-8754-9

OT 185: Vasilevski, N., Commutative Algebras of Toeplitz Operators on the Bergman Space (2008). ISBN 978-3-7643-8725-9

OT 184: Qin, Y., Nonlinear Parabolic-Hyperbolic Coupled Systems and Their Attractors (2008, to appear). Subseries **A**dvances in **P**artial **D**ifferential **E**quations. ISBN 978-3-7643-8813-3

OT 183: Nazaikinskii, V.E.. / Savin, A.Yu. / Sternin, B.Yu., Elliptic Theory and Noncommutative Geometry (2008). Subseries **A**dvances in **P**artial **D**ifferential **E**quations. ISBN 978-3-7643-8774-7

OT 182: van der Mee, C., Exponentially Dichotomous Operators and Applications (2008). Subseries **L**inear **O**perators and **L**inear **S**ystems. ISBN 978-3-7643-8732-7

OT 181: Bastos, A. / Gohberg, I. / Lebre, A.B. / Speck, F.-O. (eds.), Operator Algebras, Operator Theory and Applications (2008). ISBN 978-3-7643-8683-2

OT 180: Okada, S. / Ricker, W.J. / Sánchez Pérez, E.A., Optimal Domain and Integral Extension of Operators Acting in Function Spaces (2008). ISBN 978-3-7643-8647-4

OT 179: Ball, J.A. / Eidelman, Y. / Helton, J.W. / Olshevsky, V. / Rovnyak, J. (eds.), Recent Advances in Matrix and Operator Theory (2007). ISBN 978-3-7643-8538-5

OT 178: Bart, H. / Gohberg, I. / Kaashoek, M.A. / Ran, A.C.M., Factorization of Matrix and Operator Functions: The State Space Method (2007). ISBN 978-3-7643-8267-4

OT 177: López-Gómez, J. / Mora-Corral, C., Algebraic Multiplicity of Eigenvalues of Linear Operators (2007). ISBN 978-3-7643-8400-5

OT 176: Alpay, D. / Vinnikov, V. (Eds.), System Theory, the Schur Algorithm and Multidimensional Analysis (2007). ISBN 978-3-7643-8136-3

OT 175: Förster, K.-H. / Jonas, P. / Langer, H. / Trunk, C. (Eds.), Operator Theory in Inner Product Spaces (2007). ISBN 978-3-7643-8269-8

OT 174: Janas, J. / Kurasov, P. / Laptev, A. / Naboko, S. / Stolz, G. (Eds.), Operator Theory, Analysis and Mathematical Physics (2007). ISBN 978-3-7643-8134-9

OT 173: Emel'yanov, E.Yu., Non-spectral Asymptotic Analysis of One-Parameter Operator Semigroups (2006). ISBN 978-3-7643-8095-3

OT 172: Toft, J. / Wong, M.W. / Zhu, H. (Eds.), Modern Trends in Pseudo-Differential Operators (2007). ISBN 3-7643-8097-7

OT 171: Dritschel, M. (Ed.), The Extended Field of Operator Theory (2007). ISBN 3-7643-7979-0

OT 170: Erusalimsky, Ya.M. / Gohberg, I. / Grudsky, S.M. / Rabinovich, V. / Vasilevski, N. (Eds.), Modern Operator Theory and Applications (2007). ISBN 3-7643-7736-4